Phase Change Materials for Building Energy Applications

Phase Change Materials for Building Energy Applications

Guest Editors

Facundo Bre
Antonio Caggiano
Umberto Berardi

MDPI

Basel • Beijing • Wuhan • Barcelona • Belgrade • Novi Sad • Cluj • Manchester

Guest Editors
Facundo Bre
Environmental Research and Innovation (ERIN) Department
Luxembourg Institute of Science and Technology
Luxembourg
Luxembourg

Antonio Caggiano
Department of Civil, Chemical and Environmental Engineering
University of Genova
Genova
Italy

Umberto Berardi
ArCoD Department
Polytechnic University of Bari
Bari
Italy

Editorial Office
MDPI AG
Grosspeteranlage 5
4052 Basel, Switzerland

This is a reprint of the Special Issue, published open access by the journal *Energies* (ISSN 1996-1073), freely accessible at: https://www.mdpi.com/journal/energies/special_issues/X85KHJ2CG0.

For citation purposes, cite each article independently as indicated on the article page online and as indicated below:

Lastname, A.A.; Lastname, B.B. Article Title. *Journal Name* **Year**, *Volume Number*, Page Range.

ISBN 978-3-7258-4759-4 (Hbk)
ISBN 978-3-7258-4760-0 (PDF)
https://doi.org/10.3390/books978-3-7258-4760-0

Cover image courtesy of Umberto Berardi

Contents

About the Editors

Facundo Bre

Facundo Bre is a Senior Researcher at the Luxembourg Institute of Science and Technology (LIST), specializing in building physics, energy-efficient buildings, advanced building materials, building performance simulation, and the generation and integration of climate data for engineering applications. He holds a PhD in Engineering with a specialization in Computational Mechanics, awarded by the National Littoral University, Argentina, in 2018. He actively participates in national and European research initiatives that focus on the thermal characterization of high-performance building façades, energy modeling for urban regeneration, and the development of sustainability labeling for urban and built environments. His current research aims to advance innovative technologies, methods, and computational tools that support the design of energy-efficient, climate-resilient buildings. Dr. Bre has also contributed to several fields of material science, computational methods, and machine learning applications in science and engineering, achieving various milestones, such as scientific articles, funding, patents, software, awards, and datasets. He has received several competitive research grants, including a Marie Skłodowska-Curie Individual Fellowship, and he was awarded the Young Researcher Award in 2022 by the Argentine Association of Computational Mechanics (AMCA).

Antonio Caggiano

Antonio Caggiano is an Associate Professor at the Department of Civil, Chemical and Environmental Engineering (DICCA) of the University of Genova (UniGE), Italy and affiliated at University of Buenos Aires (Argentina). His research lies at the intersection of structural mechanics, sustainable construction materials, multiscale and multiphysics modelling, and energy efficiency in the built environment. He holds a PhD in Engineering from the Universities of Salerno (Italy) and Tucuman (Argentina) and has completed several international research stays across Europe and Latin America. Prof. Caggiano is actively engaged in EU-funded projects (e.g., Horizon Europe, MSCA, H2020), focusing on low-carbon cementitious composites, phase-change materials for thermal energy storage, and AI-driven design of building materials. His contributions span from fundamental research to applied innovation in concrete mechanics, durability, and digital twins for resilient infrastructures. He has authored over 100 peer-reviewed scientific publications, coordinated international research networks, and served on technical committees such as RILEM and fib. He is the recipient of prestigious fellowships, including the Alexander von Humboldt Fellowship, and acts as scientific coordinator of multiple interdisciplinary initiatives in energy and materials science.

Umberto Berardi

Umberto Berardi is the President of the Canadian Acoustical Association and of the International Association of Building Physics. Dr. Berardi has been Canada Research Chair in Building Science and Full Professor and Director of the BeTOP center at Toronto Metropolitan University in Toronto, Canada. Dr. Berardi has an extensive publication record, including 200 peer-reviewed journals and ten books. His awards include Frontiers in 2024, the 2023 European Thermophysical Properties ECTP-NETZSCH AWARD; the 2021 10th Canada's Clean16 award and Clean50 winner.

energies

MDPI

Editorial

Phase Change Materials for Building Energy Applications

Facundo Bre [1,*], Antonio Caggiano [2] and Umberto Berardi [3]

[1] Sustainable Urban & Built Environment Research Group, Luxembourg Institute of Science and Technology, L-4362 Esch-sur-Alzette, Luxembourg
[2] DICCA Department, University of Genova, via Montallegro 1, 16145 Genova, Italy; antonio.caggiano@unige.it
[3] ArCoD Department, Politecnico di Bari, via Orabona n.4, 71122 Bari, Italy; umberto.berardi@poliba.it
* Correspondence: facundo.bre@list.lu

Abstract: This editorial introduces the Special Issue entitled "Phase Change Materials for Building Energy Applications", which gathers nine original research articles focused on advancing thermal energy storage solutions in the built environment. The selected contributions explore the application of phase change materials (PCMs) across a range of building components and systems, including façades, flooring, glazing, and pavements, aimed at enhancing energy efficiency, reducing peak loads, and improving thermal comfort. This Special Issue highlights both experimental and numerical investigations, ranging from nanomaterial-enhanced PCMs and solid–solid PCM glazing systems to full-scale applications and the modeling of encapsulated PCM geometries. Collectively, these studies reflect the growing potential of PCMs to support sustainable, low-carbon construction and provide new insights into material design, system optimization, and energy resilience. We thank all contributing authors and reviewers for their valuable input and hope that this Special Issue serves as a resource for ongoing innovation in the field.

Keywords: phase change materials; thermal energy storage; building energy efficiency; PCM-integration; sustainable construction; building materials; experimental studies; numerical modeling; energy systems; climate-resilient buildings; carbon-neutrality

1. Introduction

The building sector is responsible for over 30% of global energy consumption and nearly 40% of total direct and indirect greenhouse gas emissions [1]. To address this, researchers in the building sector are increasingly focusing on innovative thermal management strategies [2]. Phase change materials (PCMs) have emerged as a particularly promising solution, enabling thermal energy storage and dynamic temperature regulation through latent heat processes [3]. Their integration into building components and systems, such as walls, floors, glazing, and HVAC, has been shown to improve energy efficiency [4,5], reduce peak demand [6], and enhance indoor thermal comfort [7].

Recent PCM developments have demonstrated substantial improvements in thermal storage capacity [8] and application versatility [9]. However, challenges remain in understanding their effectiveness in buildings, developing new integrated systems, and optimizing their thermal conductivity, long-term durability, and cost-effective manufacturing, thus making continued research essential.

2. Paper Contributions

This Special Issue of *Energies* presents a curated selection of nine research articles that explore the design, application, technology deployment, and performance evaluation of PCMs across a wide range of building energy systems and components. The contributions collectively highlight both experimental and theoretical advances in the field.

Some studies examine PCM integration into traditional building elements, such as façade sandwich panels [10] and energy storage flooring for rural settlements [11], to enhance thermal inertia and reduce peak loads. Others investigate more advanced and emerging applications, including the use of nanomaterials to boost PCM conductivity [12], the optimization of glazing systems with solid–solid PCMs [13], and the deployment of PCM-based technologies in off-grid environments [14].

A significant emphasis is placed on PCM numerical modeling [13], encapsulation geometry and shape optimization [15], and real-scale performance assessments, as demonstrated in the studies on precast wall panels [10] and transpired solar collectors [12]. Additionally, novel materials such as carbon aerogels derived from citrus waste [16] and PCM-enhanced asphalt for pavements [17] showcase the broadening scope of PCM applications beyond conventional building envelopes.

3. Conclusions

Together, these contributions offer a comprehensive snapshot of current research and future directions for PCMs in building energy applications. This collection will serve as a valuable reference for academics, engineers, and policymakers working to advance the design of more energy-efficient and sustainable buildings.

We sincerely thank all contributing authors for their innovative work and the reviewers for their thoughtful evaluations and feedback. We also extend our appreciation to the *Energies* editorial team for their continued support.

We curated this Special Issue to highlight the growing potential of phase change materials in addressing energy efficiency and sustainability challenges in the built environment. We hope that this collection contributes meaningfully to the development of advanced building energy technologies.

Author Contributions: Conceptualization, F.B., and A.C.; writing—original draft preparation, F.B.; writing—review and editing, A.C. and U.B.; supervision, F.B. All authors have read and agreed to the published version of the manuscript.

Funding: Bre has received funding from the European Union's Horizon 2020 research and innovation program under the Marie Skłodowska-Curie grant agreement No. 101024627, as part of the 0E-BUILDINGS action project.

Data Availability Statement: No new data were created or analyzed in this study.

Acknowledgments: We thank the *Energies* editorial team for their support.

Conflicts of Interest: The authors declare no conflicts of interest.

References

1. IEA. *Technology and Innovation Pathways for Zero-Carbon-Ready Buildings by 2030—Analysis;* IEA: Paris, France, 2022. Available online: https://www.iea.org/reports/technology-and-innovation-pathways-for-zero-carbon-ready-buildings-by-2030 (accessed on 27 May 2025).
2. Voronkova, I.; Podlasek, A. The Use of Transparent Structures to Improve Light Comfort in Library Spaces and Minimize Energy Consumption: A Case Study of Warsaw, Poland. *Energies* **2024**, *17*, 3007. [CrossRef]

3. Berardi, U.; Soudian, S. Experimental Investigation of Latent Heat Thermal Energy Storage Using PCMs with Different Melting Temperatures for Building Retrofit. *Energy Build.* **2019**, *185*, 180–195. [CrossRef]
4. Bre, F.; Caggiano, A.; Koenders, E.A.B. Multiobjective Optimization of Cement-Based Panels Enhanced with Microencapsulated Phase Change Materials for Building Energy Applications. *Energies* **2022**, *15*, 5192. [CrossRef]
5. Bre, F.; Lamberts, R.; Flores-Larsen, S.; Koenders, E.A.B. Multi-Objective Optimization of Latent Energy Storage in Buildings by Using Phase Change Materials with Different Melting Temperatures. *Appl. Energy* **2023**, *336*, 120806. [CrossRef]
6. Lei, J.; Yang, J.; Yang, E.-H. Energy Performance of Building Envelopes Integrated with Phase Change Materials for Cooling Load Reduction in Tropical Singapore. *Appl. Energy* **2016**, *162*, 207–217. [CrossRef]
7. Auzeby, M.; Wei, S.; Underwood, C.; Tindall, J.; Chen, C.; Ling, H.; Buswell, R. Effectiveness of Using Phase Change Materials on Reducing Summer Overheating Issues in UK Residential Buildings with Identification of Influential Factors. *Energies* **2016**, *9*, 605. [CrossRef]
8. Zhang, J.; Cao, Z.; Huang, S.; Huang, X.; Han, Y.; Wen, C.; Honoré Walther, J.; Yang, Y. Solidification Performance Improvement of Phase Change Materials for Latent Heat Thermal Energy Storage Using Novel Branch-Structured Fins and Nanoparticles. *Appl. Energy* **2023**, *342*, 121158. [CrossRef]
9. Wang, P.; Liu, Z.; Zhang, X.; Hu, M.; Zhang, L.; Fan, J. Adaptive Dynamic Building Envelope Integrated with Phase Change Material to Enhance the Heat Storage and Release Efficiency: A State-of-the-Art Review. *Energy Build.* **2023**, *286*, 112928. [CrossRef]
10. Niall, D.; West, R. Development of Concrete Façade Sandwich Panels Incorporating Phase Change Materials. *Energies* **2024**, *17*, 2924. [CrossRef]
11. Wang, K.; Xu, G.; Zhao, X.; Li, G.; Mai, L. Experimental Study on Phase Change Energy Storage Flooring for Low-Carbon Energy Systems in Grassland Pastoral. *Energies* **2024**, *17*, 4828. [CrossRef]
12. Croitoru, C.; Bode, F.; Calotă, R.; Berville, C.; Georgescu, M. Harnessing Nanomaterials for Enhanced Energy Efficiency in Transpired Solar Collectors: A Review of Their Integration in Phase-Change Materials. *Energies* **2024**, *17*, 1239. [CrossRef]
13. Arasteh, H.; Maref, W.; Saber, H.H. 3D Numerical Modeling to Assess the Energy Performance of Solid–Solid Phase Change Materials in Glazing Systems. *Energies* **2024**, *17*, 3759. [CrossRef]
14. Kalair, A.; Jamei, E.; Seyedmahmoudian, M.; Mekhilef, S.; Abas, N. Building the Future: Integrating Phase Change Materials in Network of Nanogrids (NoN). *Energies* **2024**, *17*, 5862. [CrossRef]
15. Gonçalves, M.; Figueiredo, A.; Vela, G.; Rebelo, F.; Almeida, R.M.S.F.; Oliveira, M.S.A.; Vicente, R. Effect of Macrocapsule Geometry on PCM Performance for Thermal Regulation in Buildings. *Energies* **2025**, *18*, 303. [CrossRef]
16. Suchorowiec, K.; Bieda, M.; Szatkowska, M.; Sieradzka, M.; Kuźnia, M.; Ziąbka, M.; Pielichowska, K. From Waste to Functional Material—Carbon Aerogels from Citrus Biomass Infiltrated with Phase Change Materials for Possible Application in Solar-Thermal Energy Conversion and Storage. *Energies* **2025**, *18*, 814. [CrossRef]
17. Korniejenko, K.; Nykiel, M.; Choinska, M.; Jexembayeva, A.; Konkanov, M.; Aruova, L. An Overview of Phase Change Materials and Their Applications in Pavement. *Energies* **2024**, *17*, 2292. [CrossRef]

energies

MDPI

Review

Harnessing Nanomaterials for Enhanced Energy Efficiency in Transpired Solar Collectors: A Review of Their Integration in Phase-Change Materials

Cristiana Croitoru [1], Florin Bode [2,*], Răzvan Calotă [1], Charles Berville [1] and Matei Georgescu [1]

1 CAMBI Research Centre, Technical University Civil Engineering Bucharest, 020396 Bucharest, Romania; cristiana.croitoru@utcb.ro (C.C.); razvan.calota@utcb.ro (R.C.); charles.berville@phd.utcb.ro (C.B.); matei.georgescu@utcb.ro (M.G.)

2 Department of Mechanical Engineering, Technical University of Cluj-Napoca, 400114 Cluj-Napoca, Romania

* Correspondence: florin.bode@termo.utcluj.ro

Abstract: The building sector plays an important role in the global climate change mitigation objectives. The reduction of CO_2 emissions and energy consumption in the building sector has been intensively investigated in the last decades, with solar thermal energy considered to be one of the most promising solutions due to its abundance and accessibility. However, the discontinuity of solar energy has led to the study of thermal energy storage to improve the thermal performance of solar thermal systems. In this review paper, the integration of various types of phase-change materials (PCMs) in transpired solar collectors (TSC) is reviewed and discussed, with an emphasis on heat transfer enhancements, including nanomaterials. Thermal energy storage applied to TSC is studied in terms of design criteria, materials technologies, and its impact on thermal conductivity. This review highlights the potential of nanomaterial technology integration in terms of thermal performance improvements. The utilization of nanomaterials in solar walls holds the potential to significantly enhance their performance. The integration of diverse materials such as graphene, graphite, metal oxides, and carbon nanoparticles can pave the way for improving thermal conductivity.

Keywords: phase-change materials; nanomaterials; nano-enhanced phase-changing materials transpired solar walls; thermal energy storage; PCM; nePCMs

1. Introduction

The buildings and construction sectors are responsible for 30% of global final energy consumption and 27% of total energy sector CO_2 emissions, being considered one of the highest energy consumers [1]. Each year, CO_2 emissions increase around the world, and global warming threats become increasingly visible. Greenhouse gas emissions are a direct consequence of abrupt industrialization and urbanization, with citizens becoming both beneficiaries as well as victims. All these concerns have led to #climatestrike movements to raise climate change awareness. Finding and implementing solutions should be priority topics for research and development in the building and construction sectors.

Building envelopes serve as crucial thermal interfaces between the external environment and indoor spaces. Ensuring optimal thermal conditions is vital for the thermal comfort and well-being of the occupants. As a result, the energy needs of buildings are inseparably tied to the thermal performance of their structural components. Inadequate thermal performance in the building envelope results in either excessive heat buildup within the premises, causing increased energy consumption for cooling, or significant thermal losses, which directly affect the energy required for heating. This challenge becomes even more pronounced when striving to enhance the energy efficiency of buildings, moving them closer to the goal of achieving zero-energy buildings (ZEB) [2]. In this context, it is mandatory to use renewable energy sources (RES) in order to ensure indoor comfort with low energy consumption.

As the most widely available of all RES, solar energy is a natural choice in the attempt to reduce energy consumption due to its abundance and accessibility. However, since solar energy exhibits intermittent availability, solar thermal energy mandates its transformation and storage to ensure its uninterrupted availability [3]. Extensive research has been conducted with a primary focus on increasing the efficiency of both solar energy storage and conversion processes, with photovoltaic and solar thermal conversion technologies as cornerstones for efficient solar energy conversion to electricity and heat. However, using this solar energy throughout extended periods requires the integration of thermal energy storage systems in the building. This integration not only ensures the sustained availability of solar-derived thermal energy but also contributes significantly to the overall energy sustainability paradigm.

The solar collector is one of the most cost-effective pieces of equipment used to harvest solar thermal energy. Two main types can be identified: (1) water-based solar collectors (WSCs), an already popular solution for preparing domestic hot water and heating, and (2) air solar collectors (ASCs), used to preheat fresh air and to dry or heat the air inside a room. According to the manufacturer data, Reichl et al. [4] found that ASCs have the advantages of no frost risk in colder regions during the winter, a lower environmental impact, and a more cost-effective investment and operation.

TSCs are typically used for low-temperature applications with low operating costs [5]. TSCs, at their simplest, are perforated metal panels installed on the exterior of a building that absorb solar radiation and transfer this absorbed heat to the air drawn through the perforations and into the building. In theory, increasing the efficiency of TSCs is based on increasing the heat transfer from the perforated panels (the simplest thermal energy storage devices) to the air [6]. Using air solar collectors, the outlet temperatures can reach values of up to 65 °C, which is suitable for a variety of building applications [7–10].

Metal panels are efficient at absorbing and transferring heat, and thus, the first studies focused on panel designs (opaque or transparent, plate or corrugated) that would increase the heat transfer efficiency [10–14]. The problem is that metal panels are poor heat storage devices, and the discontinuity of solar thermal energy limits their long term effectiveness. Combining this with the limited availability of solar thermal energy, it becomes clear that broader research into other efficiency-augmenting methods is necessary.

Thus, efforts to increase TSC efficiency have shifted to increasing their thermal energy storage (TES) capabilities through passive means, such as the implementation of phase-change materials (PCM) in the TSC [15]. PCMs have the capacity to absorb substantial heat during the daytime as they transition from a solid to a liquid state and then release this stored heat during the night as they shift from a liquid to a solid state. Thus, besides PCMs increasing the TES capabilities of TSCs, their release of thermal energy during the night also addresses the problem of intermittent solar energy availability.

PCM elements have attracted considerable attention as an efficient method for thermal energy storage due to their high thermal storage capacity, desirable fluidity, and thermal properties, allowing for efficient management of peak load demands [15] when incorporated in building envelopes. PCMs, however, are not a perfect solution. Their most important and glaring fault is their low thermal conductivity, which leads to low efficiency in transferring the considerable amount of stored heat to the airflow entering the building [15].

In order to address specific limitations associated with PCMs, the most important being low thermal conductivity, researchers have explored the integration of nanoparticles in PCM, which further elevates the thermal performance of the building envelope [16].

Thus, the implementation of nanomaterials and phase-change materials plays a vital role in enhancing the efficiency of solar energy use in the building energy balance, providing real benefits for the user's finances and health. PCM elements have attracted considerable attention as an efficient method for thermal energy storage due to their high thermal storage capacity, desirable fluidity, and thermal properties [17].

This paper presents the current state of research related to the integration of different types of nanomaterial-enhanced PCMS in TSCs for the purpose of optimizing solar thermal energy storage and usage in buildings. This paper is based on a comprehensive examination of over 100 research studies focused on increasing the energy efficiency of transpired solar collectors. Most of these studies are indexed in the Web of Science database, with publication dates spanning from 2009 to 2023. A quick review of the literature highlights the recent interest in the subject, as reflected in the growing number of articles on solar collectors, particularly after 2017 (an increase that is mirrored in this paper's bibliography).

Worldwide interest in TSC technology is centred in Asia, with Europe ranking as the second-highest contributor by a wide margin (the two continents account for more than 80% of the references examined in this paper). This statistic suggests that the majority of the solutions currently under discussion are likely geared toward urban locations in Europe and Asia and their respective climates.

An evaluation of the keywords of each research or review paper, based on their frequency of occurrence, has been undertaken to guide the paper's focus. The keywords for all references used in the present paper were extracted and processed (overly general keywords such as "buildings" or "renewable energies" were removed) and presented in a word cloud in Figure 1.

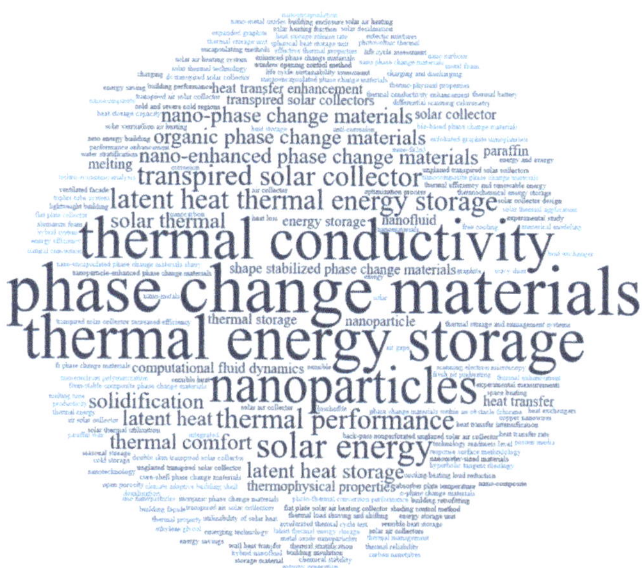

Figure 1. Keyword cloud for all the references considered in the present review article.

The results of the keyword analysis indicate that the most relevant research topics related to solar collectors concern (a) the use of PCMs as the primary means of energy storage and (b) the physics of thermal energy storage inside solar collectors through the study of different materials and techniques used to enhance materials' thermal conductivity, such as nanoparticle integration. Secondary areas of interest concern the economic viability of PCM-enhanced solar collectors or very niche methods and techniques for improving their performance.

This paper was structured in accordance with the overall objective of this work, which is to address the state of the art in TSCs using nanomaterial-enhanced PCMs (Figure 2).

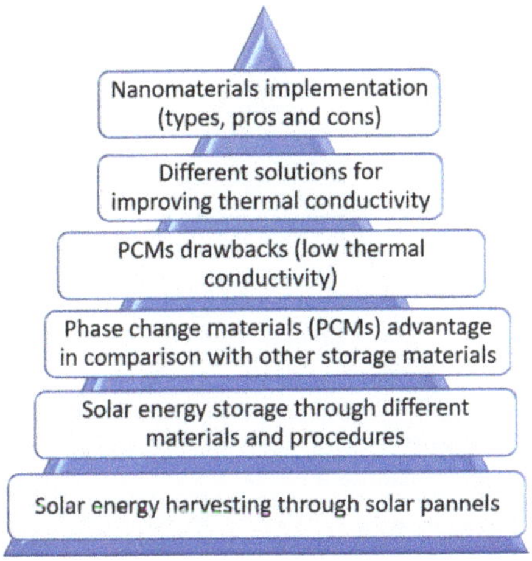

Figure 2. Paper's workflow.

The next sections will provide a qualitative description of common methods and techniques for PCM usage inside TSCs, as well as for the use of nanomaterials to enhance PCM performance. This will be followed by a quantitative evaluation of the aforementioned methods and techniques from the point of view of TES and heat transfer efficiency. Finally, the conclusion section of the present paper will summarize the most important findings and formulate recommendations for future research directions for the purpose of further enhancing TSC efficiency.

2. Thermal Energy Storage Systems Using Phase-Change Materials (PCMs)

The disparity between energy supply and consumption, along with the necessity to store excess energy that might otherwise go to waste, as well as the need to manage peak power demand, underscores the importance of thermal energy storage in various applications, including hot water, space heating, and air conditioning. Thermal energy storage systems facilitate more efficient utilization of fluctuating energy sources by aligning energy supply with demand. This has the potential to significantly reduce the reliance on fossil fuels, consequently mitigating the phenomena of Urban Heat Island and Urban Pollution Island and contributing to the development of sustainable urban areas.

There are several methods for storing thermal energy, with the main ones being sensible heat storage [18,19] and latent heat storage [20–23]. Sensible heat refers to the quantity of heat that is either added to or extracted from a material when its temperature changes. On the other hand, latent heat refers to the quantity of heat that is added to or removed from a substance during a phase change while maintaining a constant temperature. In the context of energy storage, latent heat storage represents a nearly isothermal process, offering significantly higher storage density and smaller temperature fluctuations compared to sensible heat storage systems. Furthermore, latent heat storage has the unique capability of storing the heat associated with a material's phase transition at a constant or nearly constant temperature, which corresponds to the phase transition temperature of the phase-change material.

While sensible heat storage is widely used in practical applications, latent heat storage, such as phase-change materials (PCMs), offers substantially higher energy storage density. PCMs exhibit minimal temperature variation during both charging and discharging processes, making them highly efficient for thermal energy storage [24–26].

An energy analysis for various collector types, along with the latest innovations to improve the heat collection efficiency of low-temperature solar collectors, has been provided in the study conducted by Gorjian S. et al. [27]. The methodologies discussed in this research cover a diverse set of strategies, including structural adjustments, absorber coatings, integration with reflectors, the use of alternative working fluids such as nanofluids, and the implementation of Thermal Energy Storage (TES) systems.

Instability and dependency on the outdoor climate are two of the main shortcomings when implementing RES. In order to eliminate or reduce the waste of surplus energy produced during periods with high-intensity solar radiation, the use of TES materials becomes essential [28]. In this way, the surplus energy can be stored for use during cloudy periods or during the night, thus increasing the solar collectors' overall efficiency, number of operating hours, and exterior air variation [9,29]. Moreover, incorporating PCMs into conventional building materials has shown significant promise for enhancing energy efficiency. Studies have indicated that such integration can result in a substantial 20% decrease in the cooling load for buildings, making it an interesting solution for hot climate countries, as Zahir, M.H. et al. noted in [30].

A comprehensive analysis of phase-change materials (PCMs) commonly employed in various applications was performed by Javadi, F.S. et al. in [31], revealing that paraffin waxes, salt hydrates, fatty acids, and eutectic organic/non-organic compounds are among the most frequently used PCMs. These PCMs are typically categorized into three primary groups: organic, inorganic (including salt hydrates and metallic compounds), and eutectic mixtures. Each group offers distinct advantages and disadvantages associated with their thermophysical properties.

Organic PCMs, for instance, undergo congruent melting without phase separation. Their non-corrosiveness and self-nucleation, as well as being chemically inert and thermally stable, make them desirable for certain applications like solar thermal energy applications, peak load shifting, or thermal management systems [32,33]. However, they tend to exhibit relatively low thermal conductivity, which can limit their overall efficiency. Notably, paraffin wax stands out as one of the most prevalent organic PCMs due to its narrow melting temperature range, spanning from -10 °C to 67 °C. This particular property enhances its suitability for a wide range of applications and technologies. Organic phase-change materials (PCMs) typically exhibit limited thermal conductivity, falling within the range of 0.1 to 0.4 W/mK. As an illustration, it can be considered n-octadecane, which displays a thermal conductivity of 0.35 W/mK in its solid state and 0.149 W/mK in its liquid state, a fact highlighted by Jegadheeswaran S. et al. [34].

Inorganic PCMs have gained attention for their exceptional capacity to maintain consistent thermophysical properties across numerous thermal cycles. Their specific feature is related to their increased thermal conductivity in comparison to organic PCMs, rendering them highly appealing for a wide array of applications. Furthermore, as Hinojosa J.F. et al. emphasize, inorganic PCMs boast an impressive volumetric latent heat density, typically around 350 MJ/m^3 [35]. This characteristic signifies their ability to efficiently store a substantial amount of energy. However, it is essential to recognize that inorganic PCMs are not without their challenges. One notable limitation is their corrosive nature, particularly in interactions with metals. This corrosiveness can give rise to concerns related to durability and compatibility, especially in specific applications, a problem yielded by Hua W. et al. [36]. Ushak S. et al. notice another potential issue is the occurrence of subcooling, where the PCM may fail to undergo a phase change at its designated melting point, instead remaining in a liquid state below that temperature [37]. Additionally, phase segregation, wherein components within the PCM may separate during phase transitions, can pose a significant concern [37].

In contrast, Singh P. et al. noted that eutectic PCMs present a distinct advantage by virtue of their well-defined melting and freezing points, guaranteeing a stable and predictable phase change without any component separation during transitions [38].

Cutting-edge technologies include PCMs derived from biomass, such as bio-based oils, which are being explored for their environmental sustainability. Research is underway for their potential use in TES systems, including solar thermal applications, like in the case of Jiang T. et al. [39].

Diverse melting temperatures must be tested and analyzed to cope with seasonal variability and different temperature requirements, or for applications such as drying (typically around 35 °C) or building heating (usually about 22 °C) [40].

The primary areas of focus for PCM optimization are improving thermal conductivity and heat transfer in PCM thermal energy storage (TES) systems. Various techniques, such as adding fins and increasing the thermal conductivity of PCM, have been proposed to enhance heat transfer.

However, implementing these techniques, according to Khan M.I. et al., may lead to increased system costs and added complexity [41]. Enhancing thermal conductivity can be accomplished through several methods, one of which is to incorporate PCM into high-conductivity porous materials. For example, one approach involves embedding the PCM within porous materials known for their exceptional thermal conductivity, such as metal foam and porous graphite [42–44]. Adding high-conductivity metal structures or particles is another method of increasing thermal conductivity. This technique entails introducing high-thermal conductivity metal structures or particles, such as silver, aluminium, copper, or other materials, into the PCM, as Shah, K.W. states [45]. Another method of increasing the thermal conductivity of PCM is by encapsulating it with high-conductivity shell materials for their superior thermal conductivity. According to Peng G. et al., shell materials are categorized into three distinct groups, namely organic, inorganic, and hybrid materials, combining both organic and inorganic components [46].

The thermal behavior of TSC equipped with PCM-based thermal storage was studied experimentally, and the results show that the latent heat thermal storage contributes to increased stability of the outlet air temperature of the collector and also a slight increase in the overall efficiency of the system, a fact suggested by Orzechowski T. et al. [47]. The use of PCM in solar collectors can increase the operating hours and thermal stability [3,48,49]. Recently, a team of researchers integrated thermal energy storage materials within the solar collector, and they used organic paraffin, a PCM with high latent heat suitable for building applications, obtaining a heat transfer efficiency of up to 38% and a higher number of operating hours of up to 800 min per 24 h compared to low-inertia TSC (the energy stored within the PCMs being released slowly during the night) [50]. Moreover, increasing the number of layers and introducing a separation plate influence the performance of the collector. The efficiency gradually increases when increasing the solar collector thickness until it reaches a value of 20 cm, according to Berville C. et al. [51].

Recent studies highlight the innovative integration of solar energy systems and thermal storage technologies to enhance the efficiency of domestic heating, food drying, and industrial applications, underscoring the potential for renewable energy solutions to address fuel poverty and contribute to sustainable development [52–60].

Several geometrical designs for PCM integration in TSC have lately been investigated. The incorporation of PCMs in spheres [22,51] is viewed as the optimum solution in terms of the unitary distribution of the flow and the size of the heat exchange surface. Furthermore, the TSC configuration can be refined with additional improvements by providing a mobile insulating blanket consisting of materials with very low heat conductivity, such as aerogel, as Calota R. suggests in a research grant report [61]. Throughout the day, during the cold season, the air is preheated as it passes over the transpired plate, which has a higher heat transfer rate due to the lobed perforations, and excess heat is stored in the highly efficient PCMs. During this operating mode, warm air is introduced inside the building. The dynamic insulation shifts to the exterior during the night, shielding the PCMs, which can now transfer heat to the interior via the conductive element. During the warm season, the process reverses, transferring heat out of the building, storing it in the PCM, and insulating as needed. The fresh, cool air may be used only at night and is bypassed during the day [61]. The process can be visualized in Figure 3 where one can observe the strategy for using Phase Change Materials (PCMs) arranged in a cascade, depicted in distinct colors (Figure 3a).

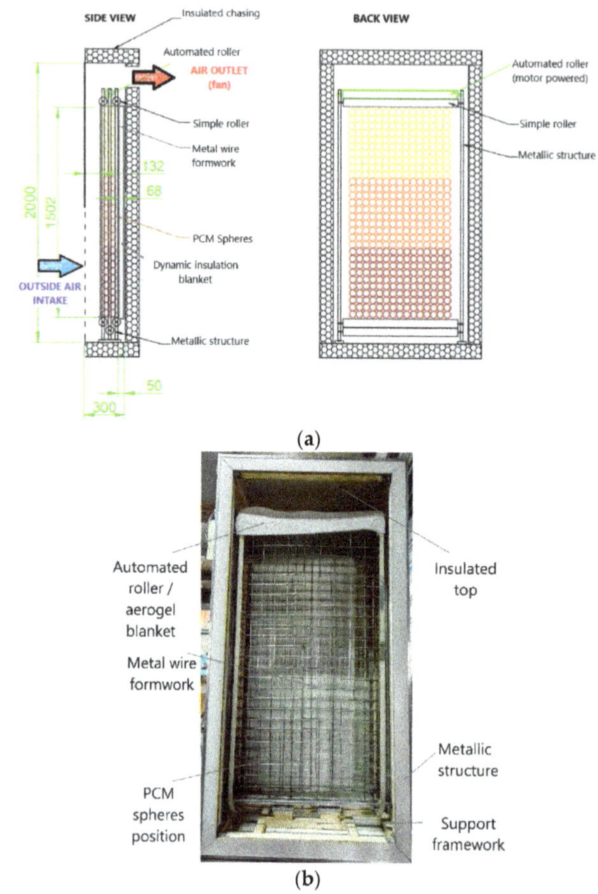

Figure 3. TSC with integrated PCMs and dynamic insulation [61]. (**a**) Side and back view; (**b**) real-scale experimental stand (with the front perforated plate removed).

Other research considered PCM incorporation in plates. According to Bejan et al. [9], after conducting experimental investigations, the performance of TSCs with lobed perforations can be substantially improved by adding 15 kg of organic PCM to the core structure. The results highlight that the maximum overall efficiency can be improved by nearly 6%, the maximum heating capacity by approximately 7.7%, and more than 17% supplementary energy was obtained during 9 h of operation.

A thermal energy storage unit made up of two plate coolers, each containing 40 kg of salt-hydrate-based PCM, was implemented in a TSC by Poole et al. [62]. The total storage capacity was 15 MJ.

Alternative geometries for PCM storage have been studied in order to identify the most energy-efficient structure. Abuska et al. [63] developed a novel solar air collector by combining the PCM with an aluminium honeycomb. The heat storage material having a mass of 26 kg was evaluated under the following conditions: PCM with honeycomb core in the first collector (type 1) and solely PCM in the second collector (type 2), as well as the third collector (type 3) with a flat absorber plate for comparison. The first one provided a usable energy output 70 min longer than the third one. Under daytime conditions, the thermal efficiencies were calculated to be 10.1%, 10.9%, and 13.6%, respectively.

Even though PCM integration in spheres, plates, and honeycomb structures is the most common, the PCMs can be embedded in complex materials' structures, such as gypsum plasterboard, like in the study performed by Bake M. [64]. The studies revealed that the construction envelope with PCM-gypsum plasterboard supplied an outlet temperature of 18 °C, which was 6 °C higher than the building envelope with simple gypsum plasterboard.

3. Integration of Nanomaterials in Thermal Energy Storage (TES) Systems

While the use of PCMs in solar collectors can increase the operating hours and their thermal stability, embedding nanoparticles in PCMs allows the improvement of thermal behavior related to melting and solidification rates as well as reducing the phase change period. Ma Z. points out that this is leading to increased benefits in the PCM's main area of weakness: their low thermal conductivity, especially in solid state [65].

3.1. Nanomaterials' Types

Existing nanomaterials used in TES have been grouped into three categories: nanometals, nanometal oxides, and nanocarbons.

Nanometals represent a class of materials wherein metallic particles are manipulated at the nanoscale, typically with dimensions below 100 nm. It is well known that metals have high thermal conductivities. This reduction in size imparts unique and enhanced properties to these materials compared to their bulk counterparts. Nanometals exhibit remarkable electrical, thermal, and optical characteristics, making them invaluable in various technological applications. They have a large surface area-to-volume ratio.

Silver exhibits the highest thermal conductivity among all metals, excelling in both heat and electricity conduction, with a thermal conductivity value of approximately 430 W/mK. Gold and copper are closely behind silver in terms of thermal conductivity, boasting values that are nearly equivalent. However, silver and gold are characterized by their high cost. In contrast, copper offers a more economical alternative and holds a competitive edge over silver and gold in various applications, as Leong K.Y. et al. notice in [66].

Nanometal oxides, composed of metal cations bonded with oxygen, are another fascinating class of nanomaterials. These materials exhibit diverse properties depending on their composition, size, and structure. They find extensive use in applications such as electronics, energy storage, environmental remediation, and catalysis. Nanometal oxides, like alumina and copper oxides, are good conductors of heat, with a thermal conductivity ranging from 30 to 40 W/mK. They are known for their exceptional surface area and reactivity, making them effective in various chemical processes.

Aluminium oxide nanoparticles (Al_2O_3) have been investigated for their potential to improve the thermal performance of phase-change materials (PCMs) used in low-temperature TES [66–68]. These nanoparticles can enhance heat transfer within the PCM, leading to more efficient energy storage and release.

Nano-titanium dioxide (TiO_2) is another metal oxide that has been studied for its applicability in low-temperature TES systems [68,69]. It can serve as a nucleating agent in PCMs, reducing subcooling effects and promoting more controlled phase transitions, thus improving overall TES efficiency. Nano-iron oxide nanoparticles (Fe_2O_3) have been explored as potential additives to enhance the thermal conductivity of PCMs used in low-temperature TES. Their presence can facilitate better heat transfer and temperature regulation within the storage medium, according to Chaichan M.T. et al. [70]. Nano-zinc oxide (ZnO) is known for its versatile properties and has been considered for low-temperature TES applications. Manoj Kumar P. et al. state that it can be integrated into PCMs to help manage temperature differentials and improve the overall thermal performance of the TES system [71]. Nano-copper oxide (CuO) particles have been investigated for their ability to enhance the thermal conductivity of PCMs in low-temperature TES applications. They can promote more efficient heat transfer and reduce temperature fluctuations [72]. Nano-cobalt oxide (Co_3O_4) is another example of a metal oxide that has shown promise in low-temperature TES systems. Its use as an additive in PCMs can contribute to better temperature control and improved energy storage capabilities.

Nanocarbons encompass a family of carbon-based nanomaterials, including carbon nanotubes (CNTs), graphene, and fullerenes. These materials have garnered immense attention due to their extraordinary electrical, mechanical, and thermal properties. Carbon nanotubes, for example, exhibit exceptional thermal conductivity, knowing that individual carbon nanotubes can range from 1000 to 3000 W/m·K or even higher and are successfully used in this matter [66]. Graphene, a single layer of carbon atoms arranged in a two-dimensional honeycomb lattice, boasts remarkable mechanical strength, electrical conductivity, and thermal properties and possesses extremely high thermal conductivity, with values exceeding 3000 W/mK. It has the potential to revolutionize various industries, including electronics and energy storage [68].

As a synthesis of the main advantages and disadvantages regarding each type of nanomaterial used in TES, nanometals have advantages such as high thermal conductivity, which leads to improved heat transfer within the storage medium, and increased heat capacity, allowing for more efficient thermal energy storage. The main disadvantages are related to agglomeration issues, which have an adverse effect on heat transfer qualities, as well as the increased cost. When compared to pure metals, nanometal oxides offer more stability and resistance to oxidation. Nevertheless, the thermal conductivity is lower, and the synthesis is more difficult, resulting in higher production costs. Nanocarbons, such as carbon nanotubes and graphene, have high heat conductivity and are lightweight, which might be useful in some applications, particularly if weight is an issue. Also, they show good chemical stability, reducing the possibility of reactions with the storage medium. Nonetheless, achieving uniform dispersion of nanocarbons in the storage medium can be challenging, affecting their performance.

The enhanced performance of nanomodified paraffins using carbon nanotubes was studied in [73,74]. These nanotubes significantly improved thermal conductivity, leading to increased efficiency in charge/discharge regimes. Moreover, the integration of carbon nanotubes enabled the development of robust thermal devices, demonstrating substantial heat flux through controlled thermal contact. Magnetically controlled heat storage materials, modified with carbon nanotubes, exhibited efficiency in practical applications like self-rescue devices.

The development of materials with thermo-optic switching properties, such as paraffin hosting carbon fillers, represents a significant advancement in the field of solar energy utilization. These materials exhibit the ability to alter their optical properties in response to temperature changes, making them valuable for applications in thermal energy storage [75,76]. The integration of nanomaterials, including carbon nanotubes and graphene, into paraffin enhances its thermal conductivity and allows for efficient absorption and release of thermal energy. Notably, nanometals, nanometal oxides, and nanocarbons bring unique characteristics to the composite material, such as enhanced thermal conductivity and adjustable properties, making them promising components for thermal energy storage systems.

The overview highlights the significant role of nanomaterials in overcoming challenges associated with traditional PCMs in solar collectors. The ability of nanomaterials to enhance thermal conductivity, regulate temperature differentials, and improve energy storage efficiency is a recurrent theme. The selection of specific nanomaterials, such as copper and graphene, reflects a consideration of both performance and cost-effectiveness.

Nanometals, nanometal oxides, and nanocarbons emerge as highly promising materials in TES applications. Their distinctive properties, including enhanced thermal conductivity and adjustable characteristics, position them as valuable components in the quest for efficient and effective thermal energy storage. This underlines the potential impact of nanomaterials on advancing the capabilities of solar collectors and expanding their applications in renewable energy systems.

3.2. Nanomaterials Integration

The present section delves into the diverse methodologies aimed at enhancing the heat transfer of phase-change materials (PCMs) in the context of thermal energy storage for buildings and renewable energies. Various techniques, ranging from macro and micro-encapsulation to the incorporation of metallic fins, materials with high thermal conductivity, and nanoparticles, are explored.

There are many methods that can be used in order to enhance the heat transfer of the PCM elements [77–79]: macro-encapsulation of the PCMs in steel cylinders or spheres, micro-encapsulation of the PCMs, the use of metallic fins to increase heat transfer surface, embedding materials with high thermal conductivity, and embedding nanoparticles with high thermal conductivity.

Nanoparticles represent a cutting-edge technological advancement aimed at enhancing heat transfer efficiency [80,81]. Nano-enhanced PCMs have been widely adopted in the fields of thermal energy storage for buildings and renewable energies [82,83]. Li [84] analyzed, via CFD simulations, the heat transfer inside a ventilation system for building heating based on paraffin RT30 integrated into sinusoidal encapsulations.

Alumina (Al_2O_3) nano-sized material was added to the paraffin with a fraction of 0.04 to enhance the conductivity of the latent heat storage. The most time-efficient method uses alumina nanopowders, with results showing a decrease in solidification time of 5.49%. The combination of Al_2O_3 nanoparticles with metallic fins in PCM storages [85,86] has shown even better results; indeed, solidification and melting time can be reduced by 12 and 6.4%, respectively. The findings from [70] indicate that the inclusion of nano-Fe_2O_3 at any mass fraction leads to a higher viscosity and density of the resulting mixture. The introduction of nano-Fe_2O_3 into paraffin wax results in a notable enhancement of thermal conductivity, with improvements of 10.04%, 57.14%, 76.19%, and 78.57% observed when mass fractions of 0.5%, 1%, 2%, and 3% are added, respectively. The outcomes of the study [71] demonstrated that the dispersion and distribution of zinc oxide nanoparticles within the paraffin did not disrupt its chemical structure. Additionally, the nanoparticles substantially improved the thermal stability of the paraffin and increased its thermal conductivity by up to 41.67% when incorporating 2% nano-ZnO particles. In [72], copper oxide (CuO) particles with a mean size of 40 nm were dispersed within the phase-change material (PCM) at three different weight percentages: 2%, 5%, and 10%. As a result of this dispersion, the thermal conductivity of the composite increased by 6%, 6.7%, and 7.8% in its liquid state. Simultaneously, the dynamic viscosity showed enhancements of 5%, 14%, and 30% as the mass fraction of CuO nanoparticles increased. During the solidification process, the heat transfer coefficient exhibited an increase of approximately 78% at the maximum flow rate [72].

Several types of materials have been investigated, but continuous research is needed. The integration of nanocarbon and nanococonut shells has been investigated by Sun Liu et al. [87]. Oleic acid and span 80 have been used to mitigate the agglomeration of nanoparticles during melting phases. The optimal concentration of nanoparticles was 0.02 wt%, and combined with oleic acid, a shorter melting time of 21% was reached compared to pure paraffin. Despite a more sustainable profile, the nanococonut shell was less effective than the nanocarbon. A challenging problem that arises in this domain is the design of more sustainable PCM and nanoparticles. Xie et al. [88] fixed one of these two issues using an organic PCM; nevertheless, they still used expanded graphite to improve the thermal characteristics of thermal energy storage. A new approach is therefore needed to develop more sustainable thermal energy storage.

Integrating nano-enhanced phase-changing materials (nePCMs) into solar walls offers a promising strategy [65]. Several studies conducted emphasize that using fins, metallic foams, or nanoparticles increases heat transfer efficiency and thermal backup time [3,48,89] in the case of integrating PCMs within solar collectors. There are several types of nanoparticles that can be used: metal, metal oxides, metal foams, carbon nanotubes, carbon nanoparticles, graphene, and graphite, and the current nanotechnology has led to the production of even more materials of various types at the nanoscale level [66]. Different types of micro/nano-PCMs for solar thermal applications are presented in research conducted by Qiu L. et al. in [90], and a comprehensive image indicating nanomaterials with high conductivity for increasing the thermal performance of latent heat thermal energy storage systems can be seen in the Jegadheeswaran S. study [91]. The main implications of using different types of nanomaterials are presented in Figure 4.

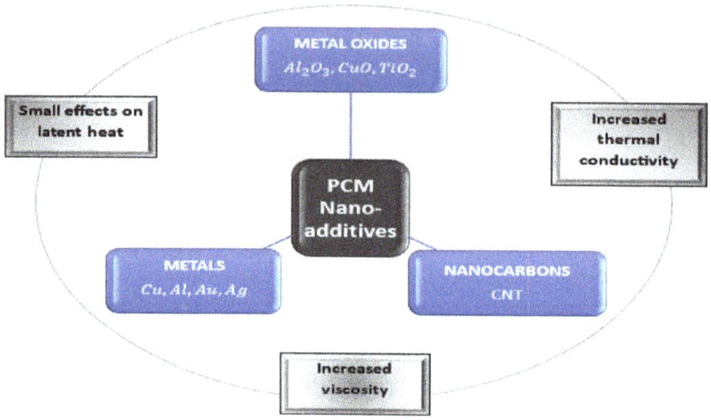

Figure 4. Nanomaterials with high conductivity for increasing the thermal performance of latent heat thermal energy storage systems.

In [92], a composite bio-phase-change material (PCM) was investigated by Yu S. et al. by incorporating carbon nanoparticles at varying mass fractions of 1%, 3%, and 5%. The findings revealed a substantial enhancement in the thermal conductivity of the composite PCM with the inclusion of carbon nanomaterials. Moreover, the composite PCM exhibited impressive latent heat properties, featuring suitable phase change temperatures and robust thermal durability. Notably, the highest increase in thermal conductivity, reaching 336%, was observed at a concentration of 5.0 wt%, while a substantial increase of 166% was recorded at 1.0 wt%.

The utilization of nanomaterials in solar walls holds the potential to significantly enhance their performance. The integration of diverse materials such as graphene, graphite, metal oxides, and carbon nanoparticles can pave the way for improved thermal conductivity. Research has shown that even small concentrations, like 3 wt% of graphene and graphite, can lead to a remarkable 101.2% increase in thermal conductivity [93], according to Dsilva Winfred Rufuss D. et al. [93].

According to the same author, the thermal conductivity of the nePCM (nano-enhanced phase-changing materials) is affected by particle concentration, dispersion, size, shape, and temperature. Moreover, using graphite nanomaterials with concentrations between 0 and 10 wt%, the thermal conductivity of PCMs can be improved by about 540 to 1000%, but the incorporation of metal foams may also be a promising technique [94]. According to Mumtaz et al. [3], nanoparticles were first used in solar applications in 2009 and can lead to an overall efficiency increase. In a study conducted by Saw et al. [95], by using nano-enhanced PCMs, a flat plate collector's efficiency was improved by 8.4%. Moreover, adding 1 wt% and 2 wt% copper nanoparticles in organic PCM improved the thermal conductivity of a pure PCM by up to 24% and 46.3%, respectively [96].

Despite advancements in developing new PCMs and improving their characteristics for use in solar walls, some limitations persist. The incorporation of nanoparticles into PCMs has shown promise in addressing these challenges. According to Kasaeian et al. [48], by embedding nanoparticles in PCM, the melting and solidification rate can improve, and the phase change time period can be reduced. As can be observed from the previous state-of-the-art survey, researchers have been concentrating on investigations mainly concerned with developing new PCMs or improving their characteristics and implementation into building components [48]. These advancements, however, have not overcome some major limitations, including the reduced potential for solidification and melting in order to discharge the accumulated heat indoors.

Nano-structuring the PCM could provide better thermal conductivity. However, other factors related to the geometry, quantity, and size of the implemented materials could have an impact on the properties of the mixture. Thus, the thermal characteristics of nePCMs should be measured, factoring in different geometries and loading ratios. By implementing various techniques, the nePCMs' thermal properties can be controlled [97,98]. However, the non-homogeneous distribution due to high concentrations of the nanoparticles in the PCMs could reduce the melting performance, as Amidu, M.A. et al. observe, and the thermal performance overall [99]. The literature data suggest that the use of PCMs and nePCMs in buildings could increase energy storage efficiency and building performance. A reduction in temperature fluctuations and internal thermal stability can also be achieved with the implementation of these materials.

On the other hand, active walls with PCM and nePCM are inefficient in very hot climates where a higher temperature gradient is required and when the differences at high temperatures are significant. Therefore, additional solutions are required.

Several studies emphasize the application of nanoparticles in solar collectors, indicating that the incorporation of PCMs within these systems can lead to increased heat transfer efficiency and extended thermal backup time. Table 1 presents a brief synthesis of the results collected from various studies about the impact of adding nanoparticles to PCM.

Table 1. Results regarding nano-enhanced phase-changing material (nePCM) properties.

Author	Investigation	Results
Keshteli et al. [85], Chen et al. [86]	Combination of Al_2O_3 nanoparticles with metallic fins in PCM storage	Solidification and melting time can be reduced by 12 and 6.4%, respectively.
Sun Liu et al. [87]	Integration of nanocarbon and nanococonut shell	Shorter melting time of 21% was reached compared to pure paraffin. The nanococonut shell was less effective than the nanocarbon.
Li W et al. [78]	Metal foam or nanoparticle integration in PCM	Increased heat transfer efficiency and thermal backup time.
Jegadheeswaran S. [91]	Nanomaterials integration in PCMs	Enhanced thermal performance of latent heat thermal energy storage systems.
Yu S. et al. [92]	Incorporate carbon nanoparticles at varying mass fractions of 1%, 3%, and 5% in PCMs	The highest increase in thermal conductivity, 336%, was observed at a concentration of 5.0 wt%, while a substantial boost of 166% was recorded at 1.0 wt%.
Dsilva Winfred Rufuss D. et al. [93]	Integration of diverse materials such as graphene, graphite, metal oxides, and carbon nanoparticles in solar walls	At concentrations of 3 wt% of graphene and graphite, a 101.2% increase in thermal conductivity was identified.
Saw et al. [95]	Use of nano-enhanced PCMs in a flat plate collector	Efficiency was improved with 8.4%
Al-Kayiem, H.H. [96]	Addition of 1 wt% and 2 wt% copper nanoparticles in organic PCM	The thermal conductivity of the pure PCM improved by up to 24%
Amidu M.A. et al. [99]	High concentrations of the nanoparticles in the PCMs	Melting performance and the thermal performance reduction

Although progress has been achieved, challenges emerge, such as the non-homogeneous distribution in high-concentration nanoparticle integration. The importance of a detailed understanding of the thermal properties of nePCMs should be investigated further, taking into account varied geometries and loading ratios.

4. Conclusions

High thermal energy density materials, such as PCMs (phase-change materials), have been widely used as storage for important quantities of energy, finding an application in the field of harvesting solar thermal energy. A PCM thermal buffer can fill the gap between solar energy availability and demand. Experimental research shows that, for transpired solar collector (TSC) systems, PCMs can increase outlet air temperature stability and overall efficiency.

This article offers both quantitative and qualitative evaluations of common methods and techniques for PCM usage inside TSCs. It not only addresses the efficiency of these methods in terms of thermal energy storage (TES) and heat transfer but also discusses the implications of incorporating nanomaterials into PCMs, providing a critical overview of the current state of research in this area.

A shortcoming of the PCMs is their low thermal conductivity, which prevents adequate heat transfer intensity. A common approach to overcoming this disadvantage is the inclusion of nanoparticles within the PCM.

The review clearly demonstrates the significant advancements in thermal energy storage (TES) and heat transfer efficiency within transpired solar collectors (TSC) through the integration of phase-change materials (PCMs) with nanomaterials. The incorporation of nanomaterials such as graphene, graphite, metal oxides, and carbon nanoparticles leads to a substantial improvement in the thermal conductivity of PCMs, thereby enhancing the overall efficiency of solar thermal systems. Researchers embedded Al_2O_3 nano-sized material in paraffin wax, decreasing solidification time by 5.49%. Other materials that increase thermal conductivity are nanocarbons, nanococonut shells, nanotubes, graphene, and graphite. These improvements determine the emergence of a new range of PCMs called nano-enhanced phase-change materials.

Several factors impact the efficiency of the use of nanoparticles: concentration [wt%], dispersion, size, shape, and temperature. Using graphite nanoparticles, the thermal conductivity can be improved by up to 1000%, depending on the concentration. In the case of copper nanoparticles, an improvement in conductivity between 24% and 46.3% was reported when incorporating the nanoparticles within a concentration range of 1–2 wt%. It was also reported that the efficiency of flat plate collectors was increased by 8.4% by incorporating nano-enhanced PCMs.

While plenty of research is dedicated to the improvement of PCMs and their thermal characteristics, as well as the development of novel PCMs, the limited potential for thermal charging and discharging, which is heavily linked to climatic conditions, should also be one of the main research topics. In the literature, concepts of dynamic systems that adapt the position of the PCM based on demand for building wall systems have been proposed.

Nanometals, nanometal oxides, and nanocarbons have emerged as exceptionally promising materials within thermal energy storage (TES) applications. Their distinctive attributes, marked by heightened thermal conductivity and the capacity for controlled adjustments, render them valuable constituents in TES systems. However, it is important to note that ongoing research and development efforts are aimed at fully unleashing their true potential. Scientists and engineers continue to explore innovative ways to harness the exceptional properties of these nanomaterials to optimize TES solutions, improve energy efficiency, and contribute to sustainable energy practices. As the field of nanotechnology advances, we can anticipate even more groundbreaking discoveries that will further elevate the role of nanometals, nanometal oxides, and nanocarbons in the realm of thermal energy storage.

The utilization of nanomaterials in TSCs (transpired solar collectors) holds the potential to significantly enhance their performance. The integration of diverse materials such as graphene, graphite, metal oxides, and carbon nanoparticles can pave the way for improved thermal conductivity. Integrating nano-enhanced Phase-Changing Materials (nePCMs) into solar walls offers a promising strategy.

Various innovative techniques for improving thermal conductivity and heat transfer in PCM-based TES systems were presented as well. Among these, the utilization of fins, high-conductivity porous materials, and the embedding of PCM in materials with high thermal conductivity stand out. These methods significantly contribute to the enhancement of heat transfer, addressing one of the critical challenges in the practical application of PCMs in TSCs.

A comprehensive analysis of different types of nanomaterials used in TES, including nanometals, nanometal oxides, and nanocarbons, is provided. This review sheds light on how these nanomaterials improve the thermal properties of PCMs, offering a detailed examination of their impact on the efficiency of solar energy storage and conversion processes.

PCMs have the capacity to contribute to increasing the overall energy efficiency of buildings that employ RESs (renewable energy sources). The development of novel PCMs with enhanced thermal properties to increase heat transfer intensity is necessary to further the knowledge in the field and facilitate practical implementation. The challenges regarding the direction of the following scientific research are related to the finding of techniques to counteract the particle agglomeration effect in the mixture with phase-change materials, followed by the problems of inhomogeneous distribution that can appear after multiple cycles of melting and solidification.

Author Contributions: Conceptualization, F.B. and C.C.; methodology, F.B. and R.C.; formal analysis, F.B. and C.B.; investigation, R.C.; writing—original draft preparation, F.B. and R.C.; writing—review and editing, F.B., M.G. and R.C.; visualization, M.G. and C.B.; supervision, F.B.; project administration, R.C. All authors have read and agreed to the published version of the manuscript.

Funding: This work was supported by the following grants of the Executive Agency for Higher Education, Research, Development, and Innovation Funding (UEFISCDI), project number PN-III-P2-2.1-PED-2021-1903, within PNCDI III; by a grant of the Ministry of Research, Innovation, and Digitization, CCCDI-UEFISCDI project number PN-III-P1-1.1-PD-2021-0342, within PNCDI III; and by a grant of the Ministry of Research, Innovation, and Digitization, CCCDI-UEFISCDI, project number PN-III-P2-2.1-PED-2021-4137, within PNCDI III CT 714PED/2022.

Data Availability Statement: No new data were created or analyzed in this study. Data sharing is not applicable to this article.

Conflicts of Interest: The authors declare no conflict of interest.

References

1. IEA. *The Energy Efficiency Policy Package: Key Catalyst for Building Decarbonisation and Climate Action*; IEA: Paris, France, 2023.
2. Rabani, M.; Bayera Madessa, H.; Nord, N. Achieving zero-energy building performance with thermal and visual comfort enhancement through optimization of fenestration, envelope, shading device, and energy supply system. *Sustain. Energy Technol. Assess.* **2021**, *44*, 101020. [CrossRef]
3. Khan, M.M.A.; Ibrahim, N.I.; Mahbubul, I.M.; Ali, H.M.; Saidur, R.; Al-Sulaiman, F.A. Evaluation of solar collector designs with integrated latent heat thermal energy storage: A review. *Sol. Energy* **2018**, *166*, 334–350. [CrossRef]
4. Reichl, C.; Kramer, K.; Thoma, C.; Benovsky, P.; Lemée, T. Comparison of modelled heat transfer and fluid dynamics of a flat plate solar air heating collector towards experimental data. *Sol. Energy* **2015**, *120*, 450–463. [CrossRef]
5. Molineaux, B.; Lachal, B.; Guisan, O. Thermal analysis of five outdoor swimming pools heated by unglazed solar collectors. *Sol. Energy* **1994**, *53*, 21–26. [CrossRef]
6. Zheng, W.; Zhang, H.; You, S.; Fu, Y. Experimental Investigation of the Transpired Solar Air Collectors and Metal Corrugated Packing Solar Air Collectors. *Energies* **2017**, *10*, 302. [CrossRef]
7. Leon, M.A.; Kumar, S. Mathematical modeling and thermal performance analysis of unglazed transpired solar collectors. *Sol. Energy* **2007**, *81*, 62–75. [CrossRef]
8. Goyal, R.K.; Tiwari, G.N.; Garg, H.P. Effect of thermal storage on the performance of an air collector: A periodic analysis. *Energy Convers. Manag.* **1998**, *39*, 193–202. [CrossRef]
9. Bejan, A.S.; Teodosiu, C.; Croitoru, C.V.; Catalina, T.; Nastase, I. Experimental investigation of transpired solar collectors with/without phase change materials. *Sol. Energy* **2021**, *214*, 478–490. [CrossRef]

10. Brown, C.; Perisoglou, E.; Hall, R.; Stevenson, V. Transpired Solar Collector Installations in Wales and England. *Energy Procedia* **2014**, *48*, 18–27. [CrossRef]
11. Bejan, A.-S.; Labihi, A.; Croitoru, C.; Catalina, T. Air solar collectors in building use—A review. *E3S Web Conf.* **2018**, *32*, 01003. [CrossRef]
12. Wang, X.; Lei, B.; Bi, H.; Yu, T. A simplified method for evaluating thermal performance of unglazed transpired solar collectors under steady state. *Appl. Therm. Eng.* **2017**, *117*, 185–192. [CrossRef]
13. Paya-Marin, M.A. Chapter 5—Solar Air Collectors for Cost-Effective Energy-Efficient Retrofitting. In *Cost-Effective Energy Efficient Building Retrofitting*; Pacheco-Torgal, F., Granqvist, C.-G., Jelle, B.P., Vanoli, G.P., Bianco, N., Kurnitski, J., Eds.; Woodhead Publishing: Cambridge, UK, 2017; pp. 141–168. [CrossRef]
14. Croitoru, C.V.; Nastase, I.; Bode, F.I.; Meslem, A. Thermodynamic investigation on an innovative unglazed transpired solar collector. *Sol. Energy* **2016**, *131*, 21–29. [CrossRef]
15. Singh, R.; Sadeghi, S.; Shabani, B. Thermal conductivity enhancement of phase change materials for low-temperature thermal energy storage applications. *Energies* **2018**, *12*, 75. [CrossRef]
16. Nematpour Keshteli, A.; Sheikholeslami, M. Nanoparticle enhanced PCM applications for intensification of thermal performance in building: A review. *J. Mol. Liq.* **2019**, *274*, 516–533. [CrossRef]
17. Lee, K.O.; Medina, M.A.; Sun, X.; Jin, X. Thermal performance of phase change materials (PCM)-enhanced cellulose insulation in passive solar residential building walls. *Sol. Energy* **2018**, *163*, 113–121. [CrossRef]
18. Lugolole, R.; Mawire, A.; Lentswe, K.A.; Okello, D.; Nyeinga, K. Thermal performance comparison of three sensible heat thermal energy storage systems during charging cycles. *Sustain. Energy Technol. Assess.* **2018**, *30*, 37–51. [CrossRef]
19. Li, G. Sensible heat thermal storage energy and exergy performance evaluations. *Renew. Sustain. Energy Rev.* **2016**, *53*, 897–923. [CrossRef]
20. Xu, J.; Wang, R.Z.; Li, Y. A review of available technologies for seasonal thermal energy storage. *Sol. Energy* **2014**, *103*, 610–638. [CrossRef]
21. Liu, W.; Bie, Y.; Xu, T.; Cichon, A.; Królczyk, G.; Li, Z. Heat transfer enhancement of latent heat thermal energy storage in solar heating system: A state-of-the-art review. *J. Energy Storage* **2022**, *46*, 103727. [CrossRef]
22. Berville, C.; Bode, F.; Croitoru, C.; Calota, R.; Nastase, I. Enhancing solar façade thermal performance with PCM spheres: A CFD investigation. *J. Build. Phys.* **2023**. [CrossRef]
23. Bejan, A.; Croitoru, C.; Sandu, M.; Nastase, I.; Bode, F. Solar ventilated façade with PCM integration for air preheating. In Proceedings of the Indoor Air Conference 2018, Philadelphia, PA, USA, 22–27 July 2018.
24. Rashid, F.L.; Al-Obaidi, M.A.; Dulaimi, A.; Mahmood, D.M.N.; Sopian, K. A Review of Recent Improvements, Developments, and Effects of Using Phase-Change Materials in Buildings to Store Thermal Energy. *Designs* **2023**, *7*, 90. [CrossRef]
25. Hayatina, I.; Auckaili, A.; Farid, M. Review on the Life Cycle Assessment of Thermal Energy Storage Used in Building Applications. *Energies* **2023**, *16*, 1170. [CrossRef]
26. Zhan, H.; Mahyuddin, N.; Sulaiman, R.; Khayatian, F. Phase change material (PCM) integrations into buildings in hot climates with simulation access for energy performance and thermal comfort: A review. *Constr. Build. Mater.* **2023**, *397*, 132312. [CrossRef]
27. Gorjian, S.; Ebadi, H.; Calise, F.; Shukla, A.; Ingrao, C. A review on recent advancements in performance enhancement techniques for low-temperature solar collectors. *Energy Convers. Manag.* **2020**, *222*, 113246. [CrossRef]
28. Arkar, C.; Šuklje, T.; Vidrih, B.; Medved, S. Performance analysis of a solar air heating system with latent heat storage in a lightweight building. *Appl. Therm. Eng.* **2016**, *95*, 281–287. [CrossRef]
29. Pop, O.; Berville, C.; Bode, F.; Croitoru, C. Numerical investigation of cascaded phase change materials use in transpired solar collectors. *Energy Rep.* **2022**, *8*, 184–193. [CrossRef]
30. Zahir, M.H.; Irshad, K.; Shafiullah, M.; Ibrahim, N.I.; Islam, A.K.; Mohaisen, K.O.; Sulaiman, F.A.A. Challenges of the application of PCMs to achieve zero energy buildings under hot weather conditions: A review. *J. Energy Storage* **2023**, *64*, 107156. [CrossRef]
31. Javadi, F.S.; Metselaar, H.S.C.; Ganesan, P. Performance improvement of solar thermal systems integrated with phase change materials (PCM), a review. *Sol. Energy* **2020**, *206*, 330–352. [CrossRef]
32. Milián, Y.E.; Gutiérrez, A.; Grágeda, M.; Ushak, S. A review on encapsulation techniques for inorganic phase change materials and the influence on their thermophysical properties. *Renew. Sustain. Energy Rev.* **2017**, *73*, 983–999. [CrossRef]
33. Zeng, J.-L.; Zhu, F.-R.; Yu, S.-B.; Zhu, L.; Cao, Z.; Sun, L.-X.; Deng, G.-R.; Yan, W.-P.; Zhang, L. Effects of copper nanowires on the properties of an organic phase change material. *Sol. Energy Mater. Sol. Cells* **2012**, *105*, 174–178. [CrossRef]
34. Jegadheeswaran, S.; Pohekar, S.D. Performance enhancement in latent heat thermal storage system: A review. *Renew. Sustain. Energy Rev.* **2009**, *13*, 2225–2244. [CrossRef]
35. Hinojosa, J.F.; Moreno, S.F.; Maytorena, V.M. Low-Temperature Applications of Phase Change Materials for Energy Storage: A Descriptive Review. *Energies* **2023**, *16*, 3078. [CrossRef]
36. Hua, W.; Xu, X.; Zhang, X.; Yan, H.; Zhang, J. Progress in corrosion and anti-corrosion measures of phase change materials in thermal storage and management systems. *J. Energy Storage* **2022**, *56*, 105883. [CrossRef]
37. Ushak, S.; Gutierrez, A.; Galleguillos, H.; Fernandez, A.G.; Cabeza, L.F.; Grágeda, M. Thermophysical characterization of a by-product from the non-metallic industry as inorganic PCM. *Sol. Energy Mater. Sol. Cells* **2015**, *132*, 385–391. [CrossRef]

38. Singh, P.; Sharma, R.K.; Ansu, A.K.; Goyal, R.; Sarı, A.; Tyagi, V.V. A comprehensive review on development of eutectic organic phase change materials and their composites for low and medium range thermal energy storage applications. *Sol. Energy Mater. Sol. Cells* **2021**, *223*, 110955. [CrossRef]
39. Jiang, T.; Zhang, Y.; Olayiwola, S.; Lau, C.; Fan, M.; Ng, K.; Tan, G. Biomass-derived porous carbons support in phase change materials for building energy efficiency: A review. *Mater. Today Energy* **2022**, *23*, 100905. [CrossRef]
40. Sharma, R.K.; Ganesan, P.; Tyagi, V.V. Long-term thermal and chemical reliability study of different organic phase change materials for thermal energy storage applications. *J. Therm. Anal. Calorim.* **2016**, *124*, 1357–1366. [CrossRef]
41. Khan, M.I.; Asfand, F.; Al-Ghamdi, S.G. Progress in research and development of phase change materials for thermal energy storage in concentrated solar power. *Appl. Therm. Eng.* **2022**, *219*, 119546. [CrossRef]
42. Oró, E.; de Gracia, A.; Castell, A.; Farid, M.M.; Cabeza, L.F. Review on phase change materials (PCMs) for cold thermal energy storage applications. *Appl. Energy* **2012**, *99*, 513–533. [CrossRef]
43. Mills, A.; Farid, M.; Selman, J.R.; Al-Hallaj, S. Thermal conductivity enhancement of phase change materials using a graphite matrix. *Appl. Therm. Eng.* **2006**, *26*, 1652–1661. [CrossRef]
44. Baumeister, J.; Weise, J.; Myslicki, S.; Kieseritzky, E.; Lindenberg, G. PCM-Based Energy Storage System with High Power Output Using Open Porous Aluminum Foams. *Energies* **2020**, *13*, 6198. [CrossRef]
45. Shah, K.W. A review on enhancement of phase change materials—A nanomaterials perspective. *Energy Build.* **2018**, *175*, 57–68. [CrossRef]
46. Peng, G.; Dou, G.; Hu, Y.; Sun, Y.; Chen, Z. Phase Change Material (PCM) Microcapsules for Thermal Energy Storage. *Adv. Polym. Technol.* **2020**, *2020*, 9490873. [CrossRef]
47. Orzechowski, T.; Stokowiec, K. Quasi-stationary phase change heat transfer on a fin. *EPJ Web Conf.* **2016**, *114*, 02086. [CrossRef]
48. Kasaeian, A.; Bahrami, L.; Pourfayaz, F.; Khodabandeh, E.; Yan, W.-M. Experimental studies on the applications of PCMs and nano-PCMs in buildings: A critical review. *Energy Build.* **2017**, *154*, 96–112. [CrossRef]
49. Bejan, A.-S.; Croitoru, C.; Bode, F.; Teodosiu, C.; Catalina, T. Experimental investigation of an enhanced transpired air solar collector with embodied phase changing materials. *J. Clean. Prod.* **2022**, *336*, 130398. [CrossRef]
50. Bejan, A.-S.; Croitoru, C.V.; Bode, F. Preliminary numerical studies conducted for the numerical model of a real transpired solar collector with integrated phase changing materials. *E3S Web Conf.* **2019**, *111*, 03047. [CrossRef]
51. Berville, C.; Bode, F.; Croitoru, C. Numerical Simulation Investigation of a Double Skin Transpired Solar Air Collector. *Appl. Sci.* **2022**, *12*, 520. [CrossRef]
52. Wang, F.; Manzanares-Bennett, A.; Tucker, J.; Roaf, S.; Heath, N. A feasibility study on solar-wall systems for domestic heating—An affordable solution for fuel poverty. *Sol. Energy* **2012**, *86*, 2405–2415. [CrossRef]
53. Lakshmi, D.; Muthukumar, P.; Nayak, P.K. Experimental investigations on active solar dryers integrated with thermal storage for drying of black pepper. *Renew. Energy* **2021**, *167*, 728–739. [CrossRef]
54. Srinivasan, G.; Rabha, D.; Muthukumar, P. A review on solar dryers integrated with thermal energy storage units for drying agricultural and food products. *Sol. Energy* **2021**, *229*, 22–38. [CrossRef]
55. Mugi, V.R.; Das, P.; Balijepalli, R.; Chandramohan, V.P. A review of natural energy storage materials used in solar dryers for food drying applications. *J. Energy Storage* **2022**, *49*, 104198. [CrossRef]
56. Koçak, B.; Fernandez, A.I.; Paksoy, H. Review on sensible thermal energy storage for industrial solar applications and sustainability aspects. *Sol. Energy* **2020**, *209*, 135–169. [CrossRef]
57. Patel, A.K. Technological Innovations in Solar Heater Materials and Manufacturing. *United Int. J. Res. Technol. (UIJRT)* **2023**, *4*, 13–24.
58. Dahash, A.; Ochs, F.; Janetti, M.B.; Streicher, W. Advances in seasonal thermal energy storage for solar district heating applications: A critical review on large-scale hot-water tank and pit thermal energy storage systems. *Appl. Energy* **2019**, *239*, 296–315. [CrossRef]
59. Mitali, J.; Dhinakaran, S.; Mohamad, A. Energy storage systems: A review. *Energy Storage Sav.* **2022**, *1*, 166–216. [CrossRef]
60. Guney, M.S.; Tepe, Y. Classification and assessment of energy storage systems. *Renew. Sustain. Energy Rev.* **2017**, *75*, 1187–1197. [CrossRef]
61. Calota, R. *Research Grant: Adaptive Air Solar Collector with Integrated Nano-Enhanced Phase Changing Materials*; UEFISCDI: Bucharest, Romania, 2021.
62. Poole, M.R.; Shah, S.B.; Boyette, M.D.; Stikeleather, L.F.; Cleveland, T. Performance of a coupled transpired solar collector—Phase change material-based thermal energy storage system. *Energy Build.* **2018**, *161*, 72–79. [CrossRef]
63. Abuşka, M.; Şevik, S.; Kayapunar, A. Experimental analysis of solar air collector with PCM-honeycomb combination under the natural convection. *Sol. Energy Mater. Sol. Cells* **2019**, *195*, 299–308. [CrossRef]
64. Bake, M. Development of PCM-gypsum Plasterboard Integrated Transpired Solar Collector for Building Envelopes. Ph.D. Thesis, Coventry University, Coventry, UK, 2021.
65. Ma, Z.; Lin, W.; Sohel, M.I. Nano-enhanced phase change materials for improved building performance. *Renew. Sustain. Energy Rev.* **2016**, *58*, 1256–1268. [CrossRef]
66. Leong, K.Y.; Abdul Rahman, M.R.; Gurunathan, B.A. Nano-enhanced phase change materials: A review of thermo-physical properties, applications and challenges. *J. Energy Storage* **2019**, *21*, 18–31. [CrossRef]
67. Singh, S.K.; Verma, S.K.; Kumar, R. Thermal performance and behavior analysis of SiO_2, Al_2O_3 and MgO based nano-enhanced phase-changing materials, latent heat thermal energy storage system. *J. Energy Storage* **2022**, *48*, 103977. [CrossRef]

68. Barthwal, M.; Dhar, A.; Powar, S. Effect of Nanomaterial Inclusion in Phase Change Materials for Improving the Thermal Performance of Heat Storage: A Review. *ACS Appl. Energy Mater.* **2021**, *4*, 7462–7480. [CrossRef]
69. Chen, H.; Zhao, R.; Wang, C.; Feng, L.; Li, S.; Gong, Y. Preparation and Characterization of Microencapsulated Phase Change Materials for Solar Heat Collection. *Energies* **2022**, *15*, 5354. [CrossRef]
70. Chaichan, M.T.; Mahdi, M.T.; Kazem, H.A.; Al-Waeli, A.H.A.; Fayad, M.A.; Al-Amiery, A.A.; Isahak, W.N.R.W.; Kadhum, A.A.H.; Takriff, M.S. Modified Nano-Fe_2O_3-Paraffin Wax for Efficient Photovoltaic/Thermal System in Severe Weather Conditions. *Sustainability* **2022**, *14*, 12015. [CrossRef]
71. Manoj Kumar, P.; Mylsamy, K.; Prakash, K.B.; Nithish, M.; Anandkumar, R. Investigating thermal properties of Nanoparticle Dispersed Paraffin (NDP) as phase change material for thermal energy storage. *Mater. Today Proc.* **2021**, *45*, 745–750. [CrossRef]
72. Jesumathy, S.; Udayakumar, M.; Suresh, S. Experimental study of enhanced heat transfer by addition of CuO nanoparticle. *Heat Mass Transf.* **2012**, *48*, 965–978. [CrossRef]
73. Karpus, N.; Kovalenko, V.; Kotok, V.; Shchegolkov, A.; Schegolkov, A. Investigation of charge and discharge regimes of nanomodified heat-accumulating materials. *East.-Eur. J. Enterp. Technol.* **2017**, *3*, 23–29.
74. Shchegolkov, A.; Shchegolkov, A.; Demidova, A. The use of nanomodified heat storage materials for thermal stabilization in the engineering and aerospace industry as a solution for economy. *MATEC Web Conf.* **2018**, *224*, 03012. [CrossRef]
75. Said, A.; Salah, A.; Fattah, G.A. Thermo-optic switching properties of paraffin-wax hosting carbon fillers. *J. Energy Storage* **2018**, *19*, 260–271. [CrossRef]
76. Said, A.; Salah, A.; Abdel Fattah, G. Enhanced thermo-optical switching of paraffin-wax composite spots under laser heating. *Materials* **2017**, *10*, 525. [CrossRef]
77. Zhou, D.; Zhao, C.Y.; Tian, Y. Review on thermal energy storage with phase change materials (PCMs) in building applications. *Appl. Energy* **2012**, *92*, 593–605. [CrossRef]
78. Waqas, A.; Ud Din, Z. Phase change material (PCM) storage for free cooling of buildings—A review. *Renew. Sustain. Energy Rev.* **2013**, *18*, 607–625. [CrossRef]
79. Memon, S.A. Phase change materials integrated in building walls: A state of the art review. *Renew. Sustain. Energy Rev.* **2014**, *31*, 870–906. [CrossRef]
80. Nazir, U.; Nawaz, M.; Alharbi, S.O. Thermal performance of magnetohydrodynamic complex fluid using nano and hybrid nanoparticles. *Phys. A Stat. Mech. Its Appl.* **2020**, *553*, 124345. [CrossRef]
81. Algarni, S.; Mellouli, S.; Alqahtani, T.; Almutairi, K.; Anqi, A. Experimental investigation of an evacuated tube solar collector incorporating nano-enhanced PCM as a thermal booster. *Appl. Therm. Eng.* **2020**, *180*, 115831. [CrossRef]
82. Rathore, P.K.S.; Shukla, S.K. Enhanced thermophysical properties of organic PCM through shape stabilization for thermal energy storage in buildings: A state of the art review. *Energy Build.* **2021**, *236*, 110799. [CrossRef]
83. Yang, L.; Huang, J.-n.; Zhou, F. Thermophysical properties and applications of nano-enhanced PCMs: An update review. *Energy Convers. Manag.* **2020**, *214*, 112876. [CrossRef]
84. Li, Z. Heat transfer during solidification of PCM layers with inclusion of nano-powders. *Int. Commun. Heat Mass Transf.* **2021**, *127*, 105518. [CrossRef]
85. Keshteli, A.N.; Sheikholeslami, M. Influence of Al2O3 nanoparticle and Y-shaped fins on melting and solidification of paraffin. *J. Mol. Liq.* **2020**, *314*, 113798. [CrossRef]
86. Chen, S.-B.; Saleem, S.; Alghamdi, M.N.; Nisar, K.S.; Arsalanloo, A.; Issakhov, A.; Xia, W.-F. Combined effect of using porous media and nano-particle on melting performance of PCM filled enclosure with triangular double fins. *Case Stud. Therm. Eng.* **2021**, *25*, 100939. [CrossRef]
87. Sun, X.; Liu, L.; Mo, Y.; Li, J.; Li, C. Enhanced thermal energy storage of a paraffin-based phase change material (PCM) using nano carbons. *Appl. Therm. Eng.* **2020**, *181*, 115992. [CrossRef]
88. Xie, M.; Huang, J.; Ling, Z.; Fang, X.; Zhang, Z. Improving the heat storage/release rate and photo-thermal conversion performance of an organic PCM/expanded graphite composite block. *Sol. Energy Mater. Sol. Cells* **2019**, *201*, 110081. [CrossRef]
89. Li, W.; Zhang, D.; Jing, T.; Gao, M.; Liu, P.; He, G.; Qin, F. Nano-encapsulated phase change material slurry (Nano-PCMS) saturated in metal foam: A new stable and efficient strategy for passive thermal management. *Energy* **2018**, *165*, 743–751. [CrossRef]
90. Qiu, L.; Ouyang, Y.; Feng, Y.; Zhang, X. Review on micro/nano phase change materials for solar thermal applications. *Renew. Energy* **2019**, *140*, 513–538. [CrossRef]
91. Jegadheeswaran, S.; Sundaramahalingam, A.; Pohekar, S.D. High-conductivity nanomaterials for enhancing thermal performance of latent heat thermal energy storage systems. *J. Therm. Anal. Calorim.* **2019**, *138*, 1137–1166. [CrossRef]
92. Yu, S.; Jeong, S.-G.; Chung, O.; Kim, S. Bio-based PCM/carbon nanomaterials composites with enhanced thermal conductivity. *Sol. Energy Mater. Sol. Cells* **2014**, *120*, 549–554. [CrossRef]
93. Dsilva Winfred Rufuss, D.; Suganthi, L.; Iniyan, S.; Davies, P.A. Effects of nanoparticle-enhanced phase change material (NPCM) on solar still productivity. *J. Clean. Prod.* **2018**, *192*, 9–29. [CrossRef]
94. Zayed, M.E.; Zhao, J.; Elsheikh, A.H.; Du, Y.; Hammad, F.A.; Ma, L.; Kabeel, A.E.; Sadek, S. Performance augmentation of flat plate solar water collector using phase change materials and nanocomposite phase change materials: A review. *Process Saf. Environ. Prot.* **2019**, *128*, 135–157. [CrossRef]

95. Saw, C.L.; Al-Kayiem, H.H.; Owolabi, A.L. Experimental Investigation on the Effect of PCM and Nano-enhanced PCM of Integrated Solar Collector Performance. In *WIT Transactions on Ecology and the Environment*; WIT Press: Boston, MA, USA, 2013; pp. 899–909.
96. Al-Kayiem, H.H.; Lin, S.C. Performance evaluation of a solar water heater integrated with a PCM nanocomposite TES at various inclinations. *Sol. Energy* **2014**, *109*, 82–92. [CrossRef]
97. Sheikholeslami, M.; Jafaryar, M.; Shafee, A.; Li, Z. Nanofluid heat transfer and entropy generation through a heat exchanger considering a new turbulator and CuO nanoparticles. *J. Therm. Anal. Calorim.* **2018**, *134*, 2295–2303. [CrossRef]
98. Fuensanta, M.; Paiphansiri, U.; Romero-Sánchez, M.D.; Guillem, C.; López-Buendía, Á.M.; Landfester, K. Thermal properties of a novel nanoencapsulated phase change material for thermal energy storage. *Thermochim. Acta* **2013**, *565*, 95–101. [CrossRef]
99. Amidu, M.A.; Ali, M.; Alkaabi, A.K.; Addad, Y. A critical assessment of nanoparticles enhanced phase change materials (NePCMs) for latent heat energy storage applications. *Sci. Rep.* **2023**, *13*, 7829. [CrossRef] [PubMed]

energies

MDPI

Review

An Overview of Phase Change Materials and Their Applications in Pavement

Kinga Korniejenko [1,*], Marek Nykiel [1], Marta Choinska [2], Assel Jexembayeva [3], Marat Konkanov [3] and Lyazat Aruova [3]

[1] Faculty of Materials Engineering and Physics, Cracow University of Technology, Warszawska 24, 31-155 Kraków, Poland; marek.nykiel@pk.edu.pl

[2] Research Institute in Civil and Mechanical Engineering GeM, UMR CNRS 6183, Nantes University—IUT Saint-Nazaire, 44035 Nantes, France; marta.choinska@univ-nantes.fr

[3] Faculty of Architecture and Civil Engineering, Gumilyov Eurasian National University, Kazhymukan Str. 13, r205, Astana 010008, Kazakhstan; dzheksembayeva_ae@mail.ru (A.J.); marcon@metrology.kz (M.K.); ecoeducation@mail.ru (L.A.)

* Correspondence: kinga.korniejenko@pk.edu.pl; Tel.: +48-60997498

Abstract: The composite of a phase change material (PCM) and bitumen or asphalt as a matrix is expected as a new, advanced material for road construction. The main motivation for this article was to show the new possibilities and perspectives of developing the pavement with the usage of PCMs. Incorporating PCMs into paving materials can improve their properties, including allowing the regulation of the pavement temperature, enhancement of the pavement durability, and avoiding the phenomenon of a heat-island on the road. The main purpose of this article was to evaluate contemporary investigations in the area of the application of PCMs in pavement materials, especially asphalt and bitumen; to summarize the advantages and disadvantages of the implementation of PCM for road construction; and to discuss further trends in this area. This manuscript explored the state of the art in this area based on research in the literature. It shows the possible material solutions, presenting their composition and discussing their key properties and the manufacturing technologies used. The possibilities for further implementations are considered, especially economic issues.

Keywords: PCM; phase change material; asphalt; bitumen; road construction; pavement; concrete

1. Introduction

A phase change material (PCM) can be defined as a substance that has the possibility to store thermal energy by absorbing or releasing a large amount of latent heat in the process of changing physical state, especially between a liquid and a solid [1,2]. These materials react to environmental changes and their properties are modified depending on the external conditions by absorbing, storing, or releasing heat without changing their temperature [3]. The first time this kind of material was discovered and successfully applied was in the 1970s [2,4]. The early applications were connected with different kinds of building materials to improve their thermal efficiency through thermal energy storage [4–6]. Since then, these materials have also found many other applications, including in cooling systems, heat transfer, and thermal protection devices [7,8]. They also find an application in foamed materials to enhance their thermal properties [9,10]. Nowadays, they also find applications in thermoplastic materials, such as asphalt [3,11,12].

The wide range of applications for these materials is possible thanks to the many advantages that PCMs have compared to traditional materials [7,13]. The most important seems to be the simplicity of application and high reliability of this technology [7]. This led to there being some commercial applications for PCMs today, including in the building, electronic, and logistics industries [13]. In other applications, such as pavements, this technology is still being developed. This group of materials is attractive because of its high energy storage density and low power use at the same time [7]. Nowadays, when the price of energy is rising, it is an additional reason to develop PCMs. It is also worth stressing that during the phase transition process, they have a nearly isothermal temperature, which gives them high energy efficiency [7]. Due to their numerous advantages, PCMs are applied in many areas, including developing energy efficiency in free cooling defrosting, thermal management, and air-conditioning [7,13,14].

One of the promising areas of application for PCMs is pavements. The basic reason for applying PCMs in this area is temperature regulation [15]. These materials can be used for the reduction of consequences of low temperature (anti-ice and snow melting) as well as high temperature (helps in avoiding heat islands). Beyond these most obvious uses, these materials are also used for the improvement of functional properties, including the reduction of oxidative aging, rutting reduction, and the minimalization of creep [16]. However, while the benefits of implementing this kind of solution are significant, implementing a PCM into asphalt and bitumen mixtures as well as concrete for pavement applications is a challenging task [15]. The main challenges in this area are connected with the material properties' deterioration as a consequence of the implementation of PCMs, including lowering the mechanical properties, increasing the cracking tendency, and others connected with chemical changes [17].

The challenges and limitations in applying PCMs in practice are connected to the temperature of the working and manufacturing technology. PCMs usually work within a temperature range between −50 °C and 150 °C and only part of them is suitable for high-temperature applications [18,19]. This is applicable in high temperatures, usually based on metals and salts, and their applications are connected with a relatively high-cost comparison with this one based on polymer materials. Therefore, it is a challenge to develop a new, low-cost PCM dedicated to higher temperatures [20]. Another challenge is connected with the application of PCMs in the manufacturing process. The usage of bulk PCMs usually negatively influences matrix material properties. Because of that, PCM applications very often require the usage of the technology of encapsulation and closed PCMs in the form of macro- or microcapsules or the infiltration of different types of aggregates by PCMs [21]. It is usually connected with additional technological operations and increasing the cost of the composition. This problem also appears in the pavement applications.

The main goal of this article was to evaluate the current research in the area of the application of PCMs into pavement materials, especially asphalt and bitumen, to summarize the advantages and disadvantages of the implementation of PCM for road construction, and to discuss further trends in this area. This manuscript is based on the latest research in the literature, especially over the last 5 years. This article shows the overall knowledge in the area of PCM. It discusses the possibility of using compounds as a PCM and their applications. Next, it presents possible material solutions, presenting compositions for pavement applications in detail, including their key properties and the manufacturing technologies used. The possibilities for further implementations are considered, especially economic issues. Eventually, the challenges for further research are analyzed.

2. Methods

The Scopus (ScienceDirect) database was used as a main tool for the preparation of a systematic review. Also, some supportive tools have been used, especially Google Scholar, Wiley Online Library, ACS Publications, and IEEE Xplore Digital Library. In the first stage, overall search in the databases was conducted using the combination of two keywords: "pavement" and "PCM". The 149 results were found in Scopus in this topic (Figure 1).

Single publications on the analyzed topic were published in 1990, 2005, and 2008. However, only since 2011 have articles been regularly published that explore topics related to the use of PCM in road applications (Figure 1a). The number of publications is changeable year by year, but it is worth noting that the topic started to be popular after 2017. Thus, this conception is very new.

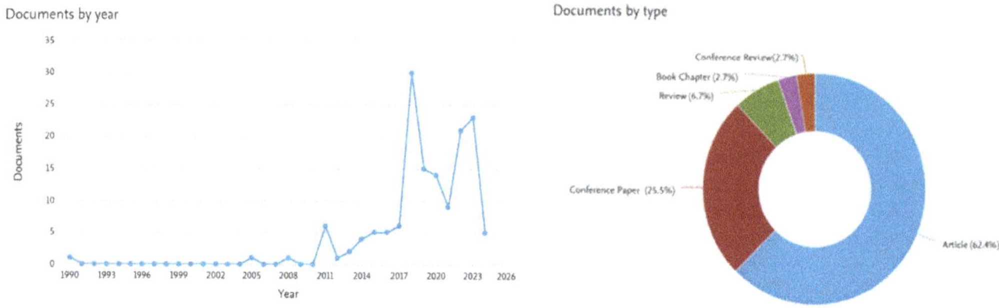

(a) (b)

Figure 1. The results of searching in the Scopus database: (**a**) numbers of publications per year; (**b**) the information about the publications by type [22].

Most of the scientific works were focused on research articles (approximately 62%) or conference papers (approximately 25%). In this area there was only a few review articles; this type of structure of the publication types is also typical for a new topic, where there is a lack of articles that can summarize the new achievement in the area. Taking this fact into consideration, this review article will be a precious supplementation to the state of the art in the application area of PCMs for pavement materials.

3. Phase Change Materials

PCMs can be divided into three main groups: organic, inorganic, and eutectic (Figure 2).

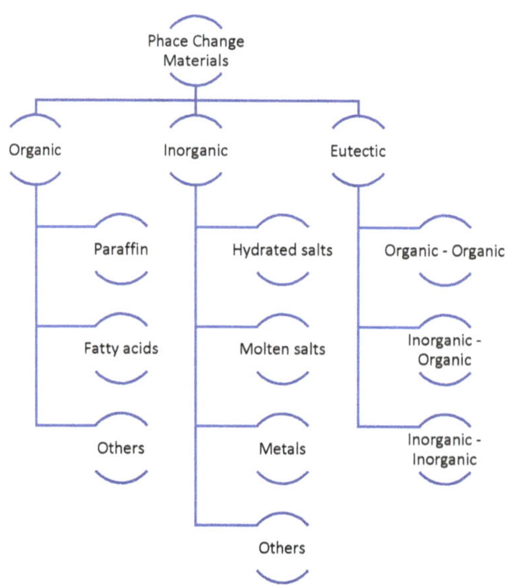

Figure 2. PCM classification according to the type of material.

3.1. Organic

Organic PCMs can be divided into paraffin, fatty acids, and others [15]. The main advantage of this group is that they do not cause corrosion problems [18]. They are usually chemically and physically stable and available for a wide range of temperatures [15]. The potential problem that touches this group is flammability [15].

Paraffin is the most widely used PCM from this group. It finds a lot of applications, especially as a component of building materials with improved thermal properties [23,24]. In the case of application in building materials, including pavements, paraffin is very often encapsulated in other organic polymers, such as a polymer shell made of polymethyl methacrylate [25]. In the case of paraffin materials, the melting point is usually between 1 and 6 °C; however, in the case of some materials it can be much higher, for example, for n-Eicosane it is 36.5 °C [15]. This range of temperature is favorable for application in construction where the temperature phase transition is near 0 °C as it is, for example, in Kazakhstan or Poland. The latent heat for this group is usually approximately 200 kJ/kg and the heat conductivity is between 0.15 and 0.45 W/mK [15]. An alternative method for the effective application of paraffin in pavements is the incorporation of it in lightweight porous aggregate [26,27].

Another wide group of organic PCMs is fatty acids. These compounds are characterized by a long aliphatic chain capped by a carboxyl group. They can have a straight or branched structure and be saturated or unsaturated [15]. The exemplary substances that are used as PCM are lauric acid, myristic acid, palmitic acid, and stearic acid [28].

Other popular organic PCMs are different kinds of polymers, especially polyethylene glycol (PEG). This substance has a lot of advantages: among others, it can be easily tunable to phase change temperatures, is characterized by a lack of toxicity, and has high melting/freezing enthalpies [29]. It is worth stressing that PEG as PCM is mostly used in asphalt mixtures [30]. Nevertheless, it has also some disadvantages, including low thermal conductivity [29].

Also, another organic PCM has been tested for pavement solutions, including eicosane, tetradecane and organic mixes [28,31]. Organic PCMs, such as PEG and tetradecane, are often applied to regulate the high and low temperatures of asphalt pavement [15].

3.2. Inorganic

The next group is inorganic PCMs. They can be divided into hydrated salts, molten salts, metals, and others. This group is characterized by high heat fusion, high thermal conductivity, and low volume changes [15]. For example, fatty salts have a more diverse melting point from 0.5 °C for RTM up to 64 °C for palmitic acid [15]. The latent heat is also in the wider range of 46–196.9 kJ/kg [15]. Another advantage is the lack of flammability. They are also relatively easily available [15]. The main problem is their sensitivity to corrosion problems [18]. Among this group, the most popular are salt hydrates and metals [32].

This group of materials was investigated for thermal energy storage applications [33]. In the area of building applications, some of them, such as hydrated salt, were also tested as a part of the composition for roof applications [1,34]. In the case of pavement applications, this group of PCMs is rarely used, although an investigation was conducted by adding NiTi alloy to asphalt mixtures [35].

3.3. Eutectic

The popular classification of this group is based on the organic and inorganic character of the compounds. It can be divided into organic–organic, inorganic–organic, and inorganic–inorganic [15]. This group is a combination of two or more components with different chemical and physical characteristics [36].

The group of eutectic PCMs is characterized by sharp melting points, and high volumetric thermal storage. The main advantage of eutectic PCM is the possibility of the customization of the desired melting temperature [36]. Compared to the two previous groups the properties of these materials are not fully investigated and still require research [15].

The eutectic PCM is also a subject for applications in the building industry. Haily et al. [37] tried to apply a eutectic mixture of lauric acid and capric acid into geopolymer materials to improve the energy efficiency of buildings [37]. They obtained satisfactory results and assessed the material possible to apply for sustainable and energy-efficient buildings. Another study was conducted by Baskar et al. [38] with the addition of lauric and palmitic as PCMs to the paints. The results showed the efficiency of the created paints for thermal regulation, especially in reducing the surface temperature of the concrete that was covered by it [38]. In the last few years, eutectic PCMs also started to be investigated as a potential additive to pavement applications [16,36].

3.4. Current Application of PCM

Nowadays, PCM materials find a wide area of applications (Figure 3). They are used in different kinds of industries for many applications; among the most popular are as follows [15,39]:

- Thermal energy storage, primarily in solar thermal applications [15,39];
- Cooling and heating applications, especially in the building industry and logistics [15,39];
- Heat dissipation of electric circuits in electronic applications and transportation areas [15,39];
- Energy-absorbing clothes (textile industry) [15,39];
- Temperature regulation in pharmaceutical and food preservation [15,39];
- Others, including energy management [15,39].

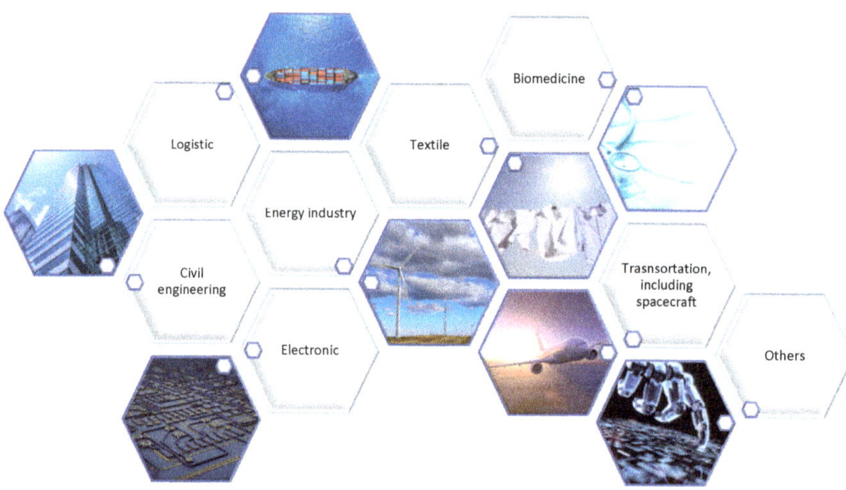

Figure 3. PCMs—main applications.

Among others, an important application is in the area of energy management, in which a PCM was also investigated as an element of Li-ion battery thermal management system to improve energy efficiency [40]. Another application in this area is a reduction of heating loads for rooms with air-conditioners [18]. It is also worth mentioning modern applications for these materials such as thermo-responsive dielectric switching/pulsing materials and temperature-sensitive electrical switching materials [41,42]. This kind of solution has potential in applications of next-generation smart electronic/electrical technology, including temperature sensors, smart switches, phase shifters, and varactors [41,42].

Among the mentioned applications, it is worth paying attention to the building industry. Li et al. [7] also explored some applications in the building industry, including the usage of PCMs for roofs, ceilings, and walls in residential houses. They showed that thanks to the application of PCM it is possible to reduce heating demand, enhance thermal comfort, and better utilize solar energy through effective storage [7]. The investigation in this area supports the development of PCMs in pavement applications [43].

4. Pavement Application of Phase Change Materials

4.1. Main Area of Investigations

There are many criteria for selection for PCMs for pavement applications. Different authors have different preferences and criteria that decide about the selection of particular substances [44–46]. The two main criteria seem to be latent heat and thermal conductivity [15]. However, in the literature there are a lot of other factors that should be taken into consideration [15]. The key properties of PCMs are connected with the following [39]:

- Thermal properties: latent heat of fusion, thermal conductivity, specific heat capacity, and phase change temperature.
- Chemical properties: corrosive, toxicity, flammability, chemical stability.
- Physical properties: volume changes, density, durability against multiple freeze and thaw cycles.
- Kinetic: nucleation rate, speed of crystal growth, and supercooling.
- Economic factors: availability and price.

Moreover, the selection of a PCM will also be dependent on the matrix material and the purpose of the pavement modification and technology of the implementation of the PCM into the material [47,48]. In Table 1, the most important research in the area of the application of PCMs in pavements is summarized based on selected literature.

Table 1. Application of PCMs in pavements.

No	Type of Application	Matrix	PCM	Main Findings	Reference
1	Snow removal system with solar thermal energy collector	Lack of typical matrix. Liquid was delivered to concrete material by the system of pipes.	Organic: n-octadecanol (agglomerated in cylindrical can made of aluminum)	PCM was used for storing thermal energy from solar collectors; it was possible to store 58 MJ. Thanks to the discharge of solar energy, the temperature of pavement rose by 30 °C. The test confirmed the possibility of effective snow removal. the pavement temperature was above 2 °C during a snowfall.	[47]
2	Reduction of the pavement surface temperature to avoid the thermal stress in high temperatures	Concrete	Organic (OM35 and OM42) encapsulated.	The most important for effective cooling are the latent heat and phase change temperature of PCM. In the night, the pavement surface temperature rose by approximately half of the reduction in temperature during the daytime. It was caused by the slower solidification rate of the PCMs.	[49]
3	Road temperature regulation	SBS-modified asphalt	HDPE, expanded graphite, and paraffin (directly mixed)	PCM addition influenced reduction consistency and temperature sensitivity. It also enhanced low-temperature performance. The deformation resistance of modified material decreased but the fatigue performance increased. The asphalt had good rutting resistance and elastic recovery ability at 64 °C.	[50]

Table 1. *Cont.*

No	Type of Application	Matrix	PCM	Main Findings	Reference
4	Capacity to store thermal energy; slow down the cooling rate; improve of the thermomechanical characteristics	Bitumen	D-Mannitol (high-speed shearing with bitumen)	The melting point of the modified material was without significant changes. PCM improved the physical characteristics of the basic material. The specific heat capacity rose gradually with the PCM content.	[51]
5	Improving rheological and thermal properties	Gilsonite-modified asphalt binder	PEG (directly mixed)	PCM balances the impacts of gilsonite. The binder PCM and gilsonite have good rutting resistance and are possible for application in low temperatures (cracking resistance).	[52]
6	Temperature control (high-temperature reduction)	Cement	(1) paraffin wax (2) myristic acid (encapsulated)	Composites had low crushing ratios during rut-forming tests. PCMs were thermally and chemically stable (minimal mass loss at 180 °C, lack of PCM leakage).	[53]
7	Cooling asphalt pavement	Asphalt	Eutectic mixture of fatty acid (palmitic acid and stearic acid), incorporated in coarse steel slags aggregate	The composition had sufficient cooling performance and durability. Additions of PCM increased the high-temperature rutting resistance of pavement by 30.7%.	[28]
8	Temperature regulation and ice-melting effects	Asphalt	PEG 800, a phase change energy storage material and polyacrylamide backbone structure (directly mixed)	Investigated composites were in line with the specification requirements. The addition of PCM enhanced mechanical properties and moisture resistance. PCM positively influenced thermal insulation performance and heat storage efficiency. PCM reduced the long-term high-temperature performance and low-temperature strength.	[54]
9	Temperature regulation; reducing the urban heat island effect	High-viscosity modified asphalt (HVMA)	(1) Paraffin/expanded graphite/high-density polyethylene composite (2) polyethylene glycol (PEG) (mixed, not explained in detail)	PCMs were uniformly distributed in HVMA. PCMs did not affect the softening points of asphalt. Composites had excellent high-temperature rutting resistance regardless of PCM addition. The effect of the regulation of temperature was visible for both PCMs.	[55]
10	Temperature regulation	Asphalt	PCM based on polyurethane (included in fine aggregate)	The viscoelastic properties of composites were related to the curing temperature, loading frequency, PCMs content, and particle sizes.	[56]

Table 1. *Cont.*

No	Type of Application	Matrix	PCM	Main Findings	Reference
11	Cooling pavement	Concrete	Organic (OM42), incorporated in expanded clay aggregate	PCM effectively decreased pavement surface temperature (2.24 °C was the annual average).	[57]
12	Cooling pavement	Concrete	Organic (OM35 and OM42), encapsulation	The cooling potential of pavements PCM improved by more than 80%. The thermal conductivity of the material increased.	[58]
13	Preventing the low-temperature impact on pavements	Asphalt	PCM based on polyurethane (directly mixed)	PCM slightly affected the high- and low-temperature performance. PCM improved the anti-aging properties. The energy storage properties of composition were found to be satisfactory.	[59]
14	Road temperature regulation	Hot-mix asphalt	Paraffin (microencapsulation)	PCM could withstand asphalt mixing and placement conditions. PCM reduced the dynamic modulus.	[60]
15	Increasing the functionality of pavements made from waste materials	OPC + waste materials (bricks)	PEG 400 Tetradecane (incorporated in recycled aggregate)	The study proved the possibility of using the waste materials as a matrix for PCMs for pavement applications.	[61]
16	Cooling pavement	Asphalt	Paraffin (mixed, not explained in detail)	PCM decreased the frequency of pavement high-temperature rutting damage. With the amount of PCM the cooling effect increased.	[62]
17	Preventing the temperature impact	Asphalt	Eutectic (solid-solid), directly mixed	PCM increased the physical properties of asphalt. PCM increased the high-temperature rutting resistance. PCM improved the low-temperature creep behavior.	[16]
18	Cooling pavement	Asphalt	Eutectic (stearic acid/palmitic acid), directly mixed	PCM application improved the rutting resistance. The structure of PCM inside the composite was stable and had a layered form. The distinguished temperature regulating property was clearly visible (more than 11 °C difference). The temperature peak was delayed 40 min.	[63]
19	Thermal stress reduction	Asphalt	Melamine–formaldehyde resin with graphene (microencapsulation)	PCM increased thermal conductivity and volume-specific heat capacity. The investigation confirmed reducing the temperature variation-induced cracking.	[64]

Table 1. *Cont.*

No	Type of Application	Matrix	PCM	Main Findings	Reference
20	Temperature regulation; avoiding urban heat island	Concrete	Organic (OM35 and OM42) incorporation in expanded clay aggregates	PCM stored latent heat at different temperatures. The material is stable up to 196.6 °C. PCM reduced the maximum pavement surface temperature by approximately 2 °C.	[65]
21	Temperature regulation	Asphalt	PEG 800, a phase change energy storage material and polyacrylamide backbone structure (directly mixed)	PCM enhanced the moisture and low-temperature cracking resistance PCM increased the thermal conductivity. PCM improved the heat preservation capacity.	[66]
22	Temperature regulation	Asphalt	PEG 800, a phase change energy storage material and polyacrylamide backbone structure (directly mixed)	PCM improved the Marshall stability and flexural–tensile strain as well as other parameters such as moisture resistance, low-temperature crack resistance, and thermal insulation properties. PCM reduced the mechanical strength and long-term high-temperature stability performance.	[67]
23	Cooling pavement	Asphalt	Eutectic (stearic acid/palmitic acid)-directly mixed	Between PCM and asphalt, no chemical reaction was detected. PCM has to be applied at higher temperatures than traditional PCM, especially organic.	[68]
24	Improvement of thermomechanical characteristics; mitigation of thermal curling.	Concrete	Paraffin incorporated in porous lightweight aggregate	The element made from composite containing PCM had lower linear strain because of the lower coefficient of thermal expansion.	[69]
25	Regulating temperature and resisting UV aging	Bitumen	PEG–PCM ZnMgAl-mixed metal oxides support (directly mixed)	ZnMgAl mixed-metal oxides as a carrier can include up to 65% of PEG. This mix has good thermal and chemical stability, sufficient phase change enthalpy, and excellent UV absorption properties.	[70]
26	Low-temperature behavior, avoiding cracking	Bitumen	Tetradecane (directly mixed)	PCM raised penetration and lowered the conventional characteristics of bitumen such as softening temperature. Direct addition of PCM also significantly influenced the rheological properties of bitumen; because of that, encapsulation is recommended.	[3]

Table 1. *Cont.*

No	Type of Application	Matrix	PCM	Main Findings	Reference
27	Temperature regulation	Asphalt	NiTi alloy (directly mixed; replacement for fine aggregate)	PCM was used as a replacement for aggregate (partially). PCM slightly reduced the water stability. PCM significantly reduced the heating rate.	[35]
28	Temperature regulation	Asphalt	Tetradecane (microencapsulation)	Different PCMs can have different thermoregulation ranges. PCM significantly improved its thermal behavior.	[71]
29	Improvement of thermophysical parameters	Asphalt	Pentadecane (microencapsulation)	The composition had good thermal stability, thermal storage performance, and mechanical properties	[72]
30	Aging	Asphalt	Tetradecane (microencapsulation)	PCM reduced temperature influence during seasonal and diurnal cycles. PCM gives only benefits in encapsulated form. The melting enthalpy decreases upon aging. PCM increased rheological properties.	[73]
31	Temperature regulation	Bitumen	Tetradecane (microencapsulation)	PCM did not affect rheological properties; it effectively regulated temperature variations.	[74]
32	Cooling pavement	Asphalt	PEG (directly mixed)	PCM complicated effect on the rheological properties. PCM harms the shear strength.	[75]
33	Temperature regulation, avoiding thermal distresses	Asphalt	PEG (microencapsulation)	Confirmation of thermal storage capacity. PCM positively influences rheological properties.	[76]
34	Freeze–thaw performance	Asphalt	Paraffin (pure and microencapsulated)	PCM helps in controlling freeze–thaw impact on subgrade soil	[77]
35	Freeze–thaw performance	Concrete	Paraffin (pure and microencapsulated)	PCM decreased the magnitude of the temperature drop. PCM deteriorated the mechanical properties.	[78]
36	Mechanical and thermal performance	Pavement (not specified)	Paraffin (macro encapsulation)	Anti-ice properties. PCM increased thermal stability and heat storage capacity.	[79]
37	Anti-freezing, temperature regulation bridge decks	Concrete	Composite organic polyol (seamless steel pipe layer with PCM)	Good effect on melting ice and snow.	[80]

The experiments in the last few years were connected with various matrices of asphalt as well as concrete. They also involve other kinds of PCMs—organic, inorganic, and eutectic. Also, different techniques of implementation were used such as encapsulation, infiltration of aggregates by PCMs, joining with backbone structure, and direct implementation. All the provided research shows a high potential of PCMs for delaying peak temperatures and accumulating heat (regulating the temperature). Depending on the application, the applied PCMs' distinguished temperature regulating was between a few and a dozen degrees Celsius. The experimental works outside the laboratory also confirmed the effectiveness of regulation between day and night. The regulation of temperature influence also affects the durability and long-term properties of the investigated materials, especially by increasing the rutting resistance. The influence of the additives on other mechanical and physical properties was not always the same and was dependent on the used material and technology. Selected research also shows that the application of PCMs has the potential for defrosting pavements, thereby increasing the safety for traffic. It is worth noticing that the effectiveness of the material, especially the latent heat effect, can influence external factors such as the solar radiation and airflow conditions [81–83].

4.2. Technologies of Manufacturing Pavement Composite

There are three main technologies used for incorporating PCM materials into pavement: immersion, impregnation, and encapsulation (macro- and microencapsulation) [32]. The other methods, such as the distribution of a PCM through a pipe system, are used quite rarely [47,80,84]. The direct addition of a PCM has a negative impact on the mechanical strength of concrete because this additive hinders the cement hydration and has a negative impact on aggregate bonding. It also affects the physical and rheological characteristics of bitumen [32,49]. Because of that, in most technologies the material is applied in the form of an encapsulated material or through impregnation into a lightweight material which is next encapsulated in the cementitious material [49]. In this case, it is also important to properly select the encapsulation medium. It has to be resistant to the stresses generated due to traffic loads. Destruction of the incorporated capsules under the influence of mechanical loads will cause PCM leakage and a loss of the material's properties [49,85].

Nowadays, the most common technology for manufacturing this kind of composite is filling the pores in the material with PCM liquid. In the first step, the matrix material is produced and the water is incorporated into the material pores. A porous matrix for filling is commonly used, such as expanded shale, clay, perlite, diatomite, and others [32,85,86]. Previous research shows that the selection of the carrier material is important for the effectiveness of the whole system [87]. In the next step, the material is drying and the usual parameters are a temperature above 100 °C and a time of approximately 24 h [49]. After this time the material is immersed in PCM liquid. To be effective, the PCM is usually heated above the melting point and the time of exposition is approximately 24 h [49]. This step caused the PCM to infiltrate the material pores. The capsules prepared this way are covered by cementitious material and the composite is eventually encapsulated [49,88].

The technologies of incorporation and encapsulation for PCMs in pavement applications are still being developed to increase the efficiency of the system and avoid potential negative influences on the matrix material [89]. The authors stressed the benefits connected with these technologies, such as preventing leakage, improving heat transfer by raising specific surface area, and protecting the PCM from the external environment [21]. Among the technologies of encapsulation, the microencapsulation of PCMs plays a special role. In the case of pavement applications, it allows for the avoidance of a huge agglomeration of the phase of the PCM in one place, which reduces the negative impact on mechanical properties.

It is also worth mentioning a new perspective for the application of PCM materials, such as fiber-based PCMs by solution spinning [90]. This method was developed over several years and involves a joint fiber reinforcement with PCMs. This kind of combination in pavement materials, besides the enhancing properties typical of a PCM application, can simultaneously improve the flexural strength and reduce brittle behavior. However, this kind of solution has not been tested in pavement applications yet.

4.3. Key Advantages

There are several reasons for incorporating PCMs in pavement materials. The most important is preventing rapid changes in the temperature by properly regulating them. Proper temperature regulation helps to improve the pavement durability, and mitigate the cracks in the material and minimize the heat island effect [32]. The unwanted phenomena in the pavement material, such as cracking and rutting, are very often connected with temperature-related distress [15]. These are caused by high solar radiation and thermal convection between the pavement surface and the air [91]. The application of a PCM helps to avoid this process by lowering the thermal conductivity of the composition and increasing heat capacity [92].

The advantages of applying this technology have also been reflected in socio-economical indicators. Avoiding the urban heat island (UHI) effect has also influenced the minimization of air pollution and greenhouse gas emissions [93,94]. The usage of PCMs helps in lowering peak energy demand and decreasing air conditioning costs [93,94].

Additionally, Sharifi et al. [95] proved the positive influence of PCMs on the reduction of fatigue-fracture damage. The implementation of a PCM reduces the stresses that have a periodic nature. This mixture reduces the cyclic flexural curling stresses that lead to the cracking of the concrete slab and are connected with a changeable temperature [95]. They modified Paris' Law and adopted it to calculate the cumulative fatigue-fracture damage of the PCM-rich concrete slab under the cyclic thermally induced curling stresses. The results showed higher resistance to the surface modified by PCM in terms of fatigue-fracture performance [95].

4.4. Main Challenges

The main challenges with the wider application of PCMs in pavements are connected with the technology of implementing it into the pavement matrix [36]. Most of the problems are connected with potential PCM leakage and the influence of the physical and rheological properties of the concrete or asphalt binder [96,97]. The effects of PCM leakage have been categorized into two categories [17,39]: the reaction with the matrix and the influence on the process of material preparation, for example, cement hydration or lubricant effect in the matrix material. Also, the PCM content could cause other problems with the material's strength, especially the soft inclusion effect, which increases the material porosity and shell material reactions [17,39]. To avoid these problems, new research needs to be conducted.

Other important points connected with PCM applications are related to the proper PCM selection, not only to ensure proper efficiency but also to avoid potential problems. A lot of PCMs are sensitive to temperature, and because of that to avoid overheating the material the proper heat transfer should be designed [94,98]. It is only one among a number of design questions such as the proper selection of the copolymerization method, problems with in situ polymerization, and the appropriate preparation of mineral-supported effective absorption into the carrier [30,99]. Last, but not least, is the problem with the high cost of investment. This kind of technology requires a high initial cost and complex construction procedures, and the effects are visible only after a longer period of time [94].

5. Further Perspectives

Today, the usage of PCM in the road industry can seem to be not economically justified, because the road industry is focused on cost minimization. However, taking into consideration wider perspectives, including the cost of maintenance and the reduction in the number of potential accidents in the long term, this investment can be beneficial [43].

The further perspective of the development of PCMs in pavements is connected with thermal conductivity investigations, increasing the efficiency of the heat transfer speed as well as the energy consumption [32]. Technological developments will also play an important role, and these may include better methods of avoiding PCM leakage in encapsulation, the improvement of pavement fatigue, and the effective limitation of the

UHI effect [32]. New methods for PCM application can also play an important role in further applications, for example, polymer fiber-based PCM by solution spinning [90].

The further perspectives seem to be especially promising for nano-PCM additives [100]. Using nanoparticles and nanofibers as shells is still a very new topic in the area of application PCMs in pavement [39]. Some early-stage research suggests that binders containing nanoparticles of PCM show lower mass during mixing and compaction [49]. This topic is additionally supported by the previous research connected with the application of microencapsulated PCM that reported a wide range of applications, not only in the building industry, but also in other branches [21].

Also, incorporating other materials, for example, mixing in carbon fibers for the further improvement of material properties, seems to be a promising topic [101]. In this area, the possibility of using waste materials or by-products to support this process has to be mentioned. They can be used especially as carriers for PCM. Some previous trials have been performed in this area, including mixing PCMs with used bricks [61] or steel slag [102], as well as their usage as a carrier for industrial by-products such as fly ash [103].

Other environmental aspects worth mentioning are recyclability issues and life-cycle assessments for this kind of composition [104]. There is a lack of work in this area in the literature. The information in the area of the social and environmental effects is focused mainly on the UHI effect [30].

Another interesting area of research work is using PCM materials as an element of the system of harvesting energy from pavements. Tahami et al. [105] tested a heat storage system using renewable energy from solar radiation generated by thermal gradients and heat flow in the pavement layers [105]. This research confirmed the possibility of harvesting heat energy from roadway pavements with the usage of PCMs [94,105]. The development of this technology is one of the interesting points for the usage of this material for an efficient supplementation of green energy [106,107].

It is also worth noticing the development of modeling methods connected with PCM applications [108,109]. In many cases the traditional methods are not sufficient for providing a full insight into the topic and the development of more advanced models is required, including 3D simulation for the particular compositions and more detailed physical modeling [110,111]. Another challenge for simulations is taking into account real conditions, including data from existing roads [112–114].

6. Conclusions

The application of PCMs in pavement materials is a new trend and a promising area for research. The provided research shows the application of PCMs in pavement as a dynamically developed area with a lot of possibilities for innovative investigations. The development of this technology to application on a full scale requires interdisciplinary knowledge from the areas of numerical modeling, technology material science, civil engineering, and environmental engineering.

The analysis of further perspectives allows us to formulate the most important areas for the nearest period:

- Improvement numerical modeling for complex problems;
- Development of modern PCM materials with wider possibilities;
- Development carriers, also with usage waste materials;
- Improvement of the technology of encapsulation and impregnation;
- Implementation of complex methods for environmental assessment.

Author Contributions: Conceptualization, K.K. and M.C.; methodology, M.N.; validation, A.J., M.K. and L.A.; formal analysis, M.C.; investigation, K.K. and M.N.; resources, L.A.; data curation, K.K.; writing—original draft preparation, K.K. and M.N.; writing—review and editing, M.C., A.J. and M.K.; supervision, L.A.; project administration, A.J. and M.K; funding acquisition, L.A. All authors have read and agreed to the published version of the manuscript.

Funding: This research was funded by the Science Committee of the Ministry of Science and Higher Education of the Republic of Kazakhstan (grant number BR18574214).

Conflicts of Interest: The authors declare no conflicts of interest. The funders had no role in the design of the study; in the collection, analyses, or interpretation of data; in the writing of the manuscript; or in the decision to publish the results.

References

1. Reddy, V.J.; Ghazali, M.F.; Kumarasamy, S. Advancements in Phase Change Materials for Energy-Efficient Building Construction: A Comprehensive Review. *J. Energy Storage* **2024**, *81*, 110494. [CrossRef]
2. Łach, M.; Pławecka, K.; Bąk, A.; Adamczyk, M.; Bazan, P.; Kozub, B.; Korniejenko, K.; Lin, W.-T. Review of Solutions for the Use of Phase Change Materials in Geopolymers. *Materials* **2021**, *14*, 6044. [CrossRef] [PubMed]
3. Kakar, M.R.; Refaa, Z.; Bueno, M.; Worlitschek, J.; Stamatiou, A.; Partl, M.N. Investigating Bitumen's Direct Interaction with Tetradecane as Potential Phase Change Material for Low Temperature Applications. *Road Mater. Pavement Des.* **2020**, *21*, 2356–2363. [CrossRef]
4. Fantini, P. Phase Change Memory Applications: The History, the Present and the Future. *J. Phys. D Appl. Phys.* **2020**, *53*, 283002. [CrossRef]
5. Kheradmand, M.; Azenha, M.; De Aguiar, J.L.B.; Castro-Gomes, J. Experimental and Numerical Studies of Hybrid PCM Embedded in Plastering Mortar for Enhanced Thermal Behaviour of Buildings. *Energy* **2016**, *94*, 250–261. [CrossRef]
6. Guarino, F.; Athienitis, A.; Cellura, M.; Bastien, D. PCM Thermal Storage Design in Buildings: Experimental Studies and Applications to Solaria in Cold Climates. *Appl. Energy* **2017**, *185*, 95–106. [CrossRef]
7. Li, Y.; Nord, N.; Xiao, Q.; Tereshchenko, T. Building Heating Applications with Phase Change Material: A Comprehensive Review. *J. Energy Storage* **2020**, *31*, 101634. [CrossRef]
8. Bouhal, T.; El Rhafiki, T.; Kousksou, T.; Jamil, A.; Zeraouli, Y. PCM Addition inside Solar Water Heaters: Numerical Comparative Approach. *J. Energy Storage* **2018**, *19*, 232–246. [CrossRef]
9. Bąk, A.; Pławecka, K.; Łach, M. Comparison of Thermal Conductivity of Foamed Geopolymers Containing Phase Change Materials. *J. Phys. Conf. Ser.* **2023**, *2423*, 012003. [CrossRef]
10. Bąk, A.; Pławecka, K.; Bazan, P.; Łach, M. Influence of the Addition of Phase Change Materials on Thermal Insulation Properties of Foamed Geopolymer Structures Based on Fly Ash. *Energy* **2023**, *278*, 127624. [CrossRef]
11. Somani, P.; Gaur, A. Evaluation and Reduction of Temperature Stresses in Concrete Pavement by Using Phase Changing Material. *Mater. Today Proc.* **2020**, *32*, 856–864. [CrossRef]
12. Partl, M.N. Quest for Improving Service Life of Asphalt Roads. *RILEM Tech. Lett.* **2020**, *4*, 154–162. [CrossRef]
13. Mehling, H.; Brütting, M.; Haussmann, T. PCM Products and Their Fields of Application—An Overview of the State in 2020/2021. *J. Energy Storage* **2022**, *51*, 104354. [CrossRef]
14. Li, F.; Zhou, S.; Chen, S.; Yang, J.; Zhu, X.; Du, Y.; Yang, Z. Low-Temperature Organic Phase Change Material Microcapsules for Asphalt Pavement: Preparation, Characterisation and Application. *J. Microencapsul.* **2018**, *35*, 635–642. [CrossRef] [PubMed]
15. Chen, Y.; Wang, H.; You, Z.; Hossiney, N. Application of Phase Change Material in Asphalt Mixture—A Review. *Constr. Build. Mater.* **2020**, *263*, 120219. [CrossRef]
16. Zhao, H.; Guo, J.; Ma, S.; Zhang, H.; Su, C.; Wang, X.; Li, Z.; Wei, J.; Cui, S. Effect of Solid-Solid Phase Change Material's Direct Interaction on Physical and Rheological Properties of Asphalt. *Coatings* **2022**, *12*, 625. [CrossRef]
17. Marani, A.; Nehdi, M.L. Integrating Phase Change Materials in Construction Materials: Critical Review. *Constr. Build. Mater.* **2019**, *217*, 36–49. [CrossRef]
18. Barreneche, C.; Navarro, M.E.; Cabeza, L.F.; Fernández, A.I. New Database to Select Phase Change Materials: Chemical Nature, Properties, and Applications. *J. Energy Storage* **2015**, *3*, 18–24. [CrossRef]
19. Farnam, Y.; Krafcik, M.; Liston, L.; Washington, T.; Erk, K.; Tao, B.; Weiss, J. Evaluating the Use of Phase Change Materials in Concrete Pavement to Melt Ice and Snow. *J. Mater. Civ. Eng.* **2016**, *28*, 04015161. [CrossRef]
20. Sharifi, N.P.; Sakulich, A. Application of Phase Change Materials to Improve the Thermal Performance of Cementitious Material. *Energy Build.* **2015**, *103*, 83–95. [CrossRef]
21. Ghasemi, K.; Tasnim, S.; Mahmud, S. PCM, Nano/Microencapsulation and Slurries: A Review of Fundamentals, Categories, Fabrication, Numerical Models and Applications. *Sustain. Energy Technol. Assess.* **2022**, *52*, 102084. [CrossRef]
22. Analyze Search Results. Available online: https://www.scopus.com/Term/Analyzer.Uri?Sort=plf-f&src=s&sid=de161c49f931b5c99e9295953af65520&sot=a&sdt=a&sl=48&s=(TITLE-ABS-KEY(pavement)+AND+TITLE-ABS-KEY(PCM))&origin=resultslist&count=10&analyzeResults=Analyze+results (accessed on 11 March 2024).
23. Akeiber, H.J.; Hosseini, S.E.; Hussen, H.M.; Wahid, M.A.; Mohammad, A.T. Thermal Performance and Economic Evaluation of a Newly Developed Phase Change Material for Effective Building Encapsulation. *Energy Convers. Manag.* **2017**, *150*, 48–61. [CrossRef]
24. Mizwar, I.K.; Napiah, M.; Sutanto, M.H. Thermal Properties of Cool Asphalt Concrete Containing Phase Change Material. *IOP Conf. Ser. Mater. Sci. Eng.* **2019**, *527*, 012049. [CrossRef]

25. Kravchenko, E.; Liu, J.; Li, X. Numerical Modeling of the Thermal Performance of Soil Containing Microencapsulated PCM. *Constr. Build. Mater.* **2021**, *298*, 123865. [CrossRef]
26. Farnam, Y.; Esmaeeli, H.S.; Zavattieri, P.D.; Haddock, J.; Weiss, J. Incorporating Phase Change Materials in Concrete Pavement to Melt Snow and Ice. *Cem. Concr. Compos.* **2017**, *84*, 134–145. [CrossRef]
27. Manning, B.J.; Bender, P.R.; Cote, S.A.; Lewis, R.A.; Sakulich, A.R.; Mallick, R.B. Assessing the Feasibility of Incorporating Phase Change Material in Hot Mix Asphalt. *Sustain. Cities Soc.* **2015**, *19*, 11–16. [CrossRef]
28. Zhang, D.; Muhammad Sani, B.; Xu, P.; Liu, K.; Gu, F. Preparation and Characterization of Binary Eutectic Phase Change Material Laden with Thermal Conductivity Enhancer for Cooling Steel Slag Asphalt Pavement. *Constr. Build. Mater.* **2023**, *388*, 131688. [CrossRef]
29. Kim, A.; Wert, N.A.; Gowd, E.B.; Patel, R. Recent Progress in PEG-Based Composite Phase Change Materials. *Polym. Rev.* **2023**, *63*, 1078–1129. [CrossRef]
30. Landi, S.; Segundo, I.R.; Homem, N.; Sousa, J.; Freitas, E.; Costa, M.F.M.; Carneiro, J. Reducing the Impact of the Sunlight in Urban Areas Using Asphalt Mixtures with Phase Change Materials: A Review in Scopus in the Last Three Years. *J. Phys. Conf. Ser.* **2022**, *2407*, 012022. [CrossRef]
31. Ma, B.; Adhikari, S.; Chang, Y.; Ren, J.; Liu, J.; You, Z. Preparation of Composite Shape-Stabilized Phase Change Materials for Highway Pavements. *Constr. Build. Mater.* **2013**, *42*, 114–121. [CrossRef]
32. Asadi, I.; Jacobsen, S.; Baghban, M.H.; Maghfouri, M.; Hashemi, M. Reviewing the Potential of Phase Change Materials in Concrete Pavements for Anti-Freezing Capabilities and Urban Heat Island Mitigation. *Buildings* **2023**, *13*, 3072. [CrossRef]
33. Rajamony, R.K.; Paw, J.K.S.; Pasupuleti, J.; Pandey, A.K.; Yaw, C.T.; Tiong, S.K.; Yusaf, T.; Samykano, M.; Sofiah, A.G.N.; Laghari, I.A.; et al. Experimental Investigation on the Performance of Binary Carbon-Based Nano-Enhanced Inorganic Phase Change Materials for Thermal Energy Storage Applications. *J. Energy Storage* **2024**, *86*, 111373. [CrossRef]
34. Bhamare, D.K.; Rathod, M.K.; Banerjee, J.; Arıcı, M. Investigation of the Effect of Air Layer Thickness on the Thermal Performance of the PCM Integrated Roof. *Buildings* **2023**, *13*, 488. [CrossRef]
35. Ma, B.; Wei, K.; Huang, X.F.; Shi, W.S.; Chen, S.S.; Hu, Y.P.; Shi, H.T. Preparation and Investigation of NiTi Alloy Phase-Change Heat Storage Asphalt Mixture. *J. Mater. Civ. Eng.* **2020**, *32*, 04020250. [CrossRef]
36. Dai, J.; Ma, F.; Fu, Z.; Liu, J.; Li, C.; Hou, Y.; Wu, H. Effectiveness of the Different Eutectic Phase-Change Materials in Cooling Asphalt Pavement. *Constr. Build. Mater.* **2023**, *407*, 133491. [CrossRef]
37. Haily, E.; Ait Ousaleh, H.; Zari, N.; Faik, A.; Bouhfid, R.; Qaiss, A. Use of a Form-Stable Phase Change Material to Improve Thermal Properties of Phosphate Sludge-Based Geopolymer Mortar. *Constr. Build. Mater.* **2023**, *386*, 131570. [CrossRef]
38. Baskar, I.; Chellapandian, M.; Jaswanth, S.S.H. Development of a Novel Composite Phase Change Material Based Paints and Mortar for Energy Storage Applications in Buildings. *J. Energy Storage* **2022**, *55*, 105829. [CrossRef]
39. Anupam, B.R.; Sahoo, U.C.; Rath, P. Phase Change Materials for Pavement Applications: A Review. *Constr. Build. Mater.* **2020**, *247*, 118553. [CrossRef]
40. Li, Y.; Du, Y.; Xu, T.; Wu, H.; Zhou, X.; Ling, Z.; Zhang, Z. Optimization of Thermal Management System for Li-Ion Batteries Using Phase Change Material. *Appl. Therm. Eng.* **2018**, *131*, 766–778. [CrossRef]
41. Li, Z.; Gong, Y.; Xu, A.; Zhao, J.; Li, Q.; Dong, L.; Xiong, C.; Jiang, M. Relaxation-Induced Significant Room-Temperature Dielectric Pulsing Effects. *Adv. Funct. Mater.* **2023**, *33*, 2301009. [CrossRef]
42. Gong, Y.; Zhao, J.; Li, Z.; Huang, J.; Zhang, Y.; Dong, L.; Xiong, C.; Jiang, M. Unparalleled Dielectric-Switching Effects Caused by Dual Polarization Synergy. *Adv. Funct. Mater.* **2023**, *33*, 2214544. [CrossRef]
43. Korniejenko, K.; Nykiel, M.; Choinska, M.; Jexembayeva, A.; Konkanov, M.; Aruova, L. An Overview of Micro- and Nano-Dispersion Additives for Asphalt and Bitumen for Road Construction. *Buildings* **2023**, *13*, 2948. [CrossRef]
44. Ricklefs, A.; Thiele, A.M.; Falzone, G.; Sant, G.; Pilon, L. Thermal Conductivity of Cementitious Composites Containing Microencapsulated Phase Change Materials. *Int. J. Heat Mass Transf.* **2017**, *104*, 71–82. [CrossRef]
45. Šavija, B.; Schlangen, E. Use of Phase Change Materials (PCMs) to Mitigate Early Age Thermal Cracking in Concrete: Theoretical Considerations. *Constr. Build. Mater.* **2016**, *126*, 332–344. [CrossRef]
46. Cerón, I.; Neila, J.; Khayet, M. Experimental Tile with Phase Change Materials (PCM) for Building Use. *Energy Build.* **2011**, *43*, 1869–1874. [CrossRef]
47. Hyun, S.W.; Kim, S.; Jeong, H.; Ko, H.S.; Shin, D.H. Development of Snow Removal System Using Embedded Piped inside Road with Solar Thermal Energy Collector and Packed Bed Latent Heat Thermal Energy Storage. *J. Energy Storage* **2024**, *83*, 110737. [CrossRef]
48. Guan, B.; Ma, B.; Qin, F. Application of Asphalt Pavement with Phase Change Materials to Mitigate Urban Heat Island Effect. In Proceedings of the 2011 International Symposium on Water Resource and Environmental Protection, Xi'an, China, 20–22 May 2011; IEEE: Piscataway, NJ, USA, 2011; pp. 2389–2392.
49. Anupam, B.R.; Sahoo, U.C.; Rath, P.; Pattnaik, S. Thermal Behavior of Phase Change Materials in Concrete Pavements: A Long-Term Thermal Impact Analysis of Two Organic Mixtures. *Int. J. Pavement Res. Technol.* **2022**, *17*, 366–378. [CrossRef]
50. Zhang, J.; Dong, Z.; Sun, G. Properties of SBS Modified Asphalt Containing Phase Change Materials. In Proceedings of the International Conference on Road and Airfield Pavement Technology, Beijing, China, 6–8 February 2023; American Society of Civil Engineers: Reston, VA, USA, 2023; pp. 512–523.

51. Najemi, L.; Belyamani, I.; Bouya, M. Effect of Blending of Medium-Temperature Phase Change Material on the Bitumen Storage Heat. *Heliyon* **2023**, *9*, e22040. [CrossRef]
52. Saberi K., F.; Wang, Y.D.; Liu, J. Improving Rheological and Thermal Performance of Gilsonite-Modified Binder with Phase Change Materials. *Constr. Build. Mater.* **2023**, *399*, 132557. [CrossRef]
53. Huang, Z.; Wei, J.; Fu, Q.; Zhou, Y.; Lei, M.; Pan, Z.; Zhang, X. Preparation and Experimental Study of Phase Change Materials for Asphalt Pavement. *Materials* **2023**, *16*, 6002. [CrossRef]
54. Cheng, C.; Liu, J.; Gong, F.; Fu, Y.; Cheng, X.; Qiao, J. Performance and Evaluation Models for Different Structural Types of Asphalt Mixture Using Shape-Stabilized Phase Change Material. *Constr. Build. Mater.* **2023**, *383*, 131411. [CrossRef]
55. Zhang, J.; Dong, Z.; Sun, G.; Qi, Y.; Zhu, X.; Li, Y. Roles of Phase Change Materials on the Morphological, Physical, Rheological and Temperature Regulating Performances of High-Viscosity Modified Asphalt. *Sci. Total Environ.* **2023**, *875*, 162632. [CrossRef] [PubMed]
56. Jiao, W.; Sha, A.; Zhang, J.; Jia, M.; Jiang, W.; Hu, L. Design and Properties of Polyurethane Solid–Solid Phase-Change Granular Temperature Regulation Asphalt Mixtures. *Sol. Energy* **2023**, *253*, 47–57. [CrossRef]
57. Anupam, B.R.; Sahoo, U.C.; Rath, P. Experimental and Numerical Investigations on an Organic Phase Change Material Incorporated Cool Concrete Pavement. *TOCIEJ* **2022**, *17*, e187414952210310. [CrossRef]
58. Anupam, B.R.; Sahoo, U.C.; Rath, P.; Bhattacharya, A. Thermal Performance Assessment of PCM Incorporated Cool Concrete Pavements Using Numerical Analysis. *Int. J. Pavement Eng.* **2023**, *24*, 2089884. [CrossRef]
59. Wang, X.; Ma, B.; Wei, K.; Si, W.; Kang, X.; Fang, Y.; Zhang, H.; Shi, J.; Zhou, X. Thermal Storage Properties of Polyurethane Solid-Solid Phase-Change Material with Low Phase-Change Temperature and Its Effects on Performance of Asphalt Binders. *J. Energy Storage* **2022**, *55*, 105686. [CrossRef]
60. Betancourt-Jimenez, D.; Montoya, M.; Haddock, J.; Youngblood, J.P.; Martinez, C.J. Regulating Asphalt Pavement Temperature Using Microencapsulated Phase Change Materials (PCMs). *Constr. Build. Mater.* **2022**, *350*, 128924. [CrossRef]
61. Ru, C.; Li, G.; Guo, F.; Sun, X.; Yu, D.; Chen, Z. Experimental Evaluation of the Properties of Recycled Aggregate Pavement Brick with a Composite Shaped Phase Change Material. *Materials* **2022**, *15*, 5565. [CrossRef] [PubMed]
62. Dai, M.; Wang, S.; Deng, J.; Gao, Z.; Liu, Z. Study on the Cooling Effect of Asphalt Pavement Blended with Composite Phase Change Materials. *Materials* **2022**, *15*, 3208. [CrossRef]
63. Dai, J.; Ma, F.; Fu, Z.; Li, C.; Wu, D.; Shi, K.; Dong, W.; Wen, Y.; Jia, M. Assessing the Direct Interaction of Asphalt Binder with Stearic Acid/Palmitic Acid Binary Eutectic Phase Change Material. *Constr. Build. Mater.* **2022**, *320*, 126251. [CrossRef]
64. Ren, Y.-X.; Hao, P.-W. Low-Temperature Performance of Asphalt Mixtures Modified by Microencapsulated Phase Change Materials with Various Graphene Contents. *Coatings* **2022**, *12*, 287. [CrossRef]
65. BR, A.; Sahoo, U.C.; Rath, P. Thermal and Mechanical Performance of Phase Change Material Incorporated Concrete Pavements. *Road Mater. Pavement Des.* **2022**, *23*, 1287–1304. [CrossRef]
66. Cheng, C.; Gong, F.; Fu, Y.; Liu, J.; Qiao, J. Effect of Polyethylene Glycol/Polyacrylamide Graft Copolymerizaton Phase Change Materials on the Performance of Asphalt Mixture for Road Engineering. *J. Mater. Res. Technol.* **2021**, *15*, 1970–1983. [CrossRef]
67. Cheng, C.; Cheng, G.; Gong, F.; Fu, Y.; Qiao, J. Performance Evaluation of Asphalt Mixture Using Polyethylene Glycol Polyacrylamide Graft Copolymer as Solid–Solid Phase Change Materials. *Constr. Build. Mater.* **2021**, *300*, 124221. [CrossRef]
68. Dai, J.; Ma, F.; Fu, Z.; Li, C.; Jia, M.; Shi, K.; Wen, Y.; Wang, W. Applicability Assessment of Stearic Acid/Palmitic Acid Binary Eutectic Phase Change Material in Cooling Pavement. *Renew. Energy* **2021**, *175*, 748–759. [CrossRef]
69. Liao, W.; Zeng, C.; Zhuang, Y.; Ma, H.; Deng, W.; Huang, J. Mitigation of Thermal Curling of Concrete Slab Using Phase Change Material: A Feasibility Study. *Cem. Concr. Compos.* **2021**, *120*, 104021. [CrossRef]
70. Zhu, S.; Ji, T.; Niu, D.; Yang, Z. Investigation of PEG/Mixed Metal Oxides as a New Form-Stable Phase Change Material for Thermoregulation and Improved UV Ageing Resistance of Bitumen. *RSC Adv.* **2020**, *10*, 44903–44911. [CrossRef] [PubMed]
71. Ma, B.; Chen, S.; Ren, Y.; Zhou, X. The Thermoregulation Effect of Microencapsulated Phase-Change Materials in an Asphalt Mixture. *Constr. Build. Mater.* **2020**, *231*, 117186. [CrossRef]
72. Ma, B.; Wang, X.; Zhou, X.; Wei, K.; Huang, W. Measurement and Analysis of Thermophysical Parameters of the Epoxy Resin Composites Shape-Stabilized Phase Change Material. *Constr. Build. Mater.* **2019**, *223*, 368–376. [CrossRef]
73. Kakar, M.R.; Refaa, Z.; Worlitschek, J.; Stamatiou, A.; Partl, M.N.; Bueno, M. Effects of Aging on Asphalt Binders Modified with Microencapsulated Phase Change Material. *Compos. Part B Eng.* **2019**, *173*, 107007. [CrossRef]
74. Kakar, M.R.; Refaa, Z.; Worlitschek, J.; Stamatiou, A.; Partl, M.N.; Bueno, M. Thermal and Rheological Characterization of Bitumen Modified with Microencapsulated Phase Change Materials. *Constr. Build. Mater.* **2019**, *215*, 171–179. [CrossRef]
75. Du, Y.; Liu, P.; Wang, J.; Wang, H.; Hu, S.; Tian, J.; Li, Y. Laboratory Investigation of Phase Change Effect of Polyethylene Glycolon on Asphalt Binder and Mixture Performance. *Constr. Build. Mater.* **2019**, *212*, 1–9. [CrossRef]
76. Kakar, M.R.; Refaa, Z.; Worlitschek, J.; Stamatiou, A.; Partl, M.N.; Bueno, M. Use of Microencapsulated Phase Change Materials in Bitumen to Mitigate the Thermal Distresses in Asphalt Pavements. In *RILEM 252-CMB Symposium*; Poulikakos, L.D., Cannone Falchetto, A., Wistuba, M.P., Hofko, B., Porot, L., Di Benedetto, H., Eds.; RILEM Bookseries; Springer International Publishing: Cham, Switzerland, 2019; Volume 20, pp. 129–135. ISBN 978-3-030-00475-0.
77. Mahedi, M.; Cetin, B.; Cetin, K.S. Freeze-Thaw Performance of Phase Change Material (PCM) Incorporated Pavement Subgrade Soil. *Constr. Build. Mater.* **2019**, *202*, 449–464. [CrossRef]

78. Yeon, J.H.; Kim, K.-K. Potential Applications of Phase Change Materials to Mitigate Freeze-Thaw Deteriorations in Concrete Pavement. *Constr. Build. Mater.* **2018**, *177*, 202–209. [CrossRef]
79. Zhou, X.; Kastiukas, G.; Lantieri, C.; Tataranni, P.; Vaiana, R.; Sangiorgi, C. Mechanical and Thermal Performance of Macro-Encapsulated Phase Change Materials for Pavement Application. *Materials* **2018**, *11*, 1398. [CrossRef] [PubMed]
80. Gao, Y.; Huang, L.; Zhang, H. Study on Anti-Freezing Functional Design of Phase Change and Temperature Control Composite Bridge Decks. *Constr. Build. Mater.* **2016**, *122*, 714–720. [CrossRef]
81. Tian, Y.; Ma, B.; Liu, F.; Li, N.; Zhou, X. Thermoregulation Effect Analysis of Microencapsulated Phase Change Thermoregulation Agent for Asphalt Pavement. *Constr. Build. Mater.* **2019**, *221*, 139–150. [CrossRef]
82. She, Z.; Wei, Z.; Young, B.A.; Falzone, G.; Neithalath, N.; Sant, G.; Pilon, L. Examining the Effects of Microencapsulated Phase Change Materials on Early-Age Temperature Evolutions in Realistic Pavement Geometries. *Cem. Concr. Compos.* **2019**, *103*, 149–159. [CrossRef]
83. Arora, A.; Sant, G.; Neithalath, N. Numerical Simulations to Quantify the Influence of Phase Change Materials (PCMs) on the Early- and Later-Age Thermal Response of Concrete Pavements. *Cem. Concr. Compos.* **2017**, *81*, 11–24. [CrossRef]
84. Kim, S.; Oh, H.J.; Han, S.J.; Ko, H.S.; Shin, Y.; Shin, D.H. Development of Black-Ice Removal System with Latent Heat Thermal Energy Storage and Solar Thermal Collectors. *Energy* **2022**, *244*, 122721. [CrossRef]
85. Pinheiro, C.; Landi, S.; Lima, O.; Ribas, L.; Hammes, N.; Segundo, I.R.; Homem, N.C.; Castelo Branco, V.; Freitas, E.; Costa, M.F.; et al. Advancements in Phase Change Materials in Asphalt Pavements for Mitigation of Urban Heat Island Effect: Bibliometric Analysis and Systematic Review. *Sensors* **2023**, *23*, 7741. [CrossRef] [PubMed]
86. Liu, Z.; Wang, Y.; Jia, J.; Sun, H.; Wang, H.; Qiao, H. Preparation and Characterization of Temperature-Adjusting Asphalt with Diatomite-Supported PEG as an Additive. *J. Mater. Civ. Eng.* **2020**, *32*, 04020019. [CrossRef]
87. Yinfei, D.; Pusheng, L.; Jiacheng, W.; Hancheng, D.; Hao, W.; Yingtao, L. Effect of Lightweight Aggregate Gradation on Latent Heat Storage Capacity of Asphalt Mixture for Cooling Asphalt Pavement. *Constr. Build. Mater.* **2020**, *250*, 118849. [CrossRef]
88. Giro-Paloma, J.; Martínez, M.; Cabeza, L.F.; Fernández, A.I. Types, Methods, Techniques, and Applications for Microencapsulated Phase Change Materials (MPCM): A Review. *Renew. Sustain. Energy Rev.* **2016**, *53*, 1059–1075. [CrossRef]
89. Somani, P.; Gaur, A. Temperature Sensitivity Analysis on Mechanical Properties of Phase Changing Material Incorporated Rigid Pavement. *Mater. Today Proc.* **2023**, *93*, 387–393. [CrossRef]
90. Zhang, Z.; Mao, H.; Kong, Y.; Niu, P.; Zheng, J.; Liu, P.; Wang, W.; Li, Y.; Yang, X. Re-Designing Cellulosic Core–Shell Composite Fibers for Advanced Photothermal and Thermal-Regulating Performance. *Small* **2024**, *20*, 2305924. [CrossRef]
91. Yuan, J.; He, P.; Li, H.; Xu, X.; Sun, W. Preparation and Performance Analysis of Ceramsite Asphalt Mixture with Phase-Change Material. *Materials* **2022**, *15*, 6021. [CrossRef]
92. Deng, Y.; Shi, X.; Kou, Y.; Chen, J.; Shi, Q. Optimized Design of Asphalt Concrete Pavement Containing Phase Change Materials Based on Rutting Performance. *J. Clean. Prod.* **2022**, *380*, 134787. [CrossRef]
93. Sha, A.; Zhang, J.; Jia, M.; Jiang, W.; Jiao, W. Development of Polyurethane-Based Solid-Solid Phase Change Materials for Cooling Asphalt Pavements. *Energy Build.* **2022**, *259*, 111873. [CrossRef]
94. Anupam, B.R.; Sahoo, U.C.; Chandrappa, A.K.; Rath, P. Emerging Technologies in Cool Pavements: A Review. *Constr. Build. Mater.* **2021**, *299*, 123892. [CrossRef]
95. Sharifi, N.P.; Askarinejad, S.; Mahboub, K.C. Fracture Performance of a PCM-Rich Concrete Pavement under Thermal Stresses. *Int. J. Pavement Eng.* **2022**, *23*, 221–230. [CrossRef]
96. Anupam, B.R.; Sahoo, U.C.; Rath, P. Effect of Two Organic Phase Change Materials on the Thermal Performance of Asphalt Pavements. *Int. J. Pavement Eng.* **2023**, *24*, 2215900. [CrossRef]
97. Liston, L.C.; Farnam, Y.; Krafcik, M.; Weiss, J.; Erk, K.; Tao, B.Y. Binary Mixtures of Fatty Acid Methyl Esters as Phase Change Materials for Low Temperature Applications. *Appl. Therm. Eng.* **2016**, *96*, 501–507. [CrossRef]
98. Young, B.A.; Falzone, G.; She, Z.; Thiele, A.M.; Wei, Z.; Neithalath, N.; Sant, G.; Pilon, L. Early-Age Temperature Evolutions in Concrete Pavements Containing Microencapsulated Phase Change Materials. *Constr. Build. Mater.* **2017**, *147*, 466–477. [CrossRef]
99. Ryms, M.; Denda, H.; Jaskuła, P. Thermal Stabilization and Permanent Deformation Resistance of LWA/PCM-Modified Asphalt Road Surfaces. *Constr. Build. Mater.* **2017**, *142*, 328–341. [CrossRef]
100. Cheng, P.; Chen, X.; Gao, H.; Zhang, X.; Tang, Z.; Li, A.; Wang, G. Different Dimensional Nanoadditives for Thermal Conductivity Enhancement of Phase Change Materials: Fundamentals and Applications. *Nano Energy* **2021**, *85*, 105948. [CrossRef]
101. Acıkök, F.; Belendir, U.; Ardoğa, M.K.; Şahmaran, M. Multi-Functional Conductive Cementitious Composites Including Phase Change Materials (PCM) with Snow/Ice Melting Capability. *Int. J. Pavement Eng.* **2023**, *24*, 2248347. [CrossRef]
102. Xu, P.; Zhang, D.; Miao, Y.; Muhammad Sani, B.; Zhang, K. Development and Characterization of a Novel Steel Slag-Based Composite Phase Change Aggregate for Snow/Ice Melting of Asphalt Pavements. *Constr. Build. Mater.* **2022**, *341*, 127769. [CrossRef]
103. Sharifi, N.P.; Jafferji, H.; Reynolds, S.E.; Blanchard, M.G.; Sakulich, A.R. Application of Lightweight Aggregate and Rice Husk Ash to Incorporate Phase Change Materials into Cementitious Materials. *J. Sustain. Cem.-Based Mater.* **2016**, *5*, 349–369. [CrossRef]
104. Athukorallage, B.; Dissanayaka, T.; Senadheera, S.; James, D. Performance Analysis of Incorporating Phase Change Materials in Asphalt Concrete Pavements. *Constr. Build. Mater.* **2018**, *164*, 419–432. [CrossRef]
105. Tahami, A.; Gholiakhani, M.; Dessouky, S.; Montoya, A.; Papagiannakis, A.T.; Fuentes, L.; Walubita, L.F. Evaluation of a Roadway Thermoelectric Energy Harvester through FE Analysis and Laboratory Tests. *Int. J. Sustain. Eng.* **2021**, *14*, 1016–1032. [CrossRef]

106. Yang, M.; Zhang, X.; Zhou, X.; Liu, B.; Wang, X.; Lin, X. Research and Exploration of Phase Change Materials on Solar Pavement and Asphalt Pavement: A Review. *J. Energy Storage* **2021**, *35*, 102246. [CrossRef]
107. Qin, Y. A Review on the Development of Cool Pavements to Mitigate Urban Heat Island Effect. *Renew. Sustain. Energy Rev.* **2015**, *52*, 445–459. [CrossRef]
108. Kant, K.; Shukla, A.; Sharma, A. Heat Transfer Studies of Building Brick Containing Phase Change Materials. *Sol. Energy* **2017**, *155*, 1233–1242. [CrossRef]
109. Li, Y.; Ding, Z.; Du, Y. Techno-Economic Optimization of Open-Air Swimming Pool Heating System with PCM Storage Tank for Winter Applications. *Renew. Energy* **2020**, *150*, 878–890. [CrossRef]
110. Mohammadnejad, F.; Hossainpour, S. A CFD Modeling and Investigation of a Packed Bed of High Temperature Phase Change Materials (PCMs) with Different Layer Configurations. *J. Energy Storage* **2020**, *28*, 101209. [CrossRef]
111. Deng, Y.; Shi, X.; Zhang, Y.; Chen, J. Numerical Modelling of Rutting Performance of Asphalt Concrete Pavement Containing Phase Change Material. *Eng. Comput.* **2023**, *39*, 1167–1182. [CrossRef]
112. Salem, T.; Bichara, L. Mitigation of UHI Effect with the Incorporation of PCM in Concrete Pavement-Case Study: City of Rotterdam. In Proceedings of the 2023 Fifth International Conference on Advances in Computational Tools for Engineering Applications (ACTEA), Zouk Mosbeh, Lebanon, 5–7 July 2023; IEEE: Piscataway, NJ, USA, 2023; pp. 93–97.
113. Si, W.; Ma, B.; Ren, J.; Hu, Y.; Zhou, X.; Tian, Y.; Li, Y. Temperature Responses of Asphalt Pavement Structure Constructed with Phase Change Material by Applying Finite Element Method. *Constr. Build. Mater.* **2020**, *244*, 118088. [CrossRef]
114. Nayak, S.; Krishnan, N.M.A.; Das, S. Microstructure-Guided Numerical Simulation to Evaluate the Influence of Phase Change Materials (PCMs) on the Freeze-Thaw Response of Concrete Pavements. *Constr. Build. Mater.* **2019**, *201*, 246–256. [CrossRef]

energies

MDPI

Article

Development of Concrete Façade Sandwich Panels Incorporating Phase Change Materials †

Dervilla Niall [1,*] and Roger West [2]

[1] School of Transport and Civil Engineering, Technological University Dublin, D01 K822 Dublin, Ireland
[2] Department of Civil, Structural and Environmental Engineering, Trinity College Dublin, D02 PN40 Dublin, Ireland; rwest@tcd.ie
* Correspondence: dervilla.niall@tudublin.ie

† This paper is an extended version of conference paper Modelling the thermal behaviour of a precast PCM enhanced concrete cladding panel, published in 2020 CERAI Conference proceedings, Cork Institute of Technology, 27–28 August 2020.

Abstract: Using the mass of a building to store or dissipate heat can reduce the demand on auxiliary heating and/or cooling systems. There is a scarcity of full-scale and full-year studies in the literature, which this study seeks to address, as it is critical to assess the performance of real phase change material (PCM) composites throughout all seasons. This study involved the design and manufacture of precast cladding sandwich panels with a PCM–concrete inner leaf used in three full-scale huts which were instrumented to record thermal data over 18 months. Analysis of these data showed that when the internal air temperature fluctuated through the phase change temperature, the PCM–concrete composite was effective at reducing the internal air temperatures by up to 16% if overnight ventilation was provided and 12% without overnight ventilation in a temperate climate. Furthermore, the PCM located deeper than 60 mm from the internal surface of the wall is ineffective at reducing internal air temperatures. The data also highlighted that the thermal conditions required to activate the PCM only occurred during 30% of the year. The thermal data were used to validate a simulation model which can be used to derive bespoke solutions for this form of technology in real scenarios in any geographical location.

Keywords: thermal energy storage; phase change material (PCM); COMSOL multiphysics; PCM–concrete composite

1. Introduction

According to the World Business Council for Sustainable Development [1], there is currently a stock of more than 80 million buildings in Europe built between 1950 and 1975, a period during which energy performance was not considered in building design. Often, these buildings are fit for purpose structurally; however, their energy performance is very poor.

In order to achieve improved energy performance while minimising associated material consumption, a proposed solution is to retain the load-bearing structure but replace the non-load-bearing façade of the building with a modern, energy-efficient building envelope. Improving the energy performance of building envelopes is a critical strategy for achieving the required reduction in space heating and cooling energy demand necessary—by at least 25% from current values—to meet the net zero carbon target for 100% of buildings by 2030 [2].

One of the most commonly proposed methods of enhancing the energy performance of a building is to use the mass of the building's envelope to store thermal energy temporarily, that is, as a thermal energy storage (TES) system. This absorption and storage of heat during the day can reduce overheating of the internal environment in a building and hence reduce the energy demand of the air conditioning system. The stored heat is then dissipated into the internal environment at night when the temperature of the building naturally reduces. This use of TES improves the thermal comfort of the occupants by moderating internal temperature fluctuations while also shifting electricity consumption to off-peak periods.

An area of research that is gaining attention is the enhancement of TESs through the addition of phase change materials (PCMs) due to their high energy storage density capabilities and their ability to store thermal energy in a constant temperature phase transition process, that is, PCMs provide an enhanced latent heat capacity (Figure 1).

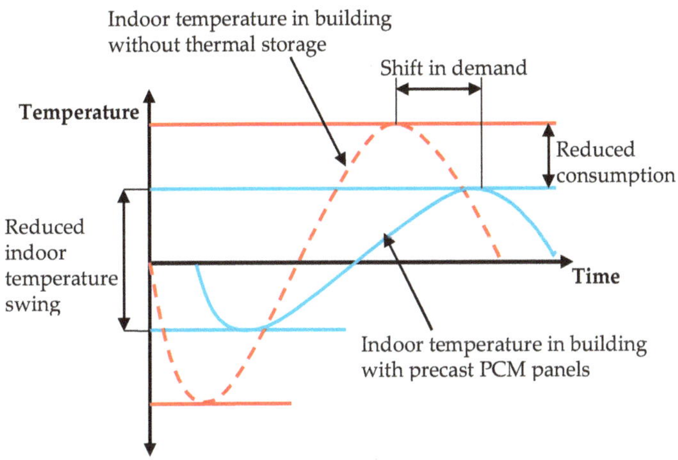

Figure 1. Schematic showing impact of increased thermal mass provided by PCM–concrete panels on internal temperature.

There are many studies in the literature which review the use of PCMs in building envelopes [3–8]. All studies reported enhanced thermal mass behaviour when PCMs are incorporated into the building envelope; however, potential applications of PCMs in building envelopes depend largely on local climate conditions, the melt temperature range of the PCM, the amount of PCM used, its thermophysical properties, the encapsulation method, and the placement location of the PCM within the building [9].

The majority of studies within the body of research into PCM-enhanced building envelopes investigate the use of PCM-integrated wallboards internally in buildings, particularly in a lightweight construction scenario. The use of PCMs in lightweight construction provides a proportionally greater enhancement of thermal mass due to the low baseline thermal mass of lightweight construction. PCM-enhanced wallboards require less space and can also be easily used in retrofitting buildings [4]. In the context of research studies, the costs and logistics of investigating PCM–concrete composites in a full-scale setting can be a significant obstacle to the exploration of the thermal performance of PCM–concrete.

Previous laboratory research carried out by the authors [10–12] has shown that the thermal mass behaviour of concrete can be enhanced by up to 50% through incorporating PCMs. This research also highlighted the fact that the effectiveness of the PCM reduces with depth into the panel. This is due to the fact that the PCM absorbs the heat as it changes phase and hinders the penetration of heat deeper into the panel. The overall thermal storage of a panel will increase as the amount of heat energy transferred to the panel increases. In a real application where a PCM–concrete composite material is used in a building to store thermal energy, the effective depth of the PCM will depend on the temperature profile of the internal environment.

The primary requirement for a PCM that is to be used in a thermal storage application is a suitable melting temperature. Research studies have concluded that the optimum melt temperature range depends on a number of variables including occupancy patterns, ventilation, local climate, and season [13–16]. The optimal phase change temperature of a

PCM is bespoke to the design and geographical location of the application. The context for this research project is to develop concrete cladding panels that will reduce overheating in buildings. From a review of the literature, a PCM with a melt temperature range of circa 19–24 °C would be appropriate for a space-cooling application in temperate climates [17].

The selected PCM must also be chemically stable and compatible with the material with which it is to be combined. From the literature review [18–24], it can be concluded that organic PCMs are the most suitable for mixing with a building material due to their suitable melt temperature ranges (0–150 °C) and their chemical stability [3,5,6]. Paraffin and various fatty acid eutectics have been successfully combined with concrete in previous research [25–27] and have appropriate melt temperature ranges for space cooling applications.

It is extensively reported in the literature that a disadvantage of organic PCMs is that they have low thermal conductivity (circa 0.2 W/mK for paraffin [9]), which can hinder their activation and hence reduce the efficiency of their application. Research studies have been carried out to explore methods of improving the heat conductivity of PCMs [28–30]. A review of the range of studied techniques for improving the thermal conductivity of PCMs can be found in Tan and Zhang [31].

In this study, it is proposed to incorporate the PCMs into concrete. The authors carried out a literature review of the various methods of incorporating PCMs into construction materials, which is included in Navarro et al. [7]. A large proportion of the research that has been carried out on PCMs incorporated into concrete has used microencapsulated organic (ME) PCMs. The distribution of the PCM containing many small capsules provides a large heat exchange surface. The capsule shell also prevents leakage and hence any chemical interaction between the PCM and the cement matrix. The capsule mitigates any issues with a volume change in the PCM material during phase change. ME PCMs also have improved chemical stability and thermal reliability as phase separation during transition is limited to microscopic distances [32]. It is important that the capsule itself is physically and chemically stable within the concrete matrix. Various studies highlighted evidence of damaged capsules within the hardened PCM–concrete composite using SEM images [33,34]. The capsule needs to be hard and durable to avoid being damaged during the concrete mixing and casting process. Tyagi et al. [35] discussed the use of zeolite and zeocarbon, which some researchers have used for reinforcing microcapsules to enable them to withstand high friction and impact during the concrete mixing process. More recently, researchers have produced more resilient PCM additives by using an emulsion polymerisation technique [36,37]. The microcapsules are relatively easy to incorporate into the concrete during the mixing process, and there is no additional site work required. There is also no need to protect the PCM capsules from destruction, such as from postfixed nails, etc.

Extensive research has been carried out on the impact of the addition of PCM on the fresh and hardened properties of concrete [38–40]. With regard to the properties of fresh concrete, it is clear that the incorporation of a PCM by any method will reduce the heat of hydration and hence slow down the early strength gain of the concrete panels. This effect must be taken into consideration particularly in the manufacture of precast cladding panels because the concrete in the panels must achieve sufficient strength to support their self-weight before they are lifted off the casting tables, which typically occurs around eighteen hours after casting in a precast factory.

Concrete strength is the most important property for concrete that is to be used in a building element. Berardi and Gallardo [41] carried out an extensive review of the properties of PCM-enhanced concrete. It was clear from this review that the limit to the quantity of ME PCM that can be incorporated into concrete while still achieving strengths that are suitable for structural applications is 5% by weight of concrete. Higher quantities of ME PCM yield impractically low concrete strengths and also cause a significant reduction in the thermal conductivity and density which tends to counteract the benefit of the increase in thermal storage capacity.

It is also clear from the literature that the addition of ME PCM reduces the workability of concrete. Most studies concluded that the maximum quantity of ME PCM that can be added to concrete while still achieving an acceptable workability is 6% by weight of concrete. It is important to note that the use of higher water-to-cement ratios to compensate for the loss of workability due to the addition of PCM microcapsules would lead to lower 28-day concrete strengths. In order to avoid compounding the strength loss caused by the addition of ME PCM, the use of a superplasticiser is a better strategy for counteracting the loss of workability, achieved without increasing the water–cement ratio.

The relatively low conductivity of the PCM also contributes to a 25 to 50% decrease in the thermal conductivity of the PCM–concrete composite [11,27,40,41]. A reduced thermal conductivity is not necessarily a problem as the desired conductivity of the PCM composite depends on the application. In a thermal storage system for a space heating application, it is required that the heat is absorbed and released gradually over a 24 h period. Hence, it is important that the heat flux characteristics of the composite are appropriate to achieve the desired thermal storage behaviour and thermal inertia. It has been shown by a number of researchers [24,42] that the range of thermal conductivities of PCM–concrete composites is suitable for a diurnal thermal storage period.

The analysis of heat transfer within PCM composite materials is challenging due to its complex thermal behaviour which is influenced by several parameters. When selecting the PCM material for a building application the literature on this topic concludes that the melting/freezing temperature of the PCM should coincide with the desired internal room temperature. However, for the PCM to have a positive effect on reducing the energy use in a building, it is critical that the air temperature in the space where the PCM is located fluctuates sufficiently within a 24 h period to ensure that the PCM material changes phase. Many factors influence this requirement including the thermophysical properties of the PCM and the material it is embedded in, along with the local climate, form of construction, building geometry, and use of the building.

For this reason, full-scale testing under real weather and internal environmental conditions through all seasons is an essential tool in the investigation into the potential of this technology as concluded by previous research studies [4–6]. There is a scarcity of full-scale research studies into the performance of PCM–concrete-enhanced building envelopes in the literature. Another gap identified in the literature is the lack of experimental data from year-long duration full-scale tests, which are required to enable the performance of the PCM composites to be assessed in all seasons [21].

The only previous study in the literature that investigated such PCM concrete walls was carried out by Cabeza et al. [43,44]. Two full-size cubicles were constructed, one with ordinary concrete and the other with concrete which contained 5% by weight of ME paraffin. The walls were instrumented with thermal sensors and internal air temperature was also recorded. Three tests, each lasting one week, were carried out under different conditions, and the study reported reductions in peak internal air temperature of 1–2 °C.

It is important to note that in this previous study, the walls consisted of a single leaf; hence, there was no layer of insulation around the outside of the cubicles. This means that the walls could more readily release heat to the external air at night and solidify, enabling it to be effective the following day. In such a scenario, the PCM would be more effective over a longer period during the year. One of the challenges of using PCMs to reduce overheating in insulated buildings is that during the summer months when the temperatures are high, the night time temperature may not drop low enough for sufficient time to allow the PCM to fully solidify. In a 'real-life' application, a panel forming the building envelope is likely to have an insulation layer outside the internal layer and also an outer layer of some form. This means that an inner layer containing PCM can only release heat from one side during the night, that is, back into the internal environment and hence there is an increased risk that the PCM will not solidify. It is important to obtain a realistic assessment of how the PCM–concrete will react in a typical form of construction, that is, in a two-wythe concrete sandwich panel which includes a layer of insulation which hinders the release of heat from the internal wall. The temperature differential between the wall and the internal environment is usually small compared with the temperature differential between the wall and the external air; hence, the release of heat from the wall at night is relatively low compared with the study carried out by Cabeza et al. [43]. Another limitation of Cabeza's study is that thermal data were only recorded over a three-week period, so the findings are limited to the environmental conditions that existed during the short test periods.

To build on the findings of Cabeza et al.'s research, in this study, three full-scale huts were constructed, two of which incorporated the novel PCM–concrete precast cladding panels (the third was a control panel hut with no PCM), which included a layer of insulation between the outer and inner leaves. All huts had one side fully double-glazed with a sliding door incorporated to allow access and solar gain. The huts were instrumented to record both internal thermal data and local climate data. Data were recorded over an 18-month period to ensure that the thermal performance of the panels could be assessed across all seasons. Thermal data were also recorded throughout the depth of the wall, which enabled the effective depth of the PCM to be determined under varying daily and seasonal thermal conditions.

As all buildings differ, each building will require a unique optimal solution for the application of a PCM composite material as a TES system. For this reason, the development of numerical simulation tools is necessary to achieve a practical and economic application of this technology. Coinciding experimental data, collected from full-scale tests within a real form of construction is required to validate a PCM numerical model. In the literature, most studies that investigated PCM composites within a full-scale setting were numerical simulations in which the model was validated using data from a laboratory study. There is a lack of modelling studies that have been validated against long-term results from full-scale experiments constructed using multiple layered walls and typical construction details [4]. In this study, the data collected in the huts are used to calibrate a simulation model that can be used to accurately predict the response of the PCM composite in any full-scale scenario and any geographical location.

This study has furthered the field of knowledge regarding the thermal performance of PCM–concrete through the provision and analysis of a thermal data set collected within a real form of construction through all seasons together with the development of a novel and validated simulation model, which facilitates further research into this form of technology and the development of optimal bespoke solutions.

This research was part of a European-funded Horizon 2020 project titled IMPRESS, the details of which can be found at https://cordis.europa.eu/project/id/636717 (accessed on 14 February 2024). The overall aim of IMPRESS was to develop innovative precast products for the renovation of existing building stock thereby improving the performance and energy efficiency of European buildings. Partners in this project include a leading concrete cladding company, Techrete Ltd. (Dublin, Ireland), and Sirus International (Dublin, Ireland), who provided the monitoring equipment in the huts.

2. Materials and Methods

2.1. PCM–Concrete Composite Material

In this study, the PCM–concrete composite was formed using an ME paraffin product called Micronal, (Microtek Laboratories Inc., Dayton, OH, USA) which was added to a self-compacting CEM I concrete mix, 5% by weight of concrete, during the mixing process. The mechanical and thermal properties of this PCM–concrete composite material were evaluated in previous research by the authors [11]. Details of the mix constituents are provided in Table 1.

Table 1. Mix design for PCM–concrete used in the inner leaf of the panel.

Quantities	Rapid Hardening CEM I Cement (kg)	Water (kg)	Fine Aggregate (kg)	Coarse 6–14 mm (kg)	Limestone Filler (kg)	SP Premier 196 (L)	Micronal (kg)
Per m^3	330	160	725	935	110	9.9	110

w/c = 0.49.

To facilitate the prediction of PCM behaviour in a thermal storage element, it is essential to have accurate information on the thermal properties of the PCM, particularly the phase change properties. Often properties provided on product data sheets lack the required precision and can be overly optimistic, as reported by a number of researchers [24,45,46]. Therefore, reliable characterisation of PCMs is essential, particularly for successful numerical simulations. When considering an overview of the literature reviewed in which PCMs were characterised, dynamic Differential Scanning Calorimetry (DSC) appears to be the most widely used method [47]. For this study, a dynamic DSC test was used to characterise the microencapsulated PCM. Small, homogenous sample sizes of less than 20 mg were used for the DSC tests along with a low-heat-rate regime, 5 °C/min as recommended by previous research studies [48,49]. The heating and cooling enthalpy curves for the ME PCM, derived from the DSC tests, are plotted in Figures 2 and 3, respectively. The DSC results show that the microencapsulated paraffin had a melt temperature range of 20.2 °C to 25.4 °C and a latent heat capacity of 91.1 J/g.

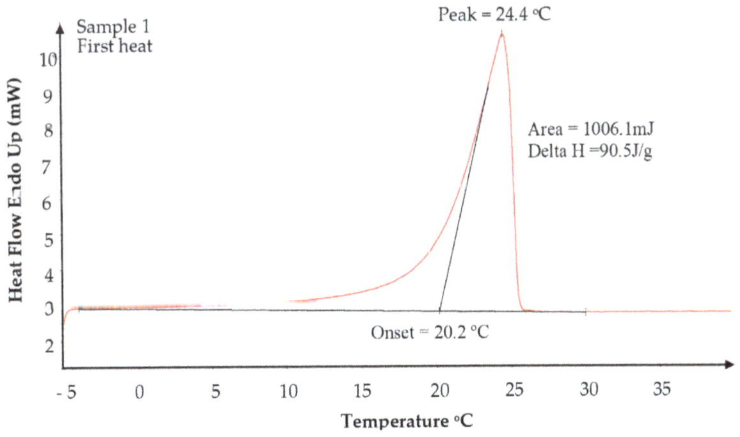

Figure 2. Heating enthalpy curve for ME paraffin measured by DSC.

It can be noted from Figures 2 and 3 that the Micronal PCM displayed thermal hysteresis behaviour during DSC testing, that is, there was a difference between the heating and cooling enthalpy curves. The onset solidification temperature (23 °C) was lower than the highest temperature of the melt temperature range (25.4 °C). This thermal hysteresis behaviour is described graphically in Figure 4. When the material is in a solid state and is being heated, the enthalpy is given by the red curve. As the material passes the highest temperature of the melting phase change temperature range, that is, 25.4 °C, the PCM is completely liquid. When the PCM is subsequently cooled, it will follow the upper blue curve; thus, the PCM remains in a completely liquid state until its temperature falls below the highest temperature of the solidification phase change temperature range, that is, 23.0 °C. During the solidification phase change, the PCM does not become completely solid until the lowest temperature of the solidification temperature range is reached, that is, at 18.2 °C. In the completely melted or completely solid state, the two enthalpy curves overlap.

With thermal hysteresis, in the absence of supercooling, there is little difference between the latent heats of solidification and melting. This hysteresis phenomenon is common in many PCM materials, and it can reduce the beneficial effect of the PCM as the temperature of the environment in which the PCM is located must drop to a lower temperature to ensure that the PCM solidifies fully and the stored heat is released effectively and hence the PCM is ready to absorb heat again when the environment overheats [50].

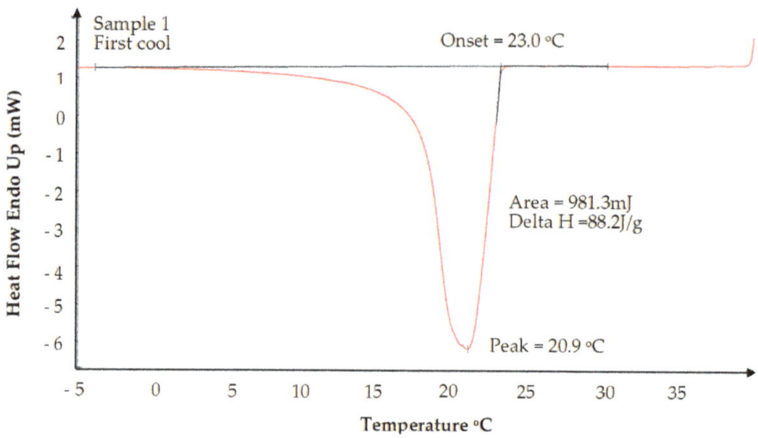

Figure 3. Cooling enthalpy curve for ME paraffin measured by DSC.

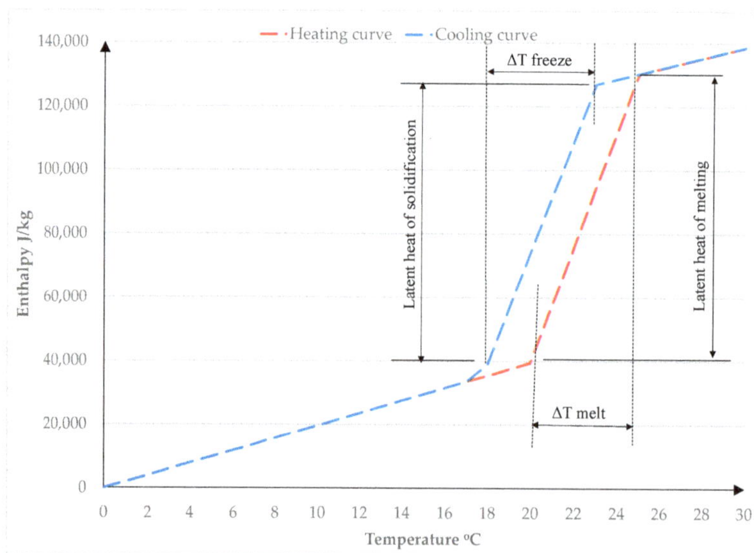

Figure 4. Heating and cooling enthalpy curves for Micronal PCM showing thermal hysteresis.

2.2. Design and Manufacture of PCM–Concrete Cladding Panels

Three different types of panels were designed and manufactured for use in three separate test huts. Each panel comprised a 70 mm thick concrete outer leaf, 120 mm insulation, and a 125 mm thick inner leaf which varies in composition (Figure 5). This study investigates the thermal behaviour of the inner leaf of the cladding panel. As there is insulation between the inner and outer leaves, the form of the outer leaf has an insignificant effect on the thermal behaviour of the inner leaf. Kingspan Kooltherm K15 Cladding Board was used for the insulation layer. This product is a high-performance rigid insulation with a thermal conductivity of 0.02 W/mK.

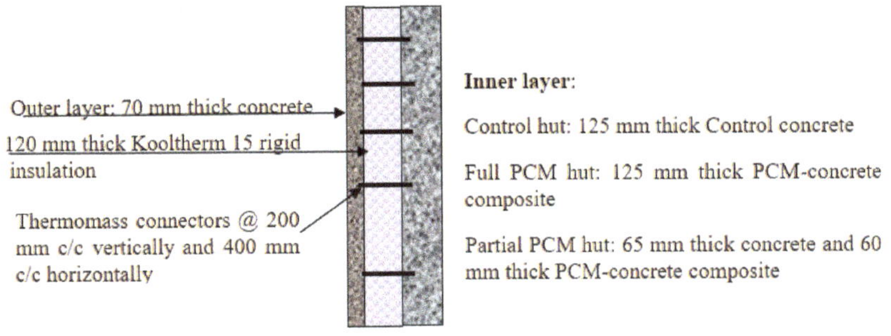

Figure 5. Summary of panel design for huts.

For the control hut, the inner leaf is constructed using a mix without any PCM. For the second hut, the inner leaf is formed using the PCM–concrete composite. This hut is referred to as the Full-PCM hut. In the third hut, the inner leaf is made up of two layers. The inner 60 mm comprises the PCM–concrete composite and the outer 65 mm of the inner leaf, adjacent to the insulation layer, comprises normal concrete without any PCM content. This hut is referred to as the Partial-PCM hut. The purpose of the Partial-PCM hut is to enable the effective depth of the PCM to be assessed. As PCM is an expensive product, it is important to only include it in locations where it can provide benefits. Previous laboratory research by the author [10] demonstrated that as the PCM absorbs heat and melts, it hinders the penetration of heat deeper into the panel, so the PCM becomes less effective with increasing depth. The amount of PCM that will melt during a diurnal period will partly depend on the intensity of heat in the environment where the panel is located. In a real building, the level of exposure to a heat source depends on heating regimes, local climate and the exposure of the wall surface to daylight. The Partial-PCM hut included less than 50% of the PCM depth in the Full-PCM hut. By comparing the thermal behaviour of the Full and Partial huts, any benefit provided by the PCM located at a depth greater than 60 mm can be observed. Also, by observing the temperatures through the walls, the effective depth of the PCM in a full scale application and real form of construction within an Irish climate scenario can be estimated.

The panels and the demonstration huts were manufactured in the Techrete Ltd. (who were partners in the IMPRESS project) manufacturing facility in Dublin. The strength of the PCM–concrete composite was sufficient to facilitate the striking of the formwork and lifting of the panels after 20 h of curing. The manufacture of the panels and erection of the huts proceeded without any problems and demonstrated that the PCM–concrete composite can be successfully scaled up from a laboratory environment to a real building scenario (Figure 6).

2.3. Design and Instrumentation of Full-Scale Demonstration Huts

In order to ensure that the data from each of the huts are comparable, all of the huts had identical design parameters including dimensions, level of insulation, air tightness, glazing, and orientation. The huts were positioned in an open area on the site to mitigate any overshadowing. All the huts are orientated with the glazed elevation facing south (Figure 7). This ensured that each hut was exposed to the same levels of solar irradiance and external air temperature.

Figure 6. PCM–concrete composite panel under construction.

Figure 7. Demonstration huts located in Techrete Ltd.

The clear internal plan dimensions of the hut were 1990 mm × 1990 mm. The main aim of the IMPRESS research study was to develop innovative building envelopes for the renovation of existing buildings. The renovation of an existing building would involve installing a new non-load-bearing façade; however, the existing structure, including the walls and floors, would remain in place and hence their thermal mass would not be enhanced. For this reason, the objective of this study was to investigate the enhanced thermal mass behaviour of just the new wall panels, so the roof slab and floor slab were thermally isolated by placing 120 mm of rigid insulation on the inner surface. This ensured that the concrete roof slab and floor slab did not provide any substantial thermal storage capacity in the hut and any difference in the internal air temperatures within the huts can be attributed to the presence of PCM within the wall panels. The junctions between the roof/floor and the cladding panel were designed to omit any thermal bridging effects.

To observe thermal behaviour, each hut was instrumented for the collection of temperature data and internal and external environmental data (Figure 8). Thermocouples were cast into the internal layer of all the panels, located at depths of 30 mm, 60 mm, and 90 mm from the inner surface. Thermocouples were also located on the internal and external surfaces of the inner leaf and also on the outer face of the insulation layer. Each set of 6 thermocouples was located at the centre of each panel on the north, east, and west elevations. There were two additional sets of 6 thermocouples located in the northern elevation panel. These 30 thermocouples allowed for through-wall temperatures to be recorded and some redundancy in case any were damaged during the manufacture or installation of the panels.

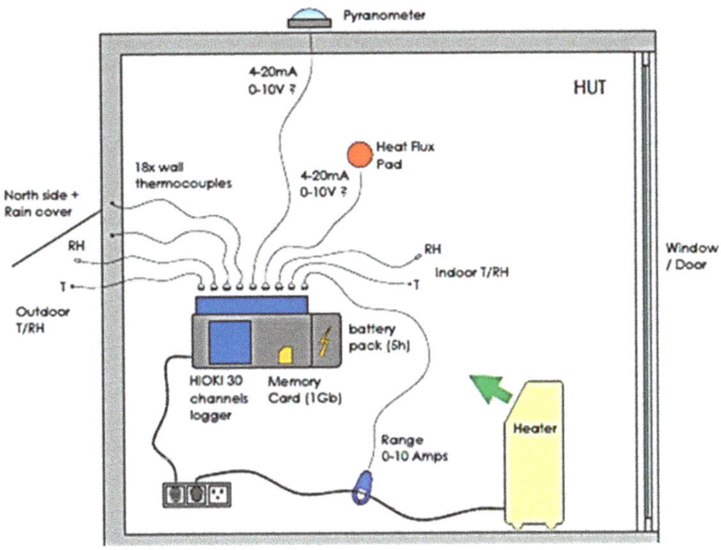

Figure 8. Schematic layout of instrumentation of each hut (courtesy of Sirus).

A heat flux pad was located on the internal face of the north wall in each hut to indicate the heat flow into and out of the wall at the surface. A type K thermocouple (Merck, Darmstadt, Germany) recorded internal air temperature and a HOIKI Z200 device (Hioki, Nangano, Japan) also recorded internal air temperature and relative humidity in each hut. The external temperature and relative humidity were also recorded and an EKO MS-802 pyranometer (Eko Instruments, Den Haag, The Netherlands) was used to record solar irradiance at the site. Interconnected programmable controls on 2 kW oil heaters with the same on/off phases were installed in each hut. The purpose of the heaters was to enable identical artificial heat load patterns to be applied to the huts that replicate a particular scenario, such as an overheating problem. All the data from the thermocouples and instruments was recorded on HOIKI portable data loggers, with one logger located in each hut (Figure 8). A summary of the sensors and monitoring equipment used in the huts is provided in Table 2.

Table 2. Summary of thermal monitoring equipment.

Sensor	Measured Parameter	Sensitivity	Location
Type K Thermocouples	Temperature	±0.86 °C	Through depth of E, W, N * walls in all huts (Refer Figure 9)
EKO MF-180	Heat flux	28.94 (μV/W·m^{-2})	Internal face of N * wall in each hut
EKO MS-802	Solar irradiance	6.98 (μV/W·m^{-2})	External
HIOKI LR8400-20	All data	Calibration certs available	All huts

* N = north; E = east; W = west.

A fundamental requirement for the design of a cladding panel for a building is that it is capable of withstanding any applied forces, in particular wind forces. Laboratory structural flexural testing was carried out on prototype panels to ensure that the proposed

PCM cladding panels would have sufficient strength and stiffness to withstand typical wind loads.

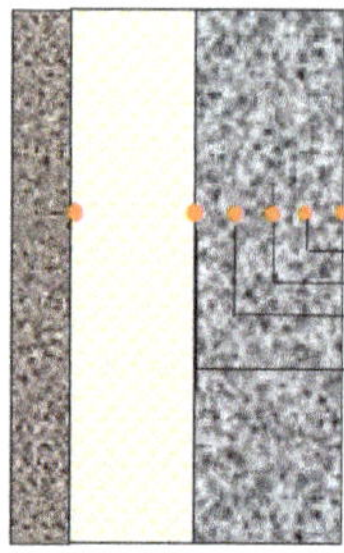

Figure 9. Schematic showing thermocouples through the depth of the wall.

2.4. Development of Simulation Model Using COMSOL Multiphysics Software

As reported widely in the literature, there is a significant number of variables that impact the potential for a PCM composite material to provide a beneficial thermal mass effect. Each individual building requires a bespoke design and hence the development of numerical simulation tools is necessary to enable practical and efficient solutions to be derived. However, as noted previously, there is a scarcity of simulation studies in the previous research that have been validated with data from 'real' full-scale experiments. In this study, COMSOL Multiphysics software version 6.1, was used to develop an accurate, experimentally verified, mathematical model for the prediction of the dynamic thermal environment of buildings which contain PCM–concrete composite material and which accounts for different types of climate and daily and seasonal variations in outdoor conditions in different geographical locations. The thermal data recorded in the full-scale huts were used to validate the model. The thermal behaviour of the PCM material was modelled using the apparent heat capacity method as described by the author in a previous paper [51].

Thermal hysteresis behaviour adds complexity to the development of numerical models for simulating the thermal behaviour of PCMs. A number of researchers have carried out experimental work and modelling studies to gain insight into the impact of thermal hysteresis in simulation models [52–55]. All these studies concluded that the hysteresis effect must be considered in modelling to achieve an accurate prediction of the thermal behaviour of PCMs. The overall heat capacity of the PCM is the derivative of the enthalpy curves with respect to temperature, and it is different depending on whether the material is being heated or cooled. To incorporate this thermal hysteresis into the COMSOL model, two separate enthalpy functions were created, one for the melting phase and one for the solidification phase. These functions were then used to define the temperature-dependent heat capacity of the PCM during phase change.

The PCM–concrete composite material was modelled using the Heat Transfer in Porous Media module in COMSOL. The CEM I control concrete mix was used as the main material in the media. The density, thermal conductivity, and specific heat capacity of this concrete were previously determined in the laboratory as 2335 kg/m^3, 1.86 W/mK, and 881 J/kgK, respectively. The volume fraction of Micronal that was added to the concrete was determined, and the volume fraction of pores filled with PCM was set as $(1 - \theta_c)$, where θ_c is the volume fraction of concrete. Therefore, the effective thermal conductivity of the media was defined as

$$k_{eff} = k_c\theta_c + k_{PCM}\,(1 - \theta_c) \quad (1)$$

The subscript c refers to concrete. Similarly, the effective heat capacity of the porous material was defined as

$$(\rho Cp)_{eff} = \rho_c C_{p,c}\, \theta_c + \rho_{PCM} Cp_{,PCM}(1 - \theta_c) \tag{2}$$

Further information on the COMSOL model can be found in the Supplemental Material—Additional Description of COMSOL model.

3. Results

Environmental data were collected in each hut over an 18-month period. To obtain an overview of when the PCM was engaged throughout the year, Figure 10 displays the temperature of the surface of the north wall in each hut over the full 18-month period. The melt temperature range of the PCM (20–24 °C) is shown as a shaded band. The plot identifies the periods during which the PCM was not active, either because the temperature of the hut walls was too high during the night to allow the PCM to solidify and release its heat, or because the wall temperatures did not get high enough during the day to cause the PCM to melt.

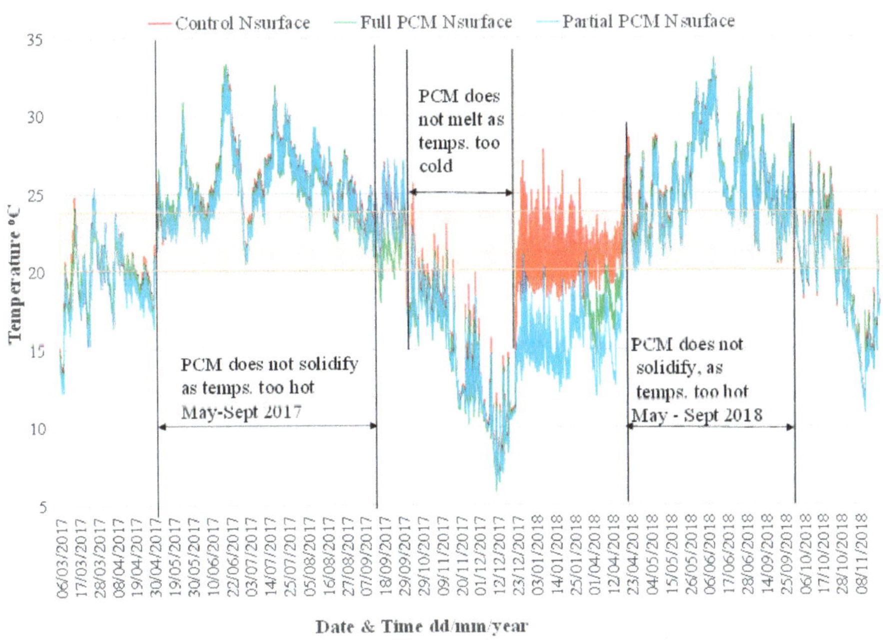

Figure 10. Overview of temperature on the surface of the north wall in each hut over an 18-month period. Nsurface indicates the internal surface of the north wall in each hut.

Overall, the data indicate that in passive conditions, that is, when the heaters are not switched on and no ventilation is provided at night, the PCM is only engaged circa 30% of the year, during the spring and autumn periods. A factor that contributes to the low engagement of the PCM is Ireland's temperate climate. The mean daily temperature in Ireland ranges from 4 °C in winter to 16 °C in summer. Hence, the periods during which the internal temperatures in a building (in the absence of artificial heating) may rise above 22 °C are shorter compared with buildings within hotter climates. Although the PCM was ineffective for circa 70% of the study period during which passive conditions were applied, the application of various heat load regimes and ventilation strategies allowed the behaviour of the PCM to be more fully observed and analysed under varying nonpassive environmental conditions. In the following sections, data for scenarios of particular interest taken over shorter periods are analysed in closer detail.

3.1. Thermal Behaviour of Panels When PCM Is Not Engaged

As the PCM within the concrete matrix is only engaged for a certain percentage of the year, it is important to observe and compare the thermal behaviour of the PCM–concrete composite and the normal concrete when the PCM is either fully solid or fully melted over a full diurnal period.

During the winter months, in passive conditions, that is, when the heaters do not come on during the day, the internal wall temperatures in the huts often remain below the lower temperature of the PCM melt temperature range (20 °C). Similarly, in summer conditions, although the PCM melted, sometimes the huts did not cool down sufficiently overnight to allow the PCM to solidify. Under these conditions, the only difference between the huts is the thermal conductivity of the inner leaf material. The impact of this difference depends on the temperature differential between the internal and external environment. The insulation layer buffers the internal environment from fluctuations in the external air temperature. The glazed elevation also buffers the internal environment from fluctuations in the external air temperature, albeit to a lesser extent due to its higher U-value.

Figure 11 shows a comparison between the internal air temperatures of the huts during a three-day winter period. The peak air temperature remains at 14 °C or below, which is well below the melt temperature of the PCM. It can be observed that there is no significant difference between the internal air temperatures of the three different types of huts when the PCM is continuously solid throughout the period. During the latter two days of said period, the internal air temperature was similar to the external air temperature, so the temperature differential between the internal and external environment was relatively small. However, during the first day of the period, the temperature differential is greater at circa 7–8 °C, and it can be noted that the internal air temperature in all the huts follows the same profile, indicating that the walls of the huts have a similar overall thermal conductivity when the PCM remains in a solid state. The sharp rise in temperature that can be observed on each day is due to radiative heat transfer from the sun showing the effect of solar gain inside the huts.

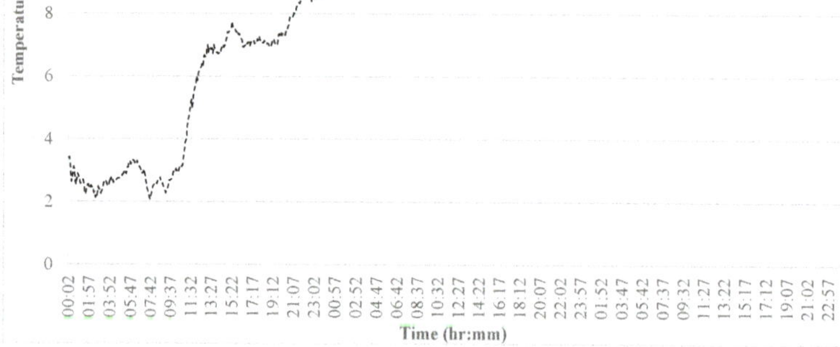

Figure 11. Comparison of internal air temperatures of the huts 19–21 November 2017—PCM fully solid.

Figure 12 displays the internal air temperatures of the huts and the external air temperature during summer. The internal air temperature does not fall below 28 °C, even though the external air temperature falls to 15 °C; hence, it can be assumed that the PCM remains in a fully melted phase during this period.

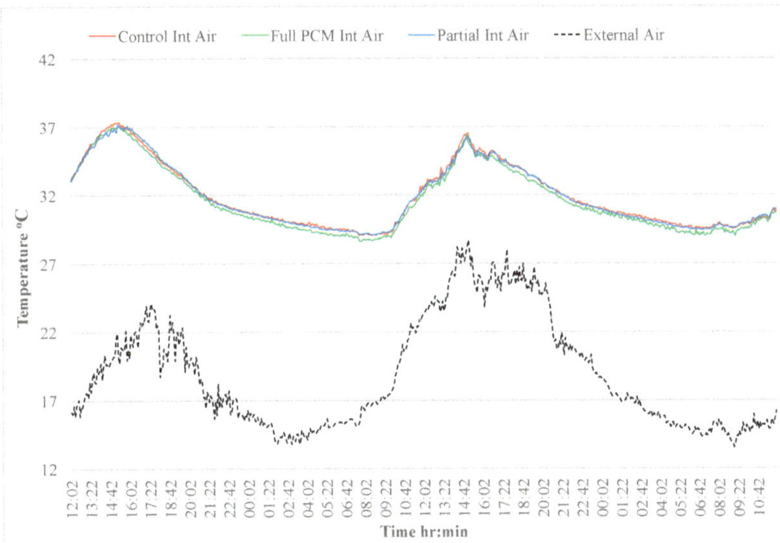

Figure 12. Comparison of internal air temperatures of the huts 20–22 June 2017—PCM not engaged.

As displayed in Figure 12, the profile of the internal air temperature follows the profile of the external air temperature. However, the amplitudes of the internal air temperature profile are lower showing the effectiveness of the insulation layer and double glazing. It can also be noted from Figure 12 that the temperature profile of the internal air in each of the huts is very similar indicating that the heat gain/loss through the walls and glazed elevations of the huts is the same despite the presence of PCM in two of them because the temperature differential between internal air temperature and the external air is the same for each hut.

It can be concluded that under the dynamic nature of the temperatures on either side of the walls and the relatively low temperature differential across the wall, the difference in overall thermal conductivity of the panels is not large enough to result in a significant difference in the rate of heat transfer through the wall and hence the thermal mass behaviour of the PCM–concrete composite panel is similar to that of the normal concrete panel when the PCM is either fully melted or fully solid.

3.2. *Thermal Behaviour of Panels When PCM Is Engaged*

To examine the influence of the presence of PCM in the walls of the huts under passive conditions, two periods were selected during which the wall temperatures fluctuated above and below the melt temperature range of the PCM, one during the winter season and one during the summer season.

3.2.1. Thermal Behaviour of Walls under Winter Passive Conditions

Data from the period from 9:00 a.m. 30 October to 8:55 a.m. on 2 November 2018 were selected and plotted to study the thermal behaviour of the walls under winter passive conditions. During this period, the temperatures in the huts increased above the melt temperature range of the PCM as a result of radiative heat from the sun, despite low external air temperatures. The internal air temperatures in each hut were similar at the start of this period, ranging from 15.1 °C to 15.5 °C. Also, the internal temperatures of the walls in each hut were very similar, both across the three huts and through the depth of each wall, ranging from 17.3 °C to 17.5 °C. As the wall temperatures at the start of the analysis are below 20 °C, it can be assumed that the PCM is in a solid phase throughout at the start of the analysis on 30 October. Figure 13 shows irradiance levels (up to 320 W/m^2) for the period together with the external air temperature (up to 10 °C).

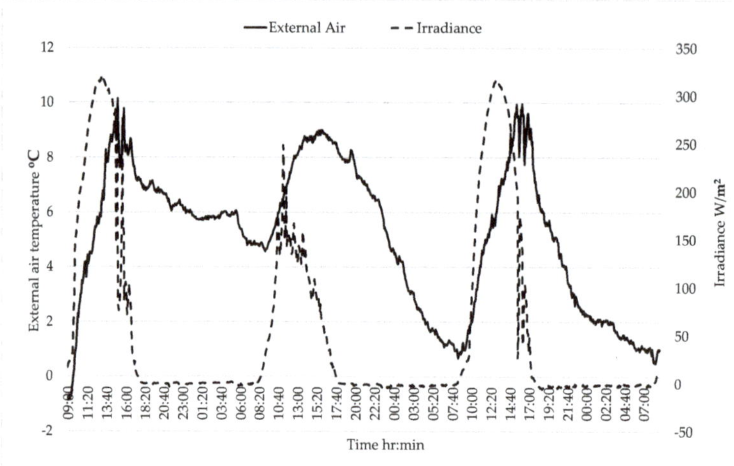

Figure 13. From 9 a.m., 30 Oct to 8:55 a.m., 2 November 2018—irradiance and external air temperature.

Figure 14 displays a comparison of the internal air temperatures of each hut over the said period. It can be observed that, although the rates of heating and cooling are similar in all the huts, the peak temperatures in the control hut on 30 October and 1 November are greater than the peak temperatures reached in the huts containing PCM. The peak internal air temperatures for each hut on the 30 October and 1 November are provided in Table 3.

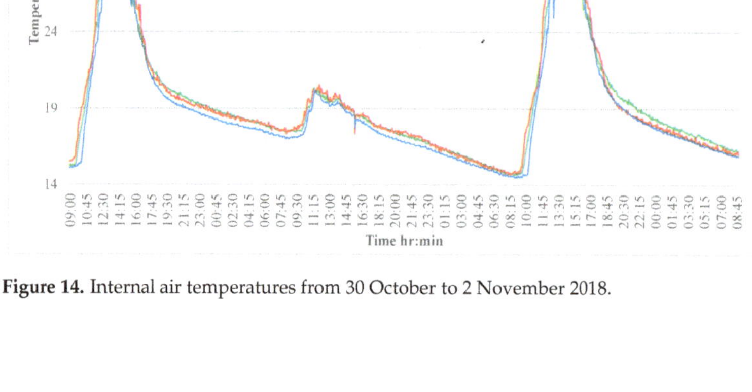

Figure 14. Internal air temperatures from 30 October to 2 November 2018.

Table 3. Peak air temperatures in huts, 30 October and 1 November 2018.

	Control Hut	Full PCM Hut	Partial PCM Hut
	Peak Temperature (°C)	Peak Temperature (°C)	Peak Temperature (°C)
30 October	37.3	32.0	31.9
1 November	34.7	31.2	31.0

There is little difference (due to experimental variability) between the peak temperatures of both huts containing PCM; however, there is a significant difference between the peak temperatures in the PCM huts and the control hut of circa 5 °C on the 30 October and circa 3.5 °C on the 1 November, which indicates a 12% reduction in the peak air temperature of the PCM huts. It is interesting to note that peak internal air temperatures of over 30 °C were reached despite external air temperatures of less than 10 °C and no heaters being on in the huts, albeit being situated in a relatively small volume room and a well-insulated space. This highlights the impact of solar gain on the internal environment particularly when ventilation is not provided. The ability of the PCM to reduce overheating effects is clearly demonstrated.

On 31 October, the irradiance levels were much lower (Figure 13), and the peak temperatures in all the huts were similar at 20.6 °C, 20.4 °C, and 20.2 °C in the control hut, Full-PCM hut and Partial-PCM hut, respectively. These temperatures are too low to activate the PCM.

3.2.2. Thermal Behaviour of Walls under Summer Passive Conditions

Data from the period from 9 a.m. on 9 June to 6 a.m. on 11 June 2017 were selected and plotted in detail to study the thermal behaviour of the walls under summer passive conditions. During this period, the air and wall temperatures in the huts fluctuated within and above the melt temperature range but did not drop low enough at night to cause the PCM to solidify completely because the external air temperatures were not low enough. However, it is of interest to examine the impact, if any, that the presence of PCM may have when the temperature is only fluctuating within the melt temperature range, so the PCM is only ever partially melted.

Figure 15 displays the external air temperatures and irradiance levels for this period. Overall, in comparison with the winter passive external conditions, the external air temperature overnight was on the order of 5 °C to 10 °C warmer, and the irradiance levels were significantly greater over a longer period during the day, not surprisingly.

Figure 16 displays the internal air temperatures in the huts on these days. Table 4 provides the peak internal air temperature reached in each hut on each day. The internal air temperature profile in each hut was very similar with a peak air temperature in the control hut of up to only 0.6 °C greater than those reached in the huts containing PCM. The air and wall temperatures in the huts at the start of the period were circa 22.3 °C, which is greater than the onset melting temperature of the PCM. In these thermal conditions, it can be assumed that a large portion of the PCM is in a melted state at the start of the heating period and hence cannot provide any further latent heat capacity.

Table 4. Peak air temperatures in huts from 9 June to 10 June 2017.

	Control Hut	Full PCM Hut	Partial PCM Hut
	Peak Temperature (°C)	Peak Temperature (°C)	Peak Temperature (°C)
9 June	28.0	27.4	27.8
10 June	29.6	29.0	28.9

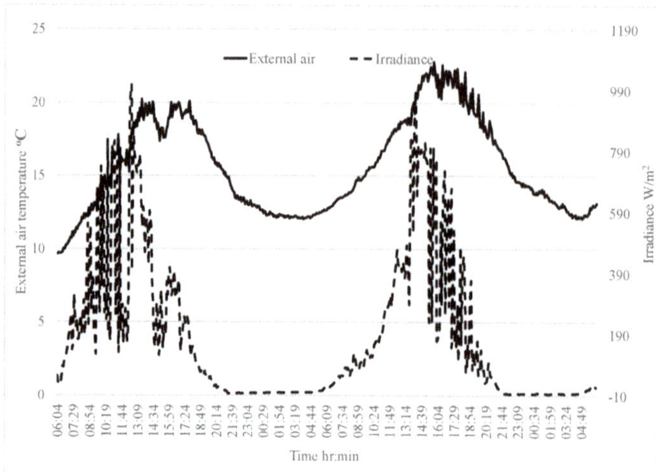

Figure 15. From 6 a.m., 9 June to 6 a.m., 11 June 2017—irradiance and external air temperature.

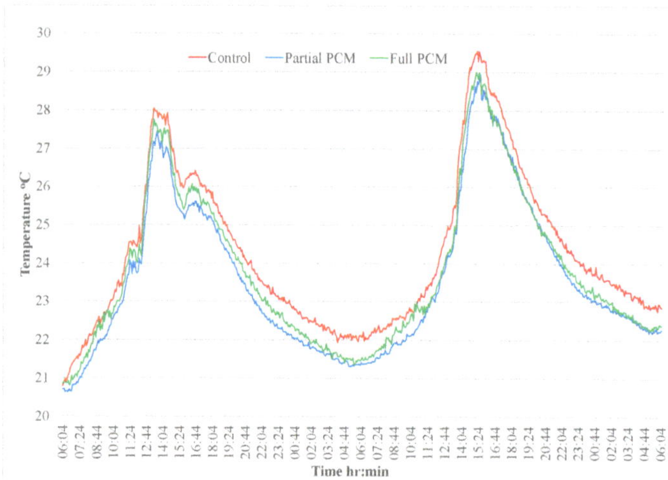

Figure 16. Internal air temperatures from 9 June to 10 June 2017.

The peak internal air temperatures achieved in the control hut during summer passive conditions at 28 °C was less than that achieved during winter passive conditions at 34 °C, despite higher levels of irradiance occurring at the site during the summer dates. The heat gain in the huts is largely due to the absorption of the incident rays by the exposed concrete surfaces. The area of concrete surface that is exposed to solar irradiance is greater in winter, primarily on the walls due to the lower solar altitude angle. In summer, less irradiance enters the hut due to higher sun elevation, and it is primarily on the floor. In this research project, the floors have been thermally isolated with a layer of insulation, so the floor does not absorb any significant incident solar radiation.

3.2.3. Thermal Behaviour of Walls under Applied Heat Load with No Overnight Ventilation

To examine the influence of the presence of PCM in the walls of the huts when an artificial heat load is applied, data from the 1st and 2nd of May 2017 are plotted in detail.

At the start of the period, the air and wall temperature in each of the huts were similar, that is, within less than 0.1 °C of each other. Also, the starting wall temperatures were circa 17 °C, so the PCM can be assumed to be solid throughout the walls. The heaters were set to come on between 8:30 a.m. and 3:30 p.m. each day, whenever the temperature fell below a set point of 20 °C, to simulate a typical heating pattern in, say, a school or working environment.

Figure 17 displays a comparison of the internal air temperatures of each hut over the two-day period. It can be noted from the figure that the temperature in the control hut increased to 27 °C when the heaters were on and then cooled to 19 °C overnight before increasing again on day two to 29 °C. During the heating period on the 2nd of May, the temperatures are slightly erratic; however, this can be attributed to variance in the solar irradiance on this day. The peak internal air temperatures for each hut are provided in Table 5.

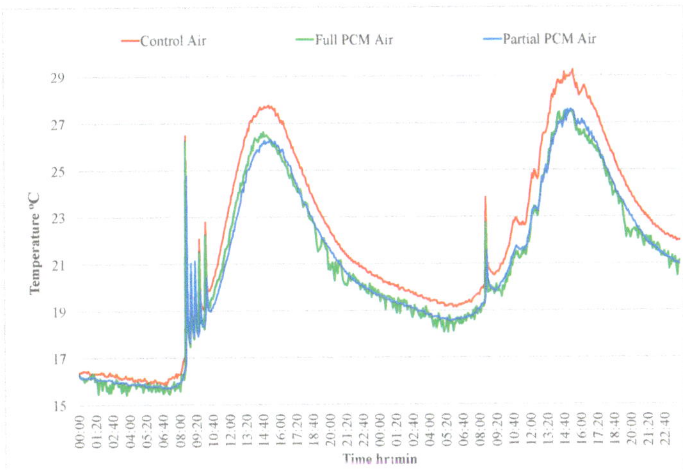

Figure 17. Internal air temperatures on 1 and 2 May 2017.

Table 5. Peak air temperatures in huts on 1 and 2 May 2017.

	Control Hut	**Full PCM Hut**	**Partial PCM Hut**
	Peak Temperature (°C)	**Peak Temperature (°C)**	**Peak Temperature (°C)**
1 May	27.7	26.6	26.3
2 May	29.2	27.5	27.6

It can be observed that when the internal air temperature increases up to circa 28 °C, the presence of PCM results in an average reduction in the internal air temperature on the order of 1.0 to 1.4 °C, a modest amount. It is also interesting to note that the reduction in the internal air temperature in the PCM huts is slightly greater on the second day when the air temperatures are also slightly higher. The higher internal air temperatures lead to higher temperatures within the wall resulting in more melting of the PCM. From these observations, it can be concluded that the PCM within the walls is changing phase during the heating periods, which are approximately 8 h long, and hence somewhat reducing the peak internal air temperatures in the huts containing PCM. The rate of cooling is similar in all the huts.

3.2.4. Thermal Behaviour of Walls When Heaters Are on and Natural Overnight Ventilation Is Provided

It is clear from the data that during the periods when the internal air temperatures of the huts increase significantly during the day, often the internal temperature of the hut at night does not fall low enough to cause the PCM to solidify and release its stored heat. This means that the PCM does not have any latent heat capacity available to absorb the excess heat again the following day. In order to extend the effective period of the PCM throughout the year, the feasibility of 'cooling' the internal environment of the huts using natural ventilation overnight was investigated. The huts were heated during the day, from 9 a.m. to 4:30 p.m., by the radiators to simulate the summer overheating conditions. Natural ventilation was provided overnight by leaving the doors open 100 mm for a fixed period. Figure 18 presents the internal air temperature in all three huts during a 24 h period from 24 to 25 October 2017.

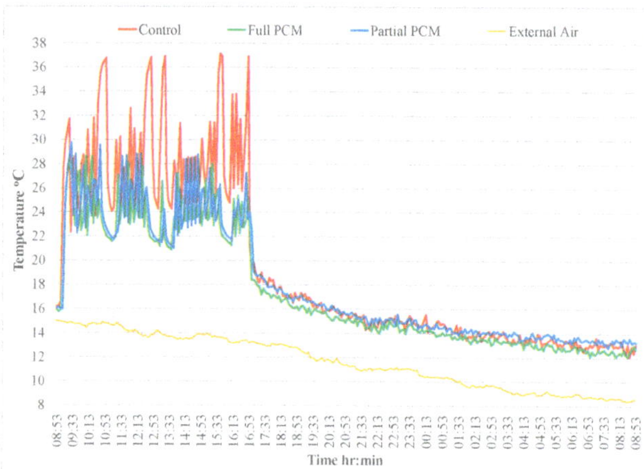

Figure 18. Internal air temperatures in the huts, 24 and 25 October 2017.

When overnight ventilation was provided, the internal air temperature dropped to below 14 °C. The internal air temperatures in all the huts remained below 18 °C from 7 p.m. on the 23rd until 9 a.m. on the 24th. The temperature of the PCM–concrete composite throughout the whole depth of the wall was below 18 °C for circa 5 h during the night. Initially, the air temperatures oscillated as the heaters turned on and off around the set point. It can be noted that the air temperature in the control hut oscillated between 24 °C and 36 °C, providing an environment in which the PCM was expected to melt. During the cooling period, the air temperature reduced below the melt temperature range, hence facilitating the solidification of the PCM and the associated release of heat.

During the heating period, both huts containing PCM were consistently cooler, and the amplitude of the oscillations in air temperature was much smaller. During the heating period, the average temperatures were 28.7 °C for the control hut, 23.9 °C for the Full-PCM hut and 24.4 °C for the Partial-PCM hut. Hence, the Full-PCM hut and Partial huts were an average of 4.7 °C and 4.2 °C cooler than the control hut, respectively, which indicates a 16% reduction in peak air temperature. The temperatures in the huts with the PCM indicated that the PCM is melting and absorbing heat. There is no difference in the rate of temperature decrease during the cooling period, which is to be expected as the doors of the huts were open. This means that the influence of the external air temperature dominated the air cooling rather than the release of heat from the internal wall. On examination of the temperatures throughout the full depth of the walls in each hut, it was noted that the provision of overnight ventilation was successful in ensuring that the temperature of the entire wall thickness dropped below the solidification temperature. It was also observed that the PCM at depths greater than 60 mm in the Full-PCM hut reached temperatures to facilitate phase change, leading to slightly lower air temperature in the Full-PCM hut; however, the difference is not significant (0.5 °C).

3.2.5. Effective Depth of PCM

Throughout the data analysis, whenever the wall temperatures reached the melt temperature range of the PCM, the engagement of the PCM and its effectiveness at reducing the internal air temperatures was evident in both the huts containing the PCM–concrete composite. It can be noted from Figures 14 and 16–18 that in all conditions there is no discernible difference between the internal air temperature profiles of the two huts containing PCM, which suggests that the additional PCM within the 60 mm to 125 mm depth of the Full-PCM hut is not significantly engaged in phase change activity. This result only applies to the environmental conditions analysed because a scenario that provided a greater temperature differential between the internal air and the wall over a longer period of time may lead to increased heat transfer into the wall and engagement of the PCM located deeper in the wall. The effective depth of PCM will be bespoke for a particular project and climate location and optimisation would require simulation modelling.

3.3. *Validation of the COMSOL Model*

In order to validate the COMSOL model, three 3D geometric models were created in COMSOL, one for each type of hut. The models replicated the geometry, orientation, and geographical location of the demonstration huts. The mesh size for the finite element model was automatically set by the COMSOL software in accordance with the physical behaviour under study. The PCM–concrete material parameters were modelled as described in Section 2.4. Virtual probes were located in the model to coincide with the location of the thermocouples in the actual huts. Simulations were carried out on each hut for each time period as analysed in Section 3.2. Conduction, radiative, and convective heat transfer conditions were included in the model. Given the number of probes and the complexity of the thermal behaviour being modelled, the computational cost of a simulation is relatively high. For this reason, the maximum time period to be simulated was restricted to 24 h. Each simulation was preconditioned by setting the initial temperatures of all the domains as per the actual temperatures recorded at the huts at the relevant point in time. The actual varying external air temperature recorded at the huts during the relevant period was imported as an interpolation function into the model and applied to the external boundaries of the hut model. Similarly, the actual irradiance recorded at the huts by the pyronometer was also applied to the model.

The wall and internal air temperatures simulated at each thermocouple location were plotted against the actual temperatures recorded at the hut for each hut during each time period. This resulted in 48 plots; hence, it is beyond the scope of this paper to review all of the simulation results. A sample of a plot comparing the actual and simulated wall temperatures in the Full-PCM hut is provided in Figure 19, and a comparison of the internal air temperature is provided in Figure 20.

To ascertain the accuracy of the simulation results, the absolute value of the simulated temperature (T_{sim}) minus the actual temperature (T_{act}), as recorded by the thermocouples, was calculated for each output at each thermocouple location. The average difference was calculated and denoted as 'Average $|T_{sim} - T_{act}|$'. To determine the precision of the Average $|T_{sim} - T_{act}|$ data, the standard deviation of each set of temperatures simulated at each thermocouple location was determined by calculating the square root of the variance of each data set. The variance of each data set was found by subtracting the Average $|T_{sim} - T_{act}|$ from each data point in the set, $|T_{sim} - T_{act}|$ and squaring the result.

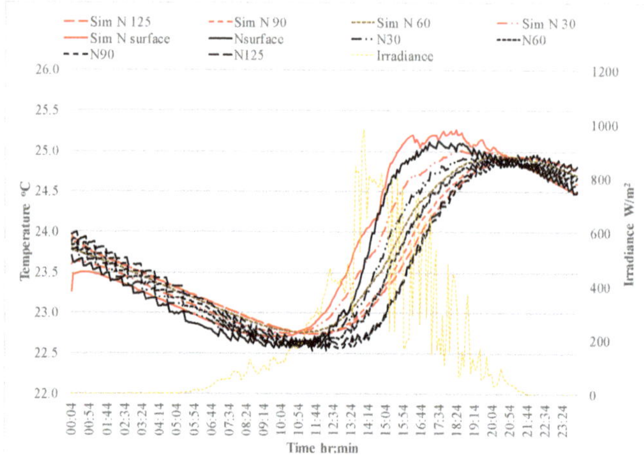

Figure 19. Data from 10 June 2017 Full-PCM hut: north wall inner leaf simulated versus actual temperatures.

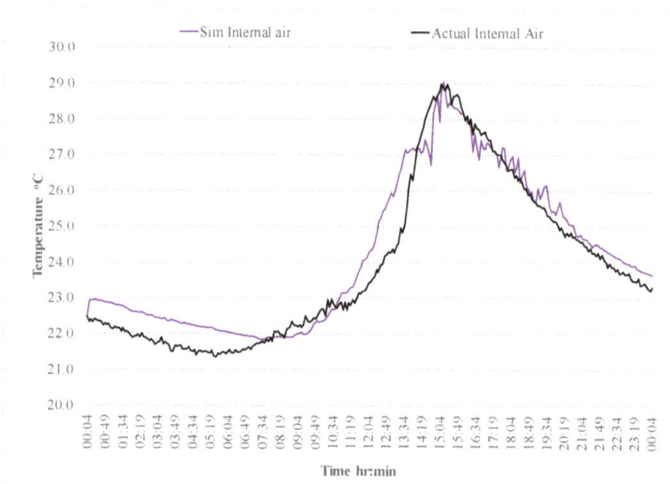

Figure 20. Data from 10 June 2017 Full-PCM hut: simulated versus actual internal air temperatures.

The variance is the average value of these squared values, that is,

$$\text{Variance} = \frac{\Sigma[|T_{sim,i} - T_{act,i}| - (\text{Average } |T_{sim} - T_{act}|)]^2}{\text{Number of data points}} \quad (3)$$

The results of the statistical analysis of the data are summarised in Table 6. The values for Average $|T_{sim} - T_{act}|$ for each wall in a particular hut were averaged. The overall change in temperature in the walls and internal air, that is, the difference between the lowest temperature and the highest temperature, that occurred during the relevant time period is also noted, as it is expected that a larger temperature difference will result in a higher value for the Average $|T_{sim} - T_{act}|$ as the difference between the simulation and actual results may accumulate.

Table 6. Summary of statistical analysis of COMSOL simulation results.

Condition	Element	Control Hut ΔT (°C)	Control Hut Average $\lvert T_{sim} - T_{act} \rvert$ (°C)	Full PCM Hut ΔT (°C)	Full PCM Hut Average $\lvert T_{sim} - T_{act} \rvert$ (°C)
PCM not engaged (solid) 19 November 2017	Wall	0.7	0.13	0.6	0.1
	Air	3.0	0.5	3.0	0.3
PCM not engaged (liquid) 21 June 2017	Wall	2.4	0.2	2.2	0.2
	Air	7.4	0.6	7.0	0.7
PCM engaged winter 1 November 2018	Wall	7.0	0.4	5.8	0.6
	Air	19.6	0.8	16.0	1.3
PCM engaged summer 10 June 2017	Wall	2.4	0.2	2.4	0.1
	Air	7.4	0.5	7.5	0.5

It can be noted from Table 6 that the accuracy of the simulations when the PCM was both engaged (10 June 2017) and not engaged (21 June 2017) was similar for both types of huts when the magnitude of the difference between the lowest and highest temperature in a particular hut, ΔT, was similar, that is, from 7.0 °C to 7.5 °C for the air and from 2.2 °C to 2.4 °C for the walls, indicating that the thermal behaviour of the PCM–concrete was being modelled with similar accuracy whether phase change occurs or not. It can also be noted that the accuracy of the simulations carried out on the same day was similar for the control concrete hut and the Full-PCM hut, indicating that the thermal behaviour of the PCM–concrete was being modelled with similar accuracy as the normal concrete.

The thermocouples used to measure temperature in the huts had a calibrated accuracy of ±0.86 °C. With the exception of the data set measuring internal air on 1 November 2018, all values for Average $\lvert T_{sim} - T_{act} \rvert$ were below 0.86, indicating that the COMSOL model simulates temperatures with similar accuracy as the thermocouples used in the huts. All simulations displayed good accuracy in the initial phase of the simulation prior to the start of the heating phase when the applied irradiance started to enter the hut. The difference between the simulation and actual temperatures increased during the heating phase and subsequent evening cooling phase, albeit while remaining acceptably low. This is expected as the complexity of the model increases with the introduction of irradiance and convection effects as the air heats up.

The graphical and statistical comparative analysis carried out on the results of the COMSOL model simulations of various environmental conditions demonstrated the validity of the developed COMSOL model. The results obtained from the simulations were aligned with the real temperatures recorded in the huts and hence the model can be used to reliably predict the thermal behaviour and impact of the PCM–concrete in a 'real-world' full-scale scenario. The model can be used to observe and quantify any effects of using the PCM–concrete in various global locations and environmental conditions and for different geometries.

4. Discussion

From the analysis of the thermal data collected at the huts presented in Section 3, one of the key findings is that the thermal conditions within the depth of the wall that would provide the potential for the PCM to change phase only occurred during 30% of the year. It is important to note that this outcome is related both to the form of construction and the local climate. The cladding panels included a layer of insulation as is typical in buildings in Ireland. This layer of insulation hinders the ability of the inner leaf to release its heat overnight, and the PCM to solidify as heat loss can only occur from the inner surface of the wall. This means that the rate of decrease in temperature of the internal air in the hut may be relatively low and hence the temperature differential between the wall and the internal air is also relatively low. A compounding factor is the temperate

Irish climate. In summer, the external temperatures, level of irradiance, and the length of the daylight period are high enough to increase the internal temperatures above the melt temperature range of the PCM; however, the overnight temperatures do not drop low enough for long enough to facilitate the solidification of the PCM throughout the walls. It is expected that if the PCM–concrete composite was located in a form of construction that included less insulation and also in a climate which provided greater fluctuation in the diurnal temperature for longer periods during the year, it would have a greater impact on reducing overheating in internal environments of buildings.

It can be concluded from the thermal data analysis that the impact that the presence of a PCM–concrete composite has on reducing overheating varies considerably between each of the scenarios analysed. In order to compare the performance of the PCM–concrete composite across the various environmental conditions, Table 7 provides a summary of key factors and outcomes under each scenario analysed. The peak temperatures reached in the control hut in each scenario are provided along with the length of the heating period. Another factor which is of interest is the temperature difference between the internal air and the internal temperature within the wall. The greater this temperature differential is, the greater the rate of heat transfer between the air and the wall. These factors are provided for in the control hut as an indication of thermal behaviour in the absence of PCM. Table 7 also summarises the key metrics of the performance of the huts containing PCM–concrete composite, that is, the reduction in peak internal air temperature relative to the control hut. The temperature range of the PCM–concrete material within the walls from the start of the heating period to the end of the heating period is also noted.

Table 7. Summary of key measures of PCM behaviour under passive and nonpassive conditions.

	Passive Conditions				**Heating Load Applied**			
	Winter		**Summer**		**No Overnight Ventilation**		**With Overnight Ventilation**	
	Control Hut		**Control Hut**		**Control Hut**		**Control Hut**	
Average difference in temperature between internal air and wall at 30 mm depth during heating phase. (°C)	9.3		1.7		5.6		7.9	
Length of heating period (h:min)	5:10		7:10		7:00		7:30	
Average peak air temperature in the Control hut (°C)	35.9		28.8		28.5		28.7	
	Full	**Partial**	**Full**	**Partial**	**Full**	**Partial**	**Full**	**Partial**
Average temperature range of wall during heating phase (°C)	17.4–22.5	17.4–23.1	22.0–24.0	21.8–23.8	16.7–20.5	16.3–20.4	17.1–21.0	17.0–21.0
Reduction in peak internal air temperature relative to Control hut (°C)	4.4	4.5	0.6	0.4	1.4	1.6	4.7	4.2

It can be noted from Table 7 that under summer passive conditions, the PCM–concrete composite has no significant impact on the internal air temperature, despite the internal air temperature in the huts reaching 28 °C. This result can be attributed to the higher temperature of the PCM–concrete composite at the start of the heating period, which would indicate that the PCM was only partially solid at the start, so its ability to absorb heat through phase change was limited. Also, the average temperature differential between the internal air of the control hut and internally within the wall was relatively low at 1.9

°C. In contrast, during winter passive conditions, the PCM can be assumed to be fully solid at the start of the heating phase, and the wall temperatures increase to 2.5–3 °C above the onset melt temperature. The average temperature differential between the wall and the internal air temperature in the control hut is significantly greater than the differential during summer passive conditions at 9.3 °C. These conditions result in a reduction in peak internal air temperature of approximately 4.5 °C—12% of the peak temperature in the control hut, which is quite significant.

It can also be noted from Table 7 that the PCM–concrete composite was equally effective during winter passive conditions and nonpassive conditions with overnight ventilation. Under winter passive conditions, the PCM–concrete composite reached higher temperatures that extended further into the melt temperature range of the PCM and did not lead to greater effectiveness at reducing internal air temperatures. This could be due to the shorter heating period under winter passive conditions and also that as the main heat source under passive conditions is solar irradiance, the volume of PCM–concrete composite that reaches the peak temperatures noted is likely to be lower as the solar irradiance does not reach all the wall surfaces. However, when an artificial heat load is applied, the primary source of heat is the internal air which is applied evenly over the full surface area of the wall and hence engages a higher volume of PCM.

With regard to the effective depth of the PCM, although the data provided some evidence of phase change in the PCM at depths greater than 60 mm under nonpassive conditions, it was relatively minor and did not impact the internal air temperature. Hence, for reasons of economy, there is a justification for specifying PCM only within the first 60 mm of the internal wall depth. This result only applies to the environmental conditions analysed as a scenario that provided a greater temperature differential between the internal air and the wall over a longer period of time, which may lead to increased heat transfer into the wall and engagement of the PCM located deeper in the wall.

The conclusions of these analyses highlight the multitude of factors that influence the effectiveness of the PCM–concrete composite in reducing internal air temperatures. The data clearly show that the inclusion of PCMs into concrete is an effective if limited strategy for reducing overheating effects in a building and hence the demand on cooling systems under certain thermal conditions. The PCM–concrete composite was effective at reducing internal air temperatures by over 4 °C during winter passive conditions and when a heating load was applied with the provision of overnight ventilation. A key factor in the effectiveness of the PCM composite is the relatively high average temperature differential between the internal air and the PCM composite material in the wall during the heating period, circa 8–9 °C. Another critical factor is the temperature differential between the internal air and the wall material during the cooling period, which must be sufficient to facilitate complete solidification of the PCM overnight. During winter passive conditions, sufficient temperature differential for cooling was provided by the low external temperatures, which were below 10 °C all the time, while the applied heating period through irradiance was only circa 5 h. Under the applied heat load, a sufficient temperature differential during the cooling period was provided through natural ventilation. It can be concluded from the analyses that the provision of a suitable overnight ventilation system would extend the annual period during which the PCM is effective.

The graphical and statistical comparative analyses carried out on the results of the COMSOL model simulations of various environmental conditions demonstrate the validity of the developed COMSOL model. The results obtained from the simulations are aligned with the real temperatures recorded in the huts, and hence, the model can be used to reliably predict the thermal behaviour and impact of the PCM–concrete in a 'real-world' full-scale scenario. The model can be used to observe and quantify any effects of using the PCM–concrete in various global locations and environmental conditions and for different geometries.

5. Conclusions and Further Research

Overall, this research study has demonstrated that a PCM–concrete composite can successfully provide enhanced thermal mass benefits but with limitations. This study has furthered the knowledge in this field of research in a number of areas. Prior to this study, there was only one previous (short-term) study [43] which assessed PCM–concrete in a full-scale scenario. In this previous study, the test huts were constructed using a single leaf wall and thermal data were only recorded over three weeks, so the findings are limited to the form of construction and the environmental conditions that existed during the limited test periods.

The key contributions of the research described in this paper are as follows:

- It has been demonstrated that PCM–concrete can be successfully upscaled for use in real construction projects using standard manufacturing methods.
- A thermal data set was collected within a 'real', full-scale form of construction throughout all seasons, which enabled the impact of this form of technology to be realistically assessed across a full annual period.
- This research study exposed the seasonal effect on the potential benefit of this technology. Analysis of the data set highlighted that the PCM only provides beneficial effects during circa 30% of the year in environmental conditions similar to an Irish climate.
- PCMs located at depths greater than 60 mm within the wall were ineffective. This finding is specific to the local climate conditions and geometry of the huts; however, it highlights the order of magnitude of the effective depth of PCM concrete and the factors that influence it.
- A simulation model was successfully developed to predict the thermal behaviour of PCM–concrete, which was validated using a data set collected from a real form of construction during all seasons.

The achievement of beneficial thermal mass effects is influenced by many variables, including geographic and building-specific characteristics; hence, it is important that the performance of PCM–concrete is assessed under full-scale and realistic building conditions. However full-scale experimental studies are significantly constrained by the cost of the necessary resources and time, and the results cannot be assumed to apply universally. To expand the understanding of this technology, calibrated modelling tools are essential. The development of the simulation model in this study facilitates bespoke solutions to be developed for any geographical location and building geometry, with optimisation of parameters including phase change temperature, latent heat capacity, effective depth, and location of PCM–concrete composite. The annual thermal data set allowed the model to be validated using data collected in a real form of construction. This has furthered the knowledge in the field of research because any models developed in previous research studies were only validated using limited data collected in a laboratory setting. The ability to simulate the thermal behaviour of a PCM–concrete composite will enhance research into this form of technology and make its use more accessible.

One of the main limitations of this technology is that the PCM may only provide beneficial effects during a portion of the year. The economic benefit of the technology is not only subject to the bespoke application of the technology for a particular building but also to the type and cost of the local energy provided to the building and hence can only be assessed on a case-by-case basis.

Another limitation that must be considered in the application of this technology is that to gain the thermal mass benefit of the PCM–concrete, the material must be left exposed within the building. This requirement may not align with architectural intent and may cause challenges in the installation of mechanical and electrical services to the building.

This research study has highlighted some areas of particular interest for further study as follows:

- The simulation model could be used to investigate the performance of PCM–concrete placed in a typical room with concrete floors, walls, and ceilings in various climatic

regions across a whole year period to determine an overview of which geographical locations and seasons would potentially benefit most from this technology.
- The performance of the PCM–concrete composite is also influenced by the ratio of the exposed surface area of the PCM–concrete elements and the enclosed volume of air within the space. Further research could be conducted to investigate if there is a limit to this ratio at which the PCM–concrete ceases to be effective.
- There are some limitations to the simulation model developed in this study, in that it does not simulate the effects of occupant behaviour, and it also assumes complete air tightness. The model could be developed further to allow other factors that influence the internal thermal environment in a building to be incorporated.

Supplementary Materials: The following supporting information can be downloaded at: https://www.mdpi.com/article/10.3390/en17122924/s1, Additional Description of COMSOL model [56,57].

Author Contributions: Writing—original draft, D.N.; Supervision, R.W. All authors have read and agreed to the published version of the manuscript.

Funding: This research received no external funding.

Data Availability Statement: The original contributions presented in the study are included in the article/Supplementary Material, further inquiries can be directed to the corresponding author.

Conflicts of Interest: The authors declare no conflict of interest.

References

1. Energy Efficiency in Buildings—Transforming the Market, World Business Council for Sustainable Development Report 2009. Available online: http://wbcsdpublications.org/project/transforming-the-market-energy-efficiency-in-buildings/ (accessed on 12 February 2024).
2. International Energy Agency (IEA). 2023. Available online: https://www.iea.org/energy-system/buildings (accessed on 12 February 2024).
3. Sharshir, S.W.; Joseph, A.; Elsharkawy, M.; Hamada, M.A.; Kandeal, A.; Elkadeem, M.R.; Thakur, A.K.; Ma, Y.; Moustapha, M.E.; Rashad, M.; et al. Thermal energy storage using phase change materials in building applications: A review of the recent development. *Energy Build.* **2023**, *285*, 112908. [CrossRef]
4. Nair, A.M.; Wilson, C.; Huang, M.J.; Griffiths, P.; Hewitt, N. Phase change materials in building integrated space heating and domestic hot water applications: A review. *J. Energy Storage* **2022**, *54*, 105227. [CrossRef]
5. Lamrani, B.; Johannes, K.; Kuznik, F. Phase change materials integrated into building walls: An updated review. *Renew. Sustain. Energy Rev.* **2021**, *140*, 110751. [CrossRef]
6. Faraj, K.; Khaled, M.; Faraj, J.; Hachem, F.; Castelain, C. A review on phase change materials for thermal energy storage in buildings: Heating and hybrid applications. *J. Energy Storage* **2021**, *33*, 101913. [CrossRef]
7. Navarro, L.; de Gracia, A.; Niall, D.; Castell, A.; Browne, M.; McCormack, S.J.; Griffiths, P.; Cabeza, L.F. Thermal energy storage in building integrated thermal systems: A review. Part 2. Integration as passive system. *Renew. Energy* **2016**, *85*, 1334–1356. [CrossRef]
8. Kosny, J.; Kossecka, E. Understanding a Potential for Application of Phase-Change Materials (PCMs) in Building Envelopes. *ASHRAE Trans.* **2013**, *119*, 1.
9. Souayfane, F.; Fardoun, F.; Biwole, P.-H. Phase change materials (PCM) for cooling applications in buildings: A review. *Energy Build.* **2016**, *129*, 396–431. [CrossRef]
10. Niall, D.; West, R.; McCormack, S. Assessment of two methods of enhancing thermal mass performance of concrete through the incorporation of phase change materials. *Sustain. Des. Appl. Res.* **2016**, *4*, 30–37.
11. Niall, D.; Kinnane, O.; West, R.; McCormack, S. Mechanical and thermal evaluation of different types of PCM-concrete composite panels. *J. Struct. Integr. Maint.* **2017**, *2*, 100–108. [CrossRef]
12. Niall, D.; Kinnane, O.; Kinnane West, R.; McCormack, S. 2016 Influence of Ground Granulated Blastfurnace Slag on the thermal properties of PCM-concrete composite panels. In Proceedings of the Advanced Building Skins Conference, Bern, Switzerland, 10–11 October 2016; pp. 963–973.
13. Ascione, F.; Bianco, N.; De Masi, R.F.; de' Rossi, F.; Vanoli, G.P. Energy refurbishment of existing buildings through the use of phase change materials: Energy savings and indoor comfort in the cooling season. *Appl. Energy* **2014**, *113*, 990–1007. [CrossRef]
14. Diaconu, B.M.; Cruceru, M. Novel concept of composite phase change material wall system for year-round thermal energy savings. *Energy Build.* **2010**, *42*, 1759–1772. [CrossRef]
15. Xiao, W.; Wang, X.; Zhang, Y. Analytical optimization of interior PCM for energy storage in a lightweight passive solar room. *Appl. Energy* **2009**, *86*, 2013–2018. [CrossRef]

16. Pasupathy, A.; Velraj, R.; Seeniraj, R. Phase change material-based building architecture for thermal management in residential and commercial establishments. *Renew. Sustain. Energy Rev.* **2008**, *12*, 39–64. [CrossRef]
17. Waqas, A.; Ud Din, Z. Phase change material (PCM) storage for free cooling of buildings—A review. *Renew. Sustain. Energy Rev.* **2013**, *18*, 607–625. [CrossRef]
18. Sarath, K.; Osman, M.F.; Mukhesh, R.; Manu, K.; Deepu, M. A review of the recent advances in the heat transfer physics in latent heat storage systems. *Therm. Sci. Eng. Prog.* **2023**, *42*, 101886. [CrossRef]
19. Zeng, C.; Liu, S.; Shukla, A. Adaptability research on phase change materials based technologies in China. *Renew. Sustain. Energy Rev.* **2017**, *73*, 145–158. [CrossRef]
20. Jelle, B.P.; Kalnæs, S.E. Chapter 3—Phase Change Materials for Application in Energy-Efficient Buildings. In *Cost-Effective Energy Ef-Ficient Building Retrofitting*; Pacheco-Torgal, F., Granqvist, C.-G., Jelle, B.P., Vanoli, G.P., Bianco, N., Kurnitski, J., Eds.; Woodhead Publishing: Cambridge, UK, 2017; pp. 57–118. [CrossRef]
21. Kenisarin, M.; Mahkamov, K. Passive thermal control in residential buildings using phase change materials. *Renew. Sustain. Energy Rev.* **2016**, *55*, 371–398. [CrossRef]
22. Osterman, E.; Butala, V.; Stritih, U. PCM thermal storage system for 'free' heating and cooling of buildings. *Energy Build.* **2015**, *106*, 125–133. [CrossRef]
23. Cui, Y.; Xie, J.; Liu, J.; Pan, S. Review of Phase Change Materials Integrated in Building Walls for Energy Saving. *Procedia Eng.* **2015**, *121*, 763–770. [CrossRef]
24. Memon, S.A. Phase change materials integrated in building walls: A state of the art review. *Renew. Sustain. Energy Rev.* **2014**, *31*, 870–906. [CrossRef]
25. Cellat, K.; Beyhan, B.; Güngör, C.; Konuklu, Y.; Karahan, O.; Dündar, C.; Paksoy, H. Thermal enhancement of concrete by adding bio-based fatty acids as phase change materials. *Energy Build.* **2015**, *106*, 156–163. [CrossRef]
26. Eddhahak-Ouni, A.; Drissi, S.; Colin, J.; Neji, J.; Care, S. Experimental and multi-scale analysis of the thermal properties of Portland cement concretes embedded with microencapsulated Phase Change Materials (PCMs). *Appl. Therm. Eng.* **2014**, *64*, 32–39. [CrossRef]
27. Fenollera, M.; Míguez, J.L.; Goicoechea, I.; Lorenzo, J.; Ángel Álvarez, M. The Influence of Phase Change Materials on the Properties of Self-Compacting Concrete. *Materials* **2013**, *6*, 3530–3546. [CrossRef] [PubMed]
28. Jin, W.; Huang, Q.; Huang, H.; Lin, Z.; Zhang, J.; Zhi, F.; Yang, G.; Chen, Z.; Wang, L.; Jiang, L. The preparation of a suspension of microencapsulated phase change material (MPCM) and thermal conductivity enhanced by MXene for thermal energy storage. *J. Energy Storage* **2023**, *73*, 108868. [CrossRef]
29. Xu, C.; Zhang, H.; Fang, G. Review on thermal conductivity improvement of phase change materials with enhanced additives for thermal energy storage. *J. Energy Storage* **2022**, *51*, 104568. [CrossRef]
30. Lin, Y.; Jia, Y.; Alva, G.; Fang, G. Review on thermal conductivity enhancement, thermal properties and applications of phase change materials in thermal energy storage. *Renew. Sustain. Energy Rev.* **2018**, *82*, 2730–2742. [CrossRef]
31. Tan, S.; Zhang, X. Progress of research on phase change energy storage materials in their thermal conductivity. *J. Energy Storage* **2023**, *61*, 106772. [CrossRef]
32. Alkan, C.; Sarı, A.; Karaipekli, A.; Uzun, O. Preparation, characterization, and thermal properties of microencapsulated phase change material for thermal energy storage. *Sol. Energy Mater. Sol. Cells* **2009**, *93*, 143–147. [CrossRef]
33. Aguayo, M.; Das, S.; Maroli, A.; Kabay, N.; Mertens, J.C.E.; Rajan, S.D.; Sant, G.; Chawla, N.; Neithalath, N. The influence of microencapsulated phase change material (PCM) characteristics on the microstructure and strength of cementitious composites: Experiments and finite element simulations. *Cem. Concr. Compos.* **2016**, *73*, 29–41. [CrossRef]
34. Lecompte, T.; Le Bideau, P.; Glouannec, P.; Nortershauser, D.; Le Masson, S. Mechanical and thermo-physical behaviour of concretes and mortars containing phase change material. *Energy Build.* **2015**, *94*, 52–60. [CrossRef]
35. Tyagi, V.; Kaushik, S.; Tyagi, S.; Akiyama, T. Development of phase change materials based microencapsulated technology for buildings: A review. *Renew. Sustain. Energy Rev.* **2011**, *15*, 1373–1391. [CrossRef]
36. Stritih, U.; Tyagi, V.; Stropnik, R.; Paksoy, H.; Haghighat, F.; Joybari, M.M. Integration of passive PCM technologies for net-zero energy buildings. *Sustain. Cities Soc.* **2018**, *41*, 286–295. [CrossRef]
37. D'Alessandro, A.; Pisello, A.L.; Fabiani, C.; Ubertini, F.; Cabeza, L.F.; Cotana, F. Multifunctional smart concretes with novel phase change materials: Mechanical and thermo-energy investigation. *Appl. Energy* **2018**, *212*, 1448–1461. [CrossRef]
38. Drissi, S.; Ling, T.-C.; Mo, K.H.; Eddhahak, A. A review of microencapsulated and composite phase change materials: Alteration of strength and thermal properties of cement-based materials. *Renew. Sustain. Energy Rev.* **2019**, *110*, 467–484. [CrossRef]
39. Snoeck, D.; Priem, B.; Dubruel, P.; De Belie, N. Encapsulated Phase-Change Materials as additives in cementitious materials to promote thermal comfort in concrete constructions. *Mater. Struct.* **2016**, *49*, 225–239. [CrossRef]
40. Hunger, M.; Entrop, A.; Mandilaras, I.; Brouwers, H.; Founti, M. The behavior of self-compacting concrete containing micro-encapsulated Phase Change Materials. *Cem. Concr. Compos.* **2009**, *31*, 731–743. [CrossRef]
41. Berardi, U.; Gallardo, A.A. Properties of concretes enhanced with phase change materials for building applications. *Energy Build.* **2019**, *199*, 402–414. [CrossRef]
42. Tang, W.; Wang, Z.; Mohseni, E.; Wang, S. A practical ranking system for evaluation of industry viable phase change materials for use in concrete. *Constr. Build. Mater.* **2018**, *177*, 272–286. [CrossRef]

43. Cabeza, L.F.; Castellón, C.; Nogués, M.; Medrano, M.; Leppers, R.; Zubillaga, O. Use of microencapsulated PCM in concrete walls for energy savings. *Energy Build.* **2007**, *39*, 113–119. [CrossRef]
44. Cabeza, L.F.; Navarro, L.; Pisello, A.L.; Olivieri, L.; Bartolomé, C.; Sánchez, J.; Álvarez, S.; Tenorio, J.A. Behaviour of a concrete wall containing micro-encapsulated PCM after a decade of its construction. *Sol. Energy* **2020**, *200*, 108–113. [CrossRef]
45. Soares, N.; Matias, T.; Durães, L.; Simões, P.; Costa, J. Thermophysical characterization of paraffin-based PCMs for low temperature thermal energy storage applications for buildings. *Energy* **2023**, *269*, 126745. [CrossRef]
46. Tyagi, V.V.; Buddhi, D. PCM thermal storage in buildings: A state of art. *Renew. Sustain. Energy Rev.* **2007**, *11*, 1146–1166. [CrossRef]
47. Klimeš, L.; Charvát, P.; Joybari, M.M.; Zálešák, M.; Haghighat, F.; Panchabikesan, K.; El Mankibi, M.; Yuan, Y. Computer modelling and experimental investigation of phase change hysteresis of PCMs: The state-of-the-art review. *Appl. Energy* **2020**, *263*, 114572. [CrossRef]
48. Memon, S.A.; Cui, H.; Zhang, H.; Xing, F. Utilization of macro encapsulated phase change materials for the development of thermal energy storage and structural lightweight aggregate concrete. *Appl. Energy* **2015**, *139*, 43–55. [CrossRef]
49. Barreneche, C.; Solé, A.; Miró, L.; Martorell, I.; Fernández, A.I.; Cabeza, L.F. Study on differential scanning calorimetry analysis with two operation modes and organic and inorganic phase change material (PCM). *Thermochim. Acta* **2013**, *553*, 23–26. [CrossRef]
50. Dincer, I.; Rosen, M.A. Energetic, environmental and economic aspects of thermal energy storage systems for cooling capacity. *Appl. Therm. Eng.* **2001**, *21*, 1105–1117. [CrossRef]
51. Niall, D.; West, R.; Kinnane, O. Modelling the thermal behaviour of a precast PCM enhanced concrete cladding panel. In Proceedings of the Civil Engineering Research Ireland Conference, Cork, Ireland, 27–28 August 2020.
52. Que, L.; Zhang, X.; Ji, J.; Gao, L.; Xie, W.; Liu, L.; Ding, X. Numerical simulation and experimental research progress of phase change hysteresis: A review. *Energy Build.* **2021**, *253*, 111402. [CrossRef]
53. Khan, R.J.; Bhuiyan, Z.H.; Ahmed, D.H. Investigation of heat transfer of a building wall in the presence of phase change material (PCM). *Energy Built Environ.* **2020**, *1*, 199–206. [CrossRef]
54. Kuznik, F.; Johannes, K.; Franquet, E.; Zalewski, L.; Gibout, S.; Tittelein, P.; Dumas, J.-P.; David, D.; Bédécarrats, J.-P.; Lassue, S. Impact of the enthalpy function on the simulation of a building with phase change material wall. *Energy Build.* **2016**, *126*, 220–229. [CrossRef]
55. Mandilaras, I.; Kontogeorgos, D.; Founti, M. A hybrid methodology for the determination of the effective heat capacity of PCM enhanced building components. *Renew. Energy* **2015**, *76*, 790–804. [CrossRef]
56. COMSOL Heat Transfer Module User's Guide. 2018. Available online: https://doc.comsol.com/5.4/doc/com.comsol.help.heat/HeatTransferModuleUsersGuide.pdf (accessed on 12 February 2024).
57. Kylili, A.; Theodoridou, M.; Ioannou, I.; Fokaides, P. Numerical heat transfer analysis of Phase Change Material (PCM)—Enhanced plasters. In Proceedings of the COMSOL Conference, Munich, Germany, 12–14 October 2016.

energies

MDPI

Article

The Use of Transparent Structures to Improve Light Comfort in Library Spaces and Minimize Energy Consumption: A Case Study of Warsaw, Poland

Ivanna Voronkova and Anna Podlasek *

Institute of Civil Engineering, Warsaw University of Life Sciences—SGGW, Nowoursynowska 159 St., 02-776 Warsaw, Poland; ivanna_voronkova@sggw.edu.pl
* Correspondence: anna_podlasek@sggw.edu.pl

Abstract: Light plays a key role in shaping the quality and atmosphere of interior spaces, and its importance and contradictions are amplified in the design of libraries. This study focuses on the problems associated with both insufficient natural light and excessive light. In both cases, visual discomfort is the result. The reason for these problems lies in the use of translucent structures with different parameters and properties in library architecture. This study analyzed the lighting environment in the main reading rooms of two university libraries in Warsaw. The research methods were based on a field survey of objects and an analysis of the architectural and construction parameters of the premises, as well as the physical and mechanical properties of various types of translucent materials used in the reading rooms. The results of this study shed light on the relationship between the interior space of the reading room and the geometry of transparent structures, as well as the effectiveness of daylighting in libraries in the natural conditions of Warsaw. The key point of the present study was to recognize the critical role that transparent materials and structures play in mitigating potential operational problems in library spaces. Wise selection of these elements at the design stage can help minimize problems related to thermal energy regulation, ventilation control, acoustic insulation, and increased visual comfort. The importance of this research area lies in its potential to optimize library buildings by increasing energy efficiency and reducing reliance on artificial lighting, heating, and air conditioning. Careful consideration of transparent materials at the design stage promises long-term benefits in the form of sustainable library spaces that not only meet functional requirements, but also contribute to a more environmentally conscious architectural landscape.

Keywords: lighting; transparent materials; translucent structures; glass façade; energy efficiency; library space

1. Introduction

The use of natural light in the design of a building's comfortable environment is a decisive aspect not only in terms of shaping the quality of its interior space, but also in terms of its energy efficiency [1,2]. Properly organized daylighting benefits people's physical and mental health, promotes efficient indoor living, and increases the energy efficiency of a building [3–5]. Using daylight to illuminate the interiors of buildings is a prerequisite for a comfortable and healthy working environment [6]. Daylight is considered the healthiest type of light for the human eye due to its color and continuous spectrum [5], while artificial lighting is unable to provide similar conditions due to flicker. The quality of lighting influences human activity-related responses: the speed, accuracy, and effort expended to perform different types of tasks [7]. However, exposure to daylight can have both positive and negative effects on health due to its dynamics and variability. As these changes are difficult to control, this creates several problems.

Too little illumination and its unevenness, excessive brightness, the appearance of glare, and the inappropriate positioning of the workplace in relation to the light source can cause visual impairment and visual fatigue, eye irritation, the appearance of blurred images, or a feeling of double vision. Prolonged exposure to poor lighting conditions can cause or exacerbate visual impairment, affect well-being, and cause accidents [4,7–9].

The optimum lighting environment for visual comfort and safety in the middle of a room depends largely on the amount of light entering the room, the intensity and brightness of the lighting, as well as the reflections that can be created on different surfaces. The lighting requirements set out in the Polish standard PN-EN 12464-1:2022-01 [10] indicate that the light intensity in a reading and writing room should be 500 lx (E_m) [11]. The homogeneity of illumination is characterized by the level of illuminance at different points of the working plane or the entire room and is determined by the ratio of the visual task area to the immediate surrounding area. According to the standard, the minimum value of uniformity in the visual task area should be 0.7 and 0.5 in the nearest surrounding area [8]. In this research study, two of these parameters are taken into account—the amount and the brightness of the daylight.

Scientific studies in recent years dedicated to investigating and discussing the quality of lighting in rooms and buildings under different conditions are mainly concerned with the lighting environment in residential buildings [12], office spaces and commercial buildings [4], waiting rooms [5], and classrooms [13,14]. However, the importance of conducting such studies in library spaces, especially in reading rooms, should not be diminished. Table 1 presents a summary of example research dedicated to library spaces conducted in the last five years.

Table 1. A summary of selected studies dedicated to the design of library spaces and buildings conducted in 2019–2024.

No.	Location	Aim of the Study	Main Findings	Reference
1	Minna, Nigeria	To assess the environmental quality conditions in three academic libraries using objective measurements.	• The illuminance levels in the three library spaces varied significantly and were also found to be unevenly distributed. • The propositions have been provided to deliver adequate daylight. • The quality of the indoor environment in the libraries was poor, indicating the need for intervention.	[15]
2	Tsukuba, Japan Hong Kong, China Lexington, USA	To examine the roles of physical academic libraries as more learning-related activities. To identify the similarities and differences in the usage of library cafés.	• The library café serves as a multi-functional center to meet different needs; it can be used as a study space or a place for social activities and interaction. • The role of light was not analyzed.	[16]
3	Portugal	To analyze light-use management in Portuguese secondary school buildings, focusing on the facilities, user behavior, and the impacts on energy consumption patterns.	• Significant improvements in the lighting conditions of school buildings are essential to improve visual comfort, energy efficiency, and environmental sustainability. • Favorable conditions exist for making natural light the main light source in Portuguese schools.	[17]
4	-	To identify relevant studies that illustrate the benefits of biophilic design specifically for urban university campuses.	• Lighting, air quality, and thermal comfort play significant roles in the quality of learning environments. • Lighting and shading levels should be able to be adjusted. • Bringing daylight into all educational areas creates a biophilic pattern of connection to natural systems, fostering a more productive working environment.	[18]

Table 1. *Cont.*

No.	Location	Aim of the Study	Main Findings	Reference
5	Pennsylvania, USA	To use methods to redesign a library facility.	• Spacious and well-designed workspaces are crucial for the users' productivity.	[19]
6	Northeast China	To identify how the indoor environment influences users' satisfaction and performance in libraries.	• The quality of the indoor environment has the greatest impact on performance in the context of a university library, with interior design playing a secondary role. • Artificial lighting can reduce student satisfaction, whereas natural daylight significantly improves mental health and productivity.	[20]
7	China	To analyze energy consumption characteristics of university facilities.	• Energy efficiency should be integrated into architectural design and layout, functional design, structural design, system design, and other aspects. • The factors that influence energy efficiency of library buildings include landscape greening design, building orientation design, building form design, daylighting and natural ventilation design, outer-building envelope design, sun-shading design, window design, and air tightness design of doors and windows.	[21]
8	Perugia, Italy	To investigate the impact of different design strategies on lighting energy use and visual comfort.	• The type of luminaire is the most critical factor in determining the quantity and quality of light in an indoor environment. • There is no single design solution to guarantee both low energy demand and high visual quality.	[22]
9	London, UK	To present an overview of current knowledge on methods for assessing daylight perception and to establish a methodology for assessing daylight perception in the context of the cultural background.	• A summary of methods for assessing daylight perception was presented and a methodology for assessing daylight perception was established. • Daylight was the key factor when choosing the best desks in the library, with privacy, outdoor views, and quietness being the next most important considerations.	[23]
10	-	To propose a comprehensive set of attributes based on user requirements for assessing higher-education buildings.	• Lighting control aims to ensure the optimal use of light to conserve energy and enhance user comfort.	[24]
11	Turkey	To assess the impacts of several lighting design parameters through a simulation case study.	• The application of LED systems with lighting energy measures can improve lighting energy performance by up to 38%. • An increase in light levels can lead to glare and higher energy consumption, while a higher color temperature can harm the psychology of occupants. • A higher melanopic illuminance and/or color temperature are required to maintain circadian rhythms under overcast skies in winter months.	[25]

A reading room is a place that requires special lighting to ensure the reader's comfort and well-being [8]. Reading rooms vary in parameters and shape, location, number and orientation of windows, as well as purpose. All of these indicators are important because they form the requirements for providing optimal lighting conditions depending on the needs of the user.

The architecture of libraries has changed considerably over the last two decades, with new, modern libraries being designed with the main planning principle of open space and free access to all kinds of information in mind. This has led to the emergence of a single flexible space in the library structure, in which one functional area flows seamlessly 'into' another [26]. In most reading rooms, in addition to areas for concentrated work which are characterized by a regular arrangement of desks, shelves with free access to books, coffee tables with soft furnishings, areas for computer work, etc., are now placed. Filling the reading room with additional functions leads to an increase in the area and volume of the room, which requires a special approach to ensure high-quality lighting for the entire room. On the one hand, it is necessary to ensure that natural light reaches the furthest corners of the room, while on the other hand, over-lighting and the appearance of glare on work surfaces close to the light source must be avoided.

Natural lighting in reading rooms can be either side or top lighting. The choice of lighting is influenced by the size of the reading rooms and their location in the library building. Small reading rooms, which are similar in size to residential units, can be illuminated by windows placed at desk level, but the number, size, and shape of the windows must meet lighting requirements that differ significantly from those for similar residential units (the horizontal lighting factor for residential units is 0.5% and the working area in reading rooms is 1.25%). As the area of the reading room increases, it is necessary to increase the level of the window openings or, if planning permits, to use double-sided raised window positions. This makes it possible to provide the necessary even lighting throughout the hall. As far as overhead lighting is concerned, it is mostly used in large reading rooms and, even then, only in those on the top floors of the building. The disadvantages of such a lighting system include structural complexity, difficulty in handling and cleaning, the possibility of condensation on the inner surface of the glass, and the danger of glare. However, despite all shortcomings, the overhead lighting system has still found its way into global practice [9].

Translucent structures and materials play an important role in shaping the architectural image of a building, influencing the esthetic perception of the building in its surroundings. They lend lightness and accessibility to the façade while protecting the interior spaces of the premises from adverse weather conditions. At the same time, the characteristics of transparent structures are important for the passage of sunlight through them. The most popular light-transmitting material in architecture is glass. Glass has allowed architects to fully express their imagination when creating a unique building image, making extensive use of the characteristics of light and space. The nature of glass is to allow sufficient daylight into a room, to create visual connections with the surroundings, and to give the room a spatial infinity and fluidity. No other material gives such an immediate visual effect [27].

Manufacturers of glass materials used in the building industry today offer a wide range of products: solar control glass, thermal insulation glass, soundproofing glass, self-cleaning glass, toughened glass, transparent photovoltaic (PV) glass, etc. Glass can be transparent, translucent, laminated, tinted, and even 'smart' [27]. Smart glass can help architects achieve their sustainability goals by reducing the amount of electricity needed for indoor lighting, lowering cooling costs, and promoting the health and well-being of building occupants [28]. In addition, Qahtan et al. [29] highlighted the benefits of smart glazing technologies, in particular, polymer-dispersed liquid crystals (PDLCs), in achieving energy efficiency, visual comfort, and environmental sustainability in building design and construction. Smart windows can reversibly change their optical properties to adapt to changing weather conditions or user preferences, thus providing selective and dynamic control over incoming solar radiation This is achieved through the incorporation of chromogenic materials such as thermochromic, photochromic, electrochromic, and gasochromic ones [30].

To choose the right type of glass and transparent design, the needs of the room as dictated by its function must be thoroughly understood. In addition to glass, other translucent materials such as polycarbonate and acrylic are also used in architecture. These are most commonly used in skylights. Polycarbonate and acrylic, due to their construction and properties, diffuse light well, which has a positive effect on reducing direct sunlight entering the room.

Over the past thirty years, trends in library design and construction have shown the widespread use of translucent structures in this type of building and space. On the one hand, this makes the library building visually attractive, open, and friendly, and its spaces bright and pleasant. On the other hand, it is a challenge to provide comfortable lighting conditions inside. Excessive use of translucent structures leads to overheating of the building during the warmer seasons and causes glare on work surfaces. The solution to the problem of overheating is to use large amounts of electricity for air conditioning, which is energy inefficient, unprofitable, and contrary to the principles of sustainable development. Moreover, the incorrect orientation of windows limits the amount of light that can enter on cloudy days. The problem of excessive light is often solved with curtains or tinted films. This contradicts the original idea of using glass in the architecture of libraries, which should be accessible and open.

For this study, it was decided to analyze the lighting environment of the main reading rooms of two university libraries in Warsaw. Both reading rooms are equipped with side and ceiling lighting. However, in both cases, different types of translucent structures are used, which affects the glazing area. Thus, the main objective of this study was to find the relationship between the parameters of the interior space of the reading room and the geometry of the translucent structures, and their influence on the efficiency of the daylighting of the libraries in the natural conditions of Warsaw.

2. Materials and Methods

This study was conducted in two university libraries in Warsaw, Poland: The Library of the Warsaw University of Life Sciences (SGGW) and the University of Warsaw Library (UW).

In the first stage of this study, data from the scientific literature on the physical properties of daylight [31] and the technical properties of translucent structures and materials [32] were analyzed and systematized. The highlighting of key information and the identification of gaps in the literature allowed for a critical look at the topic.

The next step was to analyze the technical documentation of the library buildings, which included architectural drawings of plans, façades, and cross-sections, structural diagrams, and a description of the construction and technical features of the rooms in use.

An important stage of this study was to conduct a field survey. The survey was conducted on the most cloudy and sunny days of February 2024 between 11:00 a.m. and 2:00 p.m. This study was preliminary and allowed us to become acquainted with the light environment of libraries in Poland. On-site inspections of library premises began in February 2024. The time for studying the illumination of reading rooms was chosen according to the largest angle of the solstice and according to the recommendations presented in specialized publications [33,34]. This study was performed in the main reading rooms of each of the libraries, and the subjects of the field research were the transparent structures and materials through which daylight enters the rooms. The lighting condition of the reading rooms was recorded with a Canon EOS 6D camera with a Canon TS-E 17 mm lens. The use of professional photographic equipment made it possible to capture the lighting with maximum accuracy in different weather conditions and at the chosen time of the day.

The light environment of the reading rooms was photographed when the camera was set to manual mode. The technical parameters of the camera were set in such a way as to reproduce as accurately as possible the state of illumination of the premises of the reading rooms. To photograph the lighting of the SGGW library reading room on a cloudy day, the ISO was 1000, the aperture was 8, and the shutter speed was 1/30. To capture the lighting of the UW library reading room on an overcast day, the ISO was 320, the aperture was 8, and the shutter speed was 1/30. To capture the lighting of the UW library reading room on a sunny day, ISO was 100, the aperture was 8, and the shutter speed was 1/30. A photograph of the reading room lighting of the SGGW Library on a sunny day did not show significant changes and was therefore not taken into account in the presentation of the research results in this article. All image files were made in RAW format.

2.1. Room Orientation

The illumination of a building's interior is variable and dependent on external factors such as weather, season, time of day, etc. Therefore, an important element in the design of any building is the orientation of its rooms in relation to the cardinal points. At the same time, the destination of the premises must be considered. North is generally considered the best side for a reading room, as this orientation makes it possible to provide diffused light and avoid direct sunlight entering the room. It can be seen from Figure 1 that the north façade of the building receives sunlight only in summer: in the morning and the evening.

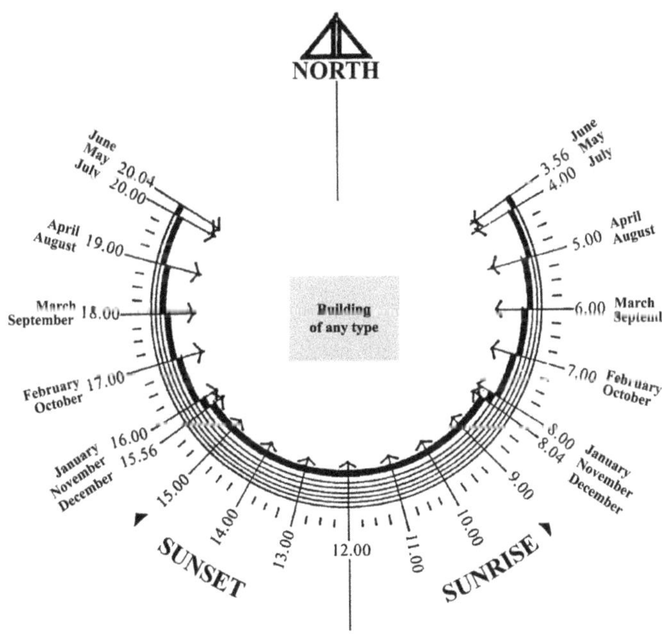

Figure 1. A diagram for determining the optimal orientation of the building according to the cardinal directions (our own interpretation based on [35]).

When determining the correct orientation of a room, the number of sunny days per year is also important. The higher the number, the more direct sunlight will enter the building, which can have both positive and negative effects on its functioning.

Warsaw is located in a temperate transitional climate zone. Throughout the year, continental and oceanic air masses flow over the city, which results in a wide variety of weather conditions within Warsaw [36]. The amount of sunlight in the city varies according to the season. This study presents indices of the number of hours and days of sunshine in Warsaw (Tables 2 and 3). From the data provided, it can be seen that the greatest amount of sunshine occurs in summer. In July, the city receives an average of 295.4 h of sunlight.

During the year in Warsaw, cloudy days far outweigh sunny days. The monthly number of sunny days ranges from 0.9 to 3.6, while the number of cloudy days varies from 3.6 to 18.7. Warsaw has an average of only 28.4 sunny days per year, which is only 8%.

Table 2. The sum of sunshine hours in Warsaw (according to [37]).

Characteristics	January	February	March	April	May	June	July	August	September	October	November	December
Maximum	71.8	109.5	214.9	317.5	356.5	390.2	437.5	345.7	298.5	196.1	90.3	67.6
Average	44.6	66.5	139.4	210.1	272.4	288.8	295.4	280.2	193.1	122.6	50.6	33.6
Minimum	12.2	16.5	74.3	96.0	134.5	164.4	148.6	187.1	79.3	47.1	22.7	4.7

Table 3. Average value of sunny and cloudy days in Warsaw (according to [37]).

Characteristics	January	February	March	April	May	June	July	August	September	October	November	December
Sunny days	1.4	1.7	3.2	3.2	2.6	1.9	2.4	3.5	3.6	2.8	0.9	1.2
Cloudy days	17.5	14.5	10.5	6.9	5.2	4.5	4.1	3.6	6.4	9.5	16.9	18.7

2.2. *Parameters of Room Geometries and Translucent Structures*

The distribution of daylight and its direction are the main determinants of visual comfort and comfort indicators, which are brightness and illumination [7]. The quality of a building's interior lighting depends on its architectural design, the geometry of the space, the size and material of the transparent structures, and the characteristics of the surfaces on which the light falls.

This study focuses on determining the geometric parameters of the main reading rooms at the SGGW and UW Libraries. The geometric parameters of the interior space of a reading room include the room's width (Wrr), depth (Drr), and height (Hrr). This study also defines the geometric parameters of the transparent structures, which include the structures' height (Hts), the structures' width (Wts), windowsill height (Hws), the number of windows (Nw), and the number of window cells (Nwc). These parameters enabled the calculation of the room's floor area (Arr), the room's volume (Vrr), the glazing area (Ag), and the ratio of the semi-transparent structure area to the floor area (RtsF).

By comparing the results of the geometric calculations with field studies, it was possible to assess the efficiency of the use of horizontal rows of windows in glass wall construction and the efficiency of the use of punctual skylights in flat roof construction.

This study also analyzed the physical and mechanical properties of various types of semi-transparent materials used in reading rooms.

3. Results

3.1. *The Main Reading Room of the SGGW Library*

The Library of the Warsaw University of Life Sciences (SGGW) is an important structural component of the university campus, which is spread across the Ursynów district in Warsaw (Poland).

The library building is adjacent to Nowoursynowska Street, surrounded by university buildings and a complex of student dormitories (Figure 2, Table 4). This location makes the library an active participant in the educational process, fostering continuous interaction with students and lecturers.

(a) (b)

Figure 2. The location of the SGGW Library: (a) a view of the library building from the southeast [38]; (b) the ground floor plan of the library building.

Table 4. Source data for SGGW library project; own research.

Description	Data
Address	Nowoursynowska St. 161, Warsaw, Poland
The distance from the city center	8.5 km, convenient connection with the city center
Location	On the campus of SGGW
Project developer	Architect: St. Fijalkowski. General Design Office "Budepol"
Implementation of the project	1996–2006
The building area	2310 m^2
The number of floors	2 above ground
The number of reading rooms	4
The total area of the reading rooms	~1200 m^2

The library building, designed by architect St. Fijałkowski, was constructed in three stages—from 1996 to 2006—as evidenced by its volume, divided into two square buildings with separate entrances. One building has three floors, while the other has two floors. The buildings are connected at the first-floor level by a covered passageway. The building has a surface area of 2310 m^2 [39]. The library structure comprises four reading rooms with a total area of approximately 1200 m^2. This study was carried out in the main reading room, located on the first floor of the two-story part of the building. The reading room has a rectangular shape, measuring 19.6 × 32.4 m with a height of 4.6 m [40,41]. The area of the room is 635 m^2 and its volume is 2921 m^3.

The exterior of the building uses a mullion–transom glazing system, which is one of the most popular methods for creating glass façades due to its technical properties and affordability [42]. The façades are decorated with glass, giving the impression that the building is made entirely of glass. However, only part of the glass acts as a semi-transparent material, allowing natural light to enter the rooms.

The rest of the glass is used as finishing material to cover the wall structures. The hall's windows decorate two façades facing northeast and northwest. This orientation maximizes the amount of direct sunlight entering the room. On the one hand, this is a good solution because it prevents glare on work surfaces and provides diffused light in the hall. On the other hand, given the size of the hall, this orientation does not provide an adequate level of interior lighting, especially in cloudy weather, despite the large number and size of the windows. Figure 3 clearly shows that the reading room is well lit near the windows, but poorly lit on the opposite side.

Figure 3. SGGW library reading room: (**a**) view from window side; (**b**) view from entrance side. Author: I. Voronkova.

There are 11 windows on the northeast side of the room and 21 windows on the northwest side. The windows are placed close together, forming two long rows on each side, taking up almost the entire wall area. Each window is 1.5 m wide and 3.05 m high. The windows are positioned 1 m above the floor. Each window consists of three sections: two narrow 0.55 m high sections (both of which can be opened) and a 1.95 m high section. The structure of the window consists of a metal frame (with a frame profile width of 50 mm and a thickness of 28 mm) into which a double-glazed window with two panes is inserted. The double glazing consists of two panes of different types of glass. The outer pane, facing the street, is made of 6 mm thick Stopsol Classic Green reflective glass, while the inner pane, facing the interior, is made of 4 mm thick Termofloat glass.

The reflective glass acts as a barrier, controlling the amount of heat energy emitted by the sun due to its low solar coefficient. The pane is coated with a special film that reflects sunlight and protects the building from excessive solar radiation, reducing cooling costs. Additionally, the highly reflective coating creates a mirror-like surface effect that provides excellent privacy, which is why it is actively used in façade glazing. From the inside, however, the glass appears transparent, allowing an undistorted view of the outside [43–45]. The main characteristics of the reflective glass are shown in Table 5.

Table 5. A summary of the light, solar, and thermal properties of the Stopsol products of 6 mm thickness (according to [44]).

Properties	Light Transmission, %	Light Reflection on the Coating Side, %	Light Reflection on the Glass Side, %	Direct Energy Transmission, %	Energy Reflection on the Coating Side, %	Energy Reflection on the Glass Side, %	Normal Emissivity
Value	31	34	20	22	28	11	0.89

Thermofloat glass, also known as low-emissivity glass, is used to retain heat in a room. The surface of Thermofloat glass is coated with a special layer that reflects long-wave heat radiation while allowing the same amount of daylight to pass through like it would through normal transparent glass. This feature increases the thermal insulation of the window and has a positive effect on both heating costs and user comfort [46].

In addition to the front windows, the reading room is equipped with skylights installed in the ceiling, allowing additional daylight to enter. This design feature is possible because there are no other rooms above the reading room and the hall's ceiling forms a flat roof. The skylights are grouped into three rows (with eight units in each row) and placed on the opposite side of the hall from the windows. The dimensions of each skylight are 1.2 × 1.2 m, with a depth of 2.5 m [41]. These proportions suggest that the skylights are insufficient to provide adequate daylight to the room, as the intensity of light decreases over such a long path, creating a "well" light effect (Figure 4).

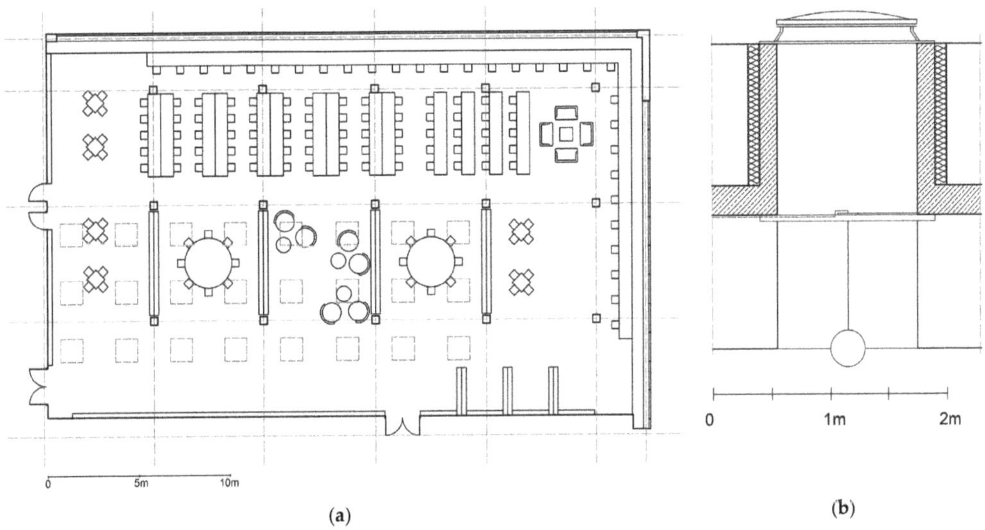

Figure 4. SGGW library reading room: (**a**) floor plan; (**b**) skylight cross-section. Author: I. Voronkova (own study based on [41]).

The top of the skylight is covered with a cellular polycarbonate sheet and a polyester panel. Cellular polycarbonate is a lightweight and practical material that is resistant to adverse weather conditions and has a special UV coating that allows only sunlight to pass through [47,48].

Cellular polycarbonate sheets and polyester panels diffuse sunlight due to their cellular structure and matt surface, providing soft and even illumination of the room. However, due to the considerable depth of the skylight opening in the ceiling of the SGGW library reading room, this function is not fully exploited. In addition, each skylight opening is fitted with an artificial light source from below a pendant lamp with a round shade that covers part of the opening, further restricting the amount of natural light entering the reading room.

Despite all of the advantages of the window design of the reading room which adorns the façade and roof of the SGGW Library, their size, number, and light and heat transmission properties still do not provide an adequate level of natural lighting. In this case, the main drawback is the orientation of the windows, which does not work well in Warsaw due to the relatively few sunny days per year.

3.2. The Main Reading Room of the University of Warsaw Library (UW)

An important functional element within the structure of the university is the University of Warsaw Library (UW), with its buildings spread across three districts in Warsaw, Śródmieście, Ochota, and Służewiec, as well as outside the city and even beyond Poland's borders [49]. The analyzed UW Library is located in the central part of the city—Śródmieście—on a 3.3 ha plot, following the development plan for Śródmieście designated for University of Warsaw buildings and green areas [50].

The area of the UW Library is bordered by Dobra, Lipowa, and Wybrzeże Kościuszkowskie Streets and is surrounded not only by university buildings but also by other types of facilities: residential buildings, shops, restaurants, beauty salons, etc. It is close to the Copernicus Science Centre and the Old Town with the Royal Castle. The UW Library's location not only makes it an important participant in the educational process, but also promotes active interaction with the city's community (Figure 5).

(a) (b)

Figure 5. The location of the UW Library: (**a**) a view of the library building from the southwest [51]; (**b**) the layout of the UW library building.

The library building at the University of Warsaw was built between 1993 and 1999 based on a design by architects Budzyński and Badowski in cooperation with the then Director of the Library, Kobierska-Maciuszka (Table 6).

Table 6. Source data from University of Warsaw Library (UW) project; own research.

Description	Data
Address	Dobra St. 56/66, Warsaw, Poland
The distance from the city center	2.1 km, convenient connection with the city center
Location	In the center of the city
Project developer	Architects: M. Budzyński, Zb. Badowski
Implementation of the project	1993–1999
The building area	~14,450 m^2
The number of floors	4 above ground and 2 below ground
The number of reading rooms	12
The total area of the reading rooms	~3200 m^2

The building is divided into two parts: the library part, where the main library operations take place, and the commercial part, where rental offices, bookshops, agencies, and cafés are located. The two parts are connected by an "alley", which serves as a public space and is used for various social events. The building has four aboveground levels and two underground levels. The floor plan of the building is approximately 129 × 112 m, with a total area of approximately 14,450 m^2 [50]. The main attraction of the library is the roof garden—one of the largest and finest roof gardens in Europe [52,53].

The UW library building is mainly made of concrete. Its façades are richly decorated with large windows, not only in the walls but also around the perimeter of the roof. In addition to the windows, one-third of the roof area is occupied by skylights which provide natural lighting for the "alley" inside the building, the library's main information hall, the main reading room, the periodicals room, the special collections room on the top floor, as well as the underground car park entrance, staircases, boiler room, kitchen, etc. [50].

The structure of the library comprises twelve zones with a total area of approximately 3200 m^2 designed for interaction with the library's collection. This study was carried out in the main reading room, located on the second floor of the building. The reading room is rectangular, measuring 19.0 × 28.8 m, with a height to the highest level of over 12.5 m [50]. The floor area of the first level of the room is approximately 550 m^2, with a volume of approximately 5965 m^3. The reading room has two levels, the second of which is a mezzanine floor with individual workstations.

The reading room has no windows. On the northwest side, the building has a semi-structural wall which, through the use of curved glass at the junction, flows seamlessly into a double-pitched glass roof. Given the amount of glazing (over 125% of the floor area) and the orientation of the transparent structure, it is clear that the reading room has a high level of natural light.

High-quality lighting is best achieved when the sky is overcast and direct sunlight is not entering the room. The room is then filled with soft, diffused light, which, given the primary function of the reading room—working with printed materials and using computers—is the most comfortable light for the human eye. In sunny weather, however, direct sunlight passes through the glass roof, causing glare not only on the walls but also on the working surfaces of the desks. This leads to overheating in the lit part of the room, creating uncomfortable conditions for concentrated mental and visual work. The difference in natural light intensity in the UW library reading room under different weather conditions is shown in Figure 6.

(a)

(b)

Figure 6. The natural light intensity in the reading room of the UW Library under different weather conditions: (**a**) on a cloudy day; (**b**) on a clear day. The photo was taken from the mezzanine of the reading room. Author: I. Voronkova.

The semi-structural wall of the reading room at Warsaw University Library consists of aluminum brackets with seals on the inside of the wall. As a result, only the glass is visible from the outside. The height of the glass wall (over 8 m from the inside) requires the use of steel columns with diameters of 150 mm and 70 mm to reinforce the wall from the inside. The glass wall rises 0.80 m above the floor and is divided into 53 sections by a frame. Of these, 27 sections are double-glazed windows measuring 1.85 × 2.10 m, 6 sections are double-glazed windows measuring 0.95 × 2.10 m, and 20 sections are curved glass measuring 0.95 × 2.20 m (Figure 7). The glass used is reflective with a low solar energy transmittance (g = 24%). Information on the use of low-emissivity glass in the double-glazed windows to retain heat in the room is not available to the authors of this study.

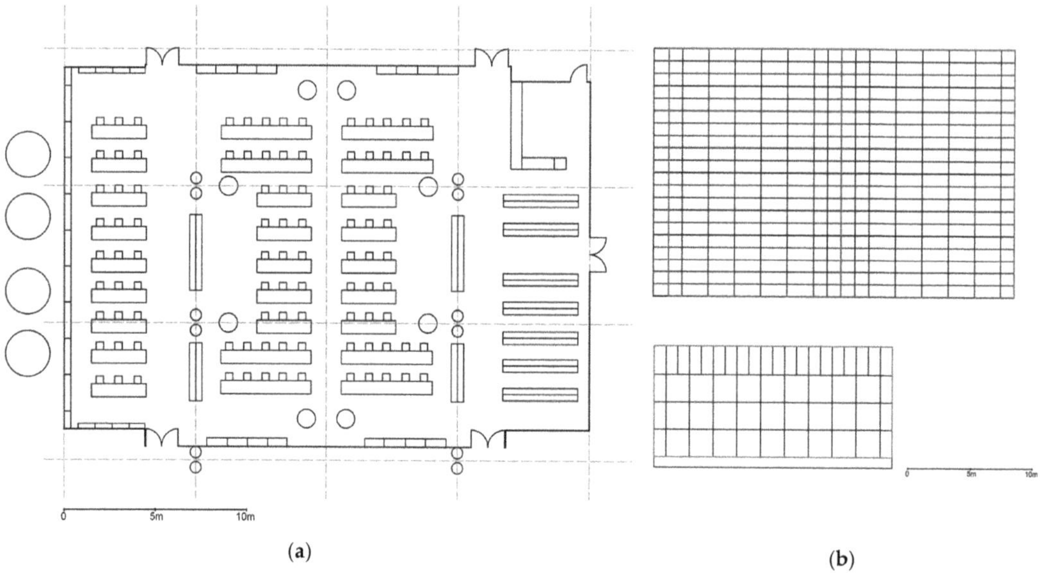

Figure 7. UW library reading room: (**a**) floor plan; (**b**) view of semi-transparent structures: glass roof above and glass wall below. Author: I. Voronkova.

As mentioned above, in addition to the design of the glass wall structure, the reading room is also illuminated by a skylight in the form of a glass roof that covers the entire area of the room. The roof is supported by a specially designed structure of steel tubes coated with fire-resistant protective coatings. The design is "organic" in style, resembling the shape of branching trees. The glazing of the skylight meets the basic requirements of reliability, light transmission, and thermal insulation. The skylight structure uses double-glazed windows with two layers of grey-green glass.

The outer glass is tempered (ESG), while the inner glass is laminated (VSG). ESG-tempered glass has increased impact resistance, reducing the risk of breakage. It is five to seven times stronger than ordinary glass and shatters into small, safe pieces when damaged. In addition, the upper pane is coated with a special film that reflects sunlight, giving the skylight a mirrored appearance from the outside. Laminated VSG glass is also designed with durability in mind. It consists of at least two layers of glass bonded together with a polyvinyl butyral (PVB) film, which gives it increased resistance to breakage and allows it to retain its shape in the event of damage [54].

The use of durable glass on the roof makes it safe to walk on. At the same time, manufacturers of such double-glazed windows claim that their product effectively retains solar energy, which has a positive effect on the air conditioning of the room. Given the physical properties of the glazing unit, characterized by indicators such as a U-value of 1.1 W/m^{2}K, a solar factor (g) of 24%, and a light transmission (LT) of 42%, direct sunlight entering the room should be minimal. However, these indicators do not seem to hold up in the actual situation shown in Figure 6b.

Table 7 shows the main architectural and structural parameters of the reading rooms studied. From the table, it can be seen that with a slight difference in floor area between the two rooms, the lighting levels vary significantly. This is mainly because the glazing area in the reading room of the UW Library is almost four times larger than the glazing area in the reading room of the SGGW Library. Another no less important criterion for increasing the lighting level is the height of the room, which has a direct effect on the volume of the room. The height of the UW reading room is almost three times greater than the height of the SGGW library reading room, and the volume is twice as large.

Table 7. A comparison of architectural and construction indicators for the SGGW and UW reading rooms; source: own research.

N	SGGW Reading Room	UW Reading Room
Wrr	19.6 m	28.8 m
Drr	32.4 m	19.0 m
Hrr	4.6 m	12.5 m
Hws	1.0 m	0.8 m
Arr	~635 m^2	~550 m^2
Vrr	~2921 m^3	~5965 m^3
Nw	Windows: 32 Skylights: 24	Glass wall: 53 glass units Glass roof: 320 glass units
Ag	~172 m^2	~700 m^2
RtsF	~27%	~127%

Table 8 presents the main architectural and construction indicators of translucent constructions of reading rooms. The table indicates the type of translucent structure (Tts), number of windows (Nw), window height (Hw), window width (Ww), window length (Hw) (for skylights and glass roof), glazing area (Ag), and translucent material (Mt).

Table 8. Comparative table of architectural and construction indicators of translucent structures of SGGW's and UW's Library; source: own research.

N	SGGW Reading Room	UW Reading Room
	Type-I translucent structures	
Tts	Windows	Glass wall
Nw	32	53 glass units
Hw	2.95 m	2.10 m (27 units) 2.10 m (6 units) 2.20 m (20 units)
Ww	1.45 m	1.85 m (27 units) 0.95 m (6 units) 0.95 m (20 units)
Ag	137 m^2	159 m^2
Mt	Reflective glass and low-emissivity glass	Reflective glass
	Type-II translucent structures	
Tts	Skylights	Glass roof
Nw	24	320 glass units
Lw	1.20 m	1.13 m (120 units) 2.16 m (200 units)
Ww	1.20 m	0.95 m
Ag	35 m^2	540 m^2
Mt	Cellular polycarbonate sheet and polyester panel	Tempered glass and laminated glass

4. Discussion

Natural lighting is an integral part of the physical environment of any space. Providing high-quality daylight helps to create a comfortable environment for productive work and effective relaxation. Therefore, providing optimal levels of natural light in library reading rooms requires increased attention.

To date, many studies on lighting in different types of buildings have focused mainly on improving environmental quality and increasing energy savings [55]. Most existing research emphasizes artificial lighting and energy consumption [56]. Separate studies focus on evaluating and improving natural lighting in residential and educational spaces.

However, few studies assess the lighting environment in libraries, in particular, reading rooms [57,58]. Therefore, studies on natural lighting in reading rooms need to be further investigated and refined. In addition, architectural norms and standards do not provide specific suggestions for the organization of daylight in libraries, especially in reading rooms. The few guidelines that do address reading room lighting often overlook the physical parameters of the space and the natural conditions in which these rooms are located. Therefore, to improve the efficiency of natural lighting in reading rooms and save energy costs due to heating and air conditioning, it is crucial to study the specifics of organizing daylight in this type of space, based not only on an understanding of the physical properties of light, but also on taking into account the physical parameters of the spaces, the natural conditions in which they are located, and the physical–technical characteristics of different types of glass and semi-transparent structures [59].

Given the important role of the library in the educational process, bright and spacious reading rooms are essential for its proper functioning. If they are well organized and well lit, they can be actively used as a primary space for individual and group study, for working with library collections and personal materials, for concentrated work, and for meeting friends. Therefore, the reading room is (or should be) the most desirable type of space in the library.

The principle of participatory design emphasizes involving stakeholders, including future users, in the design process [60]. Integrating user feedback during library design is crucial for creating spaces that truly meet the expected needs. A library designed with participatory principles creates an environment that may be not only visually appealing but also conducive to learning, relaxation, and social interaction. Nevertheless, some studies indicate that there are notable differences between the views of experts (architects and civil engineers) and the experiences of actual building occupants. Bridging the gap between expert perspectives and the real-world experiences of occupants could help guide the development of building environments that are more effective, sustainable, and user-friendly [61].

The quality of natural lighting is crucial to meeting users' needs, maintaining physical and mental health, and reducing the overall energy consumption of the building. Therefore, this study considered the physical–technical parameters of the reading rooms, namely the room dimensions (width, length, height, area, volume), type of natural lighting (side, top), type of translucent structure (window, wall, skylight), sizes of translucent structure (width, length, height, area), ratio of glazed area to room area, and physical–technical properties of different types of semi-transparent materials (glass and fiberglass). All of these parameters have a significant impact on increasing the level and improving the quality of natural lighting in reading rooms.

For example, this study suggests that to improve the effectiveness of natural lighting in the SGGW library reading room, it would be necessary to increase the height of the room and the height of the windows. Given the orientation of the windows, increasing their height would allow more diffused daylight to penetrate deeper into the room without causing problems with excessive brightness or glare on work surfaces. This would minimize the need for artificial lighting. Another solution to improve the lighting conditions in the SGGW library reading room could be to reduce the depth of the skylight. This would maintain the intensity of natural light as it passes through the translucent material into the room, making it more effective. In the case of the reading room at the UW Library, the main problem is reflection on the desktops. This is because direct sunlight enters the room through the double-sloped glass roof, specifically the sloping roof surface on the southeast side. If the roof was single-sloped and facing north, there would be no problem with glare.

Glare is a critical problem in many spaces, often caused by natural lighting. However, reflections on the glass façades of buildings can be extremely dangerous for the climate and for people. The sun's scorching rays, combined with the concave shape of buildings and the reflective coating of their façades, can turn into death rays that burn everything in their path. For example, the glass coating on the façades of the Vdara Hotel in Las Vegas and the Skyscraper in London reflects the sun's rays and the concave shape of the buildings directs the rays to a single point. The surface at this point is heated to an extremely high temperature, melting not only plastic but also metal parts [62]. Unfortunately, there are many such dangerous objects in architectural practice. In addition, the sun's rays reflected from glass façades heat the surrounding air, contributing to global warming and moving away from sustainable development.

The effect of solar glare can be minimized by using a special type of translucent material. An alternative to ordinary glass is highly reflective glass. However, as this study shows, not all is clear-cut. As mentioned earlier, the most popular material for solar shading is reflective glass, which is used in both libraries studied. Given that the reading room windows in both libraries face north, the use of reflective glass is not functionally justified. Instead, a more decorative approach to façade design was taken, focusing on the reflective effect of the surroundings. This enhances the modernity of the architectural image and gives it a sense of mystery.

On the other hand, the use of glass with low light transmission does not prevent bright reflections on the work surfaces in the reading room of the University of Warsaw Library. This suggests that the technology for producing solar control glass needs further research and improvement. An alternative is smart glass (or electrochromic glass), which changes from semi-transparent to transparent and vice versa when an electric current is applied, without changing the light transmission [27]. Electrochromic glass can be part of a healthy building strategy, providing access to daylight throughout the day without glare or excessive heat [36]. By optimizing heat exchange, smart glass reduces the load on heating, ventilation, and air conditioning systems, contributing to energy savings and building energy efficiency [57].

This study has shown that in Warsaw's weather conditions, where there are only a few sunny days a year, the optimal solution for the design of library reading rooms is to increase the ratio of glazed area to room area. The most popular and reliable semi-transparent material is glass. To maintain comfortable thermal conditions, the best approach is to use double-glazed windows with different types of glass; to avoid overheating, reflective glass is recommended, and to retain heat in the room, low-emissivity glass is suggested. Both types of glass can be fire-resistant and, if necessary, toughened for added durability.

Visual comfort and energy efficiency are therefore critical elements in the lighting design of any space, including library reading rooms. Several factors, highlighted in this study, should be considered to achieve optimum natural lighting levels and reduce the energy costs of running a building.

5. Conclusions

Daylight is natural and does not produce a stroboscopic effect, making it particularly valuable for reading room lighting. It has a positive effect on the physical and mental health of users. Natural light has unique physical properties that offer significant advantages over various artificial light sources. At the same time, natural lighting has considerable variability, which can pose a challenge when organizing it within a building. The right architectural parameters and the geometry of transparent structures can be effectively integrated into the design process, contributing positively to the creation of a comfortable lighting environment through architecture and design. In addition, improving natural lighting conditions can improve the performance of a space and reduce electricity consumption.

Providing visually comfortable lighting for reading rooms should become a priority task during the design of library buildings. At the same time, it is crucial not only to anticipate users' qualitative and quantitative needs, such as the type of activity (focused work with printed text, computer work, or quiet reading), work duration (long-term or short-term), work time (morning, noon, or evening), but also to consider the architectural image of the library (building and room geometry), as well as the natural and climatic conditions of the location (air temperature, amount of solar radiation, and day length). Taking all of these factors into account will help make the right decision regarding the building's orientation and application of the most efficient type of transparent structures and materials.

There is no doubt about the need to install transparent structures in reading rooms. They are an effective way of letting daylight into the room and maintaining visual contact with the surroundings. In addition, transparent structures are an integral part of the library building envelope and are the main points of heat accumulation and loss. The choice of the right type of transparent structure and the material to fill it should therefore be based primarily on consideration of the needs of the reading room in terms of filling its spaces with soft and diffused light to ensure high-quality functionality, as well as on an analysis of the natural and climatic factors and the physical–technical indicators described in this study.

Modern library buildings can serve as examples of sustainable development and energy efficiency, actively using natural light to illuminate spaces, consuming energy with minimal environmental impact, and making a positive contribution to human life.

Author Contributions: Conceptualization, I.V.; methodology, I.V.; software, I.V.; validation, I.V. and A.P.; formal analysis, I.V. and A.P.; investigation, I.V.; resources, I.V.; data curation, I.V.; writing—original draft preparation, I.V. and A.P.; writing—review and editing, I.V. and A.P.; visualization, I.V.; project administration, I.V.; funding acquisition, I.V. and A.P. All authors have read and agreed to the published version of the manuscript.

Funding: This study was performed within the project MINIATURA 7 titled "The role of light in shaping a comfortable environment in Polish libraries, taking into account energy efficiency", funded by the National Science Centre, Poland, under grant number DEC-2023/07/X/ST8/01209.

Data Availability Statement: The data are contained within this article.

Acknowledgments: The authors would like to thank Wojciech Wozniak—Director of the SGGW Library, Piotr Bien—Head of the Economic Section of the SGGW, Jolanta Talbierska—acting Director of BUW Library, and Marlena Nawrocka—Deputy Head of the Administrative Section of BUW for their assistance in data collection and data curation.

Conflicts of Interest: The authors declare no conflicts of interest.

References

1. Šujanová, P.; Rychtáriková, M.; Sotto Mayor, T.; Hyder, A. A Healthy, Energy-Efficient and Comfortable Indoor Environment, a Review. *Energies* **2019**, *12*, 1414. [CrossRef]
2. Soares, N.; Bastos, J.; Pereira, L.D.; Soares, A.; Amaral, A.R.; Asadi, E.; Rodrigues, E.; Lamas, F.B.; Monteiro, H.; Lopes, M.A.R.; et al. A Review on Current Advances in the Energy and Environmental Performance of Buildings towards a More Sustainable Built Environment. *Renew. Sustain. Energy Rev.* **2017**, *77*, 845–860. [CrossRef]
3. Yun, G.; Yoon, K.C.; Kim, K.S. The influence of shading control strategies on the visual comfort and energy demand of office buildings. *Energy Build.* **2014**, *84*, 70–85. [CrossRef]
4. Boyce, P.; Hunter, C.; Howlett, O. *The Benefits of Daylight through Windows. Report*; Lighting Research Center, Rensselaer Polytechnic Institute: Troy, NY, USA, 2003.
5. Xie, F.; Song, H.; Zhang, H. Research on Light Comfort of Waiting Hall of High-Speed Railway Station in Cold Region Based on Interpretable Machine Learning. *Buildings* **2023**, *13*, 1105. [CrossRef]
6. Lipowicz-Budzyńska, A. Art glass as an external partition—Its impact on visual and functional properties. *Architectus* **2019**, *3*, 69–82. [CrossRef]
7. Knera, D.; Szczepańska, E.; Heim, D. Providing an interior daylight environment through the use of light pipes. *Tech. Trans. Civ. Eng.* **2014**, 195–203. [CrossRef]

8. Sitek, M.; Malinowska-Borowska, J. Conditions of lighting in reading rooms located in public and school libraries in Silesian Voivodeship. *Environ. Med.* **2013**, *16*, 38–44.
9. Cherpinska, I. The natural lighting of library rooms and buildings. *Archit. Bull. Lviv. Polytech. Natl. Univ.* **2007**, *585*, 168–173. (In Ukrainian)
10. *PN-EN 12464-1:2022-01;* Light and Lighting. Workplace Lighting. Part 1: Indoor Workplaces. Polish Committee for Standardization: Warsaw, Poland, 2022.
11. Pawlak, A. Illumination of indoor workplaces—A new lighting standard. *Work Saf.* **2004**, *10*, 6–10. (In Polish)
12. Muñoz-González, C.; Ruiz-Jaramillo, J.; Cuerdo-Vilches, T.; Joyanes-Díaz, M.D.; Montiel Vega, L.; Cano-Martos, V.; Navas-Martín, M.Á. Natural lighting in historic houses during times of pandemic. The case of housing in the Mediterranean climate. *Int. J. Environ. Res. Public Health* **2021**, *18*, 7264. [CrossRef] [PubMed]
13. Michael, A.; Heracleous, C. Assessment of natural lighting performance and visual comfort of educational architecture in Southern Europe: The case of typical educational school premises in Cyprus. *Energy Build.* **2017**, *140*, 443–457. [CrossRef]
14. Jia, Y.; Liu, Z.; Fang, Y.; Zhang, Y.; Zhao, C.; Cai, X. Effect of interior space and window geometry on daylighting performance for terrace classrooms of universities in severe cold regions: A case study of Shenyang, China. *Buildings* **2023**, *13*, 603. Available online: https://www.mdpi.com/2075-5309/13/3/603 (accessed on 26 February 2024). [CrossRef]
15. Akanmu, W.P.; Nunayon, S.S.; Eboson, U.C. Indoor Environmental Quality (IEQ) Assessment of Nigerian University Libraries: A Pilot Study. *Energy Built Environ.* **2021**, *2*, 302–314. [CrossRef]
16. Deng, Q.; Allard, B.; Lo, P.; Chiu, D.K.; See-To, E.W.; Bao, A.Z. The Role of the Library Café as a Learning Space: A Comparative Analysis of Three Universities. *J. Librariansh. Inf. Sci.* **2019**, *51*, 823–842. [CrossRef]
17. Lourenço, P.; Pinheiro, M.D.; Heitor, T. Light Use Patterns in Portuguese School Buildings: User Comfort Perception, Behaviour, and Impacts on Energy Consumption. *J. Clean. Prod.* **2019**, *228*, 990–1010. [CrossRef]
18. Peters, T.; D'Penna, K. Biophilic Design for Restorative University Learning Environments: A Critical Review of Literature and Design Recommendations. *Sustainability* **2020**, *12*, 7064. [CrossRef]
19. Borrelli, S.; Chao, Z.; Su, C. Reflecting the Voice of the Student: A Case Study from The Pennsylvania State University Using Mixed-Methods in Assessing Space. *Libr. Manag.* **2019**, *40*, 121–127. [CrossRef]
20. Zhang, Z. The Effect of Library Indoor Environments on Occupant Satisfaction and Performance in Chinese Universities Using SEMs. *Build. Environ.* **2019**, *150*, 322–329. [CrossRef]
21. Liu, Q.; Ren, J. Research on the Building Energy Efficiency Design Strategy of Chinese Universities Based on Green Performance Analysis. *Energy Build.* **2020**, *224*, 110242. [CrossRef]
22. Makaremi, N.; Schiavoni, S.; Pisello, A.L.; Cotana, F. Effects of Surface Reflectance and Lighting Design Strategies on Energy Consumption and Visual Comfort. *Indoor Built Environ.* **2019**, *28*, 552–563. [CrossRef]
23. Izmir Tunahan, G.; Altamirano, H.; Teji, J.U.; Ticleanu, C. Evaluation of Daylight Perception Assessment Methods. *Front. Psychol.* **2022**, *13*, 805796. [CrossRef] [PubMed]
24. Abisuga, A.O.; Wang, C.C.; Sunindijo, R.Y. A Holistic Framework with User-Centred Facilities Performance Attributes for Evaluating Higher Education Buildings. *Facilities* **2019**, *38*, 132–160. [CrossRef]
25. Kaymaz, E.; Manav, B. A Proposal on Residential Lighting Design Considering Visual Requirements, Circadian Factors, and Energy Performance of Lighting. *J. Asian Archit. Build. Eng.* **2023**, *22*, 2851–2866. [CrossRef]
26. Samotyj, R.; Voronkova, I. *Forming Modern Architectural Space in a University Library in the Example of Renovation of Some Libraries "4th Wroclaw Meetings of Librarians"*; Oficyna Wydawnicza Politechniki Wrocławskiej: Wroclaw, Poland, 2014; pp. 383–393. (In Polish)
27. Voronkova, I. Types of architectural glass to improve the energy efficiency of buildings. *Acta Sci. Pol. Archit.* **2021**, *3*, 77–83. [CrossRef]
28. Sottile, G.M. Cleantech Daylighting Using Smart Glass: A Survey of LEED® Accredited Professionals. *Clean Technol. Proc.* **2008**, 201–204.
29. Qahtan, A.M.; Shaik, S.; Alwadai, Y.A.; Alhamami, S.; Alawlaki, A.S.; Alyami, M.H.; Kodati, J.A. Smart Glazing Systems vs. Conventional Glazing: A Comprehensive Study on Temperature Control, Daylighting, and Sustainability. *Int. J. Sustain. Dev. Plan.* **2024**, *19*, 939–947. [CrossRef]
30. Liu, X.; Wu, Y. A review of advanced architectural glazing technologies for solar energy conversion and intelligent daylighting control. *Archit. Intell.* **2022**, *1*, 10. [CrossRef]
31. Central Institute for Labor Protection—National Research Institute. Lighting Parameters Affecting Work Safety and Visual Comfort. Available online: https://www.ciop.pl/CIOPPortalWAR/appmanager/ciop/pl?_nfpb=true&_pageLabel=P30001831335539182278&html_tresc_root_id=23200&html_tresc_id=23217&html_klucz=19558&html_klucz_spis (accessed on 26 February 2024). (In Polish)
32. Losyk, M.; Zvir, O. *Technological and Physical and Chemical Properties of Glass: Education, Manual;* LNAM: Lviv, Ukraine, 2018; p. 40. (In Ukrainian)
33. Weissman, D. Heliotropic shading: Daylighting a rare books reading room with electrochromic glass and parametric analysis. *Leukos J. Illum. Eng. Soc.* **2024**, *20*, 148–171. Available online: https://www.tandfonline.com/doi/epdf/10.1080/15502724.2023.2219019?needAccess=true (accessed on 26 February 2024). [CrossRef]
34. Twarowski, M. The sun in architecture. *Arkady* **1970**, 296. (In Polish)

35. Gnes, I.P.; Rudyk, R.A.; Yakubovskyi, V.B.; Yakubovskyi, I.V. *Single-Family Residential Building: Study Guide*; Lviv Polytechnic National University Publishing House: Lviv, Ukraine, 2007; pp. 43–45. (In Ukrainian)
36. Pawlak, J.; Zalupka, M. Warsaw Climate. Office of Architecture and Spatial Planning Capital City Warsaw. Available online: https://architektura.um.warszawa.pl/documents/12025039/22491719/klimat_broszura.pdf/b654b97c-8c88-dbf2-0168-38fa1971c4ca?t=1634497936327 (accessed on 26 February 2024). (In Polish)
37. KLIMAT IMGW-PIB: Climate Standards 1991–2020. Available online: https://klimat.imgw.pl/pl/climate-normals/D_POGODNE (accessed on 26 February 2024). (In Polish)
38. Brach, M.; Jaworska, A.; Sarnowski, M.; Kalinowska-Szymczak, A. The Library under the Microscope, i.e., a New Dimension of the SGGW Main Library in GIS Technology. Available online: https://www.arcanagis.pl/biblioteka-pod-lupa-czyli-nowy-wymiar-biblioteki-glownej-sggw-w-technologii-gis/ (accessed on 26 February 2024). (In Polish)
39. Main Library of SGGW—Stage I, Warsaw, Poland. Architectural and Construction Design: Arch. Fijałkowski, S. Unpublished work. 1996. (In Polish)
40. Main Library of SGGW—Stage III, Warsaw, Poland. Electrical Installations: Post-Completion Documentation. Unpublished work. 2004. (In Polish)
41. Main Library of SGGW—Stage III, Warsaw, Poland. Construction Works, Part 1: Post-Completion Documentation. Unpublished work. 2005. (In Polish)
42. GreenSWIFT. Rack-and-Transom Facades. Available online: https://greenswift.com.ua/uk/stijkovo-rigelni-fasadi/ (accessed on 26 February 2024). (In Ukrainian)
43. GFO Gdansk Glass Factory. Available online: https://www.gfo.com.pl/szyby/szyby-refleksyjne-stopsol/ (accessed on 26 February 2024). (In Polish)
44. AGC. Technical Data Sheet—Stopsol, 11/2011. Available online: https://www.agc-yourglass.com/sites/default/files/technical_documents/original/stopsol-tds.pdf (accessed on 26 February 2024).
45. Glass, Properties of Glass for Various Purposes, an Assortment of Glass Products. Available online: https://vuzlit.com/282143/vikoristannya_skla_suchasnomu_budivnitstvi#533 (accessed on 26 February 2024). (In Ukrainian)
46. INTERNORM. Low-Emission Glass—Energy-Saving Windows. Available online: https://okna.waw.pl/poradnik/szklo-niskoemisyjne-okna-energooszczedne (accessed on 26 February 2024). (In Polish)
47. MERCOR. Technical and Operational Documentation. Fixed Skylights, Ventilation Flaps, mcr PROLIGHT Roof Hatches. Available online: https://mercor.com.pl/upload/product/123/attachments/65/DTR_MCR_PROLIGHT_%C5%9Awietliki,%20klapy%20wentylacyjne,%20wy%C5%82azy%20dachowe%20(HO.21.01285%20rev.B).pdf (accessed on 26 February 2024). (In Polish)
48. GREENHOUSE. Available online: https://szklarnia.pro/pl/poliweglan-komorowy/237-poliweglan-komorowy.html (accessed on 26 February 2024). (In Polish)
49. University of Warsaw. Visit Us. Available online: https://www.uw.edu.pl/odwiedz-nas/ (accessed on 26 February 2024).
50. Construction Warehouse. *Monographic Issue Devoted to the Presentation of the New Headquarters of the University Library in Warsaw*; Bertelsmann, Professional Publishing House: Gütersloh, Germany, 1999; Volume 7. (In Polish)
51. Architektura.Info. BUW—Building Festival. Available online: https://architektura.info/wiadomosci/wydarzenia_architektoniczne/buw_festiwal_budynkow (accessed on 26 February 2024). (In Polish)
52. Cherpinska, I. Architecture of library areas. *Tradit. Innov. High. Archit. Artist. Educ.* **2008**, *1–3*, 313–317. (In Ukrainian)
53. Voronkova, I.; Gumennyk, I. "Green" roofs as a crucial element for energy efficiency. Cases of library buildings. In Proceedings of the 5th International Scientific and Practical Conference "Innovative Technology in Architecture and Design" (ITAD-2021), Kharkiv, Ukraine, 20–21 May 2021; Volume 2490.
54. SageGlass Saint-Gobain. Available online: https://www.sageglass.com/smart-windows/how-electrochromic-glass-works (accessed on 26 February 2024).
55. Alsabry, A.; Szymański, K.; Backiel-Brzozowska, B. Analysis of the Energy, Environmental and Economic Efficiency of Multi-Family Residential Buildings in Poland. *Energies* **2024**, *17*, 2057. [CrossRef]
56. Prasad, R.D. School Electricity Consumption in a Small Island Country: The Case of Fiji. *Energies* **2024**, *17*, 1727. [CrossRef]
57. Shen, J.; Zhang, X.; Mylly, N.; Lin, J. A critical review of lighting design and asset management strategies. Illuminating practices and lessons learned for Swedish public libraries. *J. Phys. Conf. Ser.* **2023**, *2654*, 012139. Available online: https://iopscience.iop.org/article/10.1088/1742-6596/2654/1/012139/pdf (accessed on 26 February 2024). [CrossRef]
58. Garcia-Fernandez, B.; Omar, O. Integrated innovative solar lighting system for optimization of daylight utilization for public library in Alexandria, Egypt. *Ain Shams Eng. J.* **2023**, *14*, 101819. Available online: https://www.sciencedirect.com/science/article/pii/S2090447922001307 (accessed on 26 February 2024). [CrossRef]
59. Djoković, J.M.; Nikolić, R.R.; Bujnak, J.; Hadzima, B.; Pastorek, F.; Dwornicka, R.; Ulewicz, R. Selection of the optimal window type and orientation for the two cities in Serbia and one in Slovakia. *Energies* **2022**, *15*, 323. Available online: https://www.mdpi.com/1996-1073/15/1/323 (accessed on 26 February 2024). [CrossRef]
60. Kołacz, K.; Podlasek, A. Pro-Social Solutions in Residential Environments Created as a Result of Participatory Design. *Sustainability* **2024**, *16*, 510. [CrossRef]

61. Mokhtariyan Sorkhan, F.; Roumi, S.; Soltanzadeh Zarandi, M.; Ashraf Ganjouei, M.A. The Impact of Indoor Environmental Quality on Occupant Satisfaction in Commercial Buildings: A Comparison of Building Expert Opinions and Residents' Experiences. *Energies* **2024**, *17*, 1473. [CrossRef]
62. Learn from London's "Solar Death Ray"Tower. Available online: https://www.greenprophet.com/2013/09/learn-from-londons-solar-death-ray-tower/ (accessed on 26 February 2024).

energies

MDPI

Article

3D Numerical Modeling to Assess the Energy Performance of Solid–Solid Phase Change Materials in Glazing Systems

Hossein Arasteh [1], Wahid Maref [1,*] and Hamed H. Saber [2]

[1] Department of Construction Engineering, École de Technologie Supérieure (ÉTS), University of Quebec, Montreal, QC H3C 1K3, Canada; hossein.arasteh.1@ens.etsmtl.ca

[2] Deanship of Research and Industrial Development, Mechanical Engineering Department, Jubail Industrial College, Royal Commission of Jubail and Yanbu, Jubail Industrial City 31961, Saudi Arabia; saberh@rcjy.edu.sa

* Correspondence: wahid.maref@etsmtl.ca

Abstract: This research investigates the energy efficiency of a novel double-glazing system incorporating solid–solid phase change materials (SSPCMs), which offer significant advantages over traditional liquid–solid phase change materials. The primary objective of this study is to develop a 3D numerical model using the finite volume method, which will be followed by a parametric study under real climatic boundary conditions. A proposed double-glazing setup featuring a 2 mm layer of SSPCM applied on the inner glass pane within the air gap is modeled and analyzed. The simulations consider various transient temperatures and ranges of the SSPCM to evaluate the energy performance of the system under different weather conditions of Miami, FL during the coldest and hottest days of the year, both in sunny and cloudy conditions. The results demonstrate a notable improvement in energy performance compared to standard double-glazing windows (DGWs), with the most efficient SSPCM configuration exhibiting a phase transition temperature and range of 25 °C and 1 °C, respectively. This configuration achieved energy savings of 24%, 26%, and 23% during summer sunny, winter sunny, and winter cloudy days, respectively, relative to DGWs during cooling and heating degree hours. However, a 3% energy loss was observed during summer cloudy days. Overall, the findings of this study have shown the potential for energy savings by incorporating SSPCM with suitable thermophysical properties into double-glazing systems.

Keywords: carbon neutrality; building energy; solid–solid phase change material; glazing system; zero energy buildings; computational fluid dynamics; finite volume method

1. Introduction

According to the International Energy Agency (IEA) and its current projections of energy efficiency, achieving carbon neutrality in buildings by 2050 is anticipated to be challenging [1]. Notably, approximately one-third of the world's final energy consumption and nearly 15% of direct CO_2 emissions (comprising 40–48% of both direct and indirect CO_2 emissions) stem from the combined sectors of residential and building construction [1]. The building envelope, specifically fenestration systems (e.g., windows, curtain walls, skylight devices), is identified as the primary source of heat loss in building construction, responsible for approximately 60% of the overall heat loss [2]. Consequently, research efforts have intensified in this area to alleviate the energy demand attributed to heating, ventilation, and air conditioning (HVAC) systems. Overall, advancements in enhancing the thermal performance of transparent components of the building envelope can be categorized into three main areas: solar control, thermal resistance, and thermal inertia.

The improvement of solar control systems has been the subject of extensive research involving the integration of smart glazing technologies, such as chromogenic materials (thermochromism [3], thermotropism [4], etc.), photovoltaic modules [5], switchable systems [6], coating techniques (such as low-e coatings [7] and laminated coatings [8]), and so forth. Similarly, efforts to improve the thermal resistance of transparent components in building envelopes have resulted in reduced heat loss. Techniques, such as the use of inert gases within the gap space of multi-glazing systems [9] as well as the incorporation of low-thermal-conductivity materials like aerogels [10] and hydrogels [11], are recognized methods for enhancing thermal resistance. Another approach involves increasing the thermal mass of glazing systems by utilizing thermal energy storage materials, such as phase change materials (PCMs), which can store and release heat during day and night, thereby improving not only energy savings but also indoor thermal management. These techniques have been extensively explored in recent research and were comprehensively reviewed in [12]. Furthermore, nanoscale analysis of thermal storage materials is also of vital importance. Methods, such as Molecular Dynamics and Monte Carlo simulations, can provide deep insights into the nanoscale thermal capabilities of these materials [13].

This research primarily focuses on enhancing thermal inertia/mass within glazing systems, specifically by employing phase change materials (PCMs). The PCMs undergo heat absorption and release through phase change cycles, which can occur as a solid-to-liquid transition, known as solid–liquid phase change material (SLPCM), or a solid-to-solid transition, known as solid–solid phase change material (SSPCM), as applicable in glazing systems. The SSPCMs offer several advantages over SLPCMs, including no leakage, less phase segregation, low subcooling, small volume variation, encapsulation-free implementation, extended durability upon thermal cycling (high thermal stability), more consistent optical properties, and less material degradation [14]. Furthermore, SSPCMs, remaining solid (except when reaching their melting temperature), can be directly adhered to a surface (such as the interior glass pane in the present study) with specific thickness, while the air or inert gas (possessing low thermal conductivity) remains between the indoor and outdoor environments. This arrangement is advantageous because when the SLPCM completely fills the air gap, it diminishes the thermal resistance of the glazing systems to some extent, as the thermal conductivity of SLPCM is higher than that of air unless it has been encapsulated, which will act like an SSPCM. Despite the extensive literature on SLPCM application in glazing units, studies on SSPCMs in smart glazing are limited. Additionally, to the best of the authors' knowledge, there is no direct experimental research incorporating SSPCMs into glazing systems [15].

SSPCMs experience a phase transition between opaque (semi-crystalline) and transparent (amorphous) states, wherein only the soft segments melt, supported by the hard segment (polymeric backbone) with a significantly higher melting temperature. Consequently, the SSPCM remains solid during phase transition by melting and freezing the soft segments, which are anchored by the hard segments. The illustration for the phase transition mechanism of Polymeric Graft SSPCM [15] and the transparency during phase transitioning [16] are depicted in Figure 1. Figure 2 presents the phase transition temperatures and enthalpies for the primary types of SL-PCMs and SS-PCMs. As shown in this figure, the SS-PCMs generally display lower phase transition enthalpies and transition at lower temperatures in relation to the SL-PCMs. The lower enthalpy can be explained by the restricted mobility within SS-PCMs, which hinders crystallization and limits packing efficiency [14].

As mentioned earlier, a limited number of studies have explored the application of SSPCMs in glazing systems. Fallahi et al. [14] conducted a review study focusing on the molecular properties and thermal characteristics of SSPCMs for thermal energy storage. That study examined the relationship between molecular structure, phase transition mechanisms, and thermal properties of the four main categories of SSPCMs: polymeric, organic, organometallic, and inorganic. The authors provided guidance for selecting an appropriate SSPCM for various applications based on desired physical, thermal, and mechanical properties, offering a comprehensive list of SSPCMs within each main category. Another review study [17] discussed the applications of SSPCMs and recent advancements in their thermophysical properties. That study compiled a comprehensive list of organic, polymeric, organometallic, and commercial SSPCMs, along with their thermophysical properties, phase transition temperatures, melting temperatures, molecular characteristics, and thermal behavior. The review serves as a valuable resource for researchers and practitioners interested in utilizing SSPCMs in various applications.

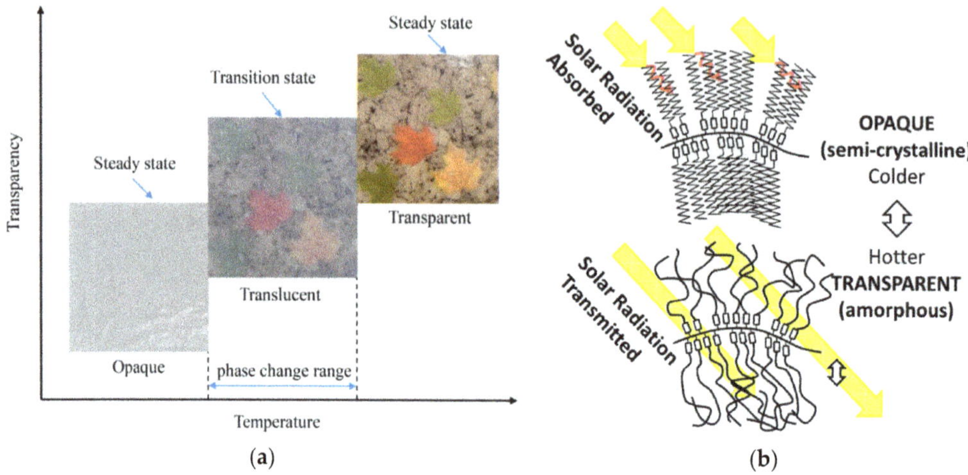

Figure 1. Schematics for the principal operation of SSPCM: (**a**) phase transition process of Polymeric Graft SSPCM [15] and (**b**) transparency of an SSPCM's phase transitioning [16].

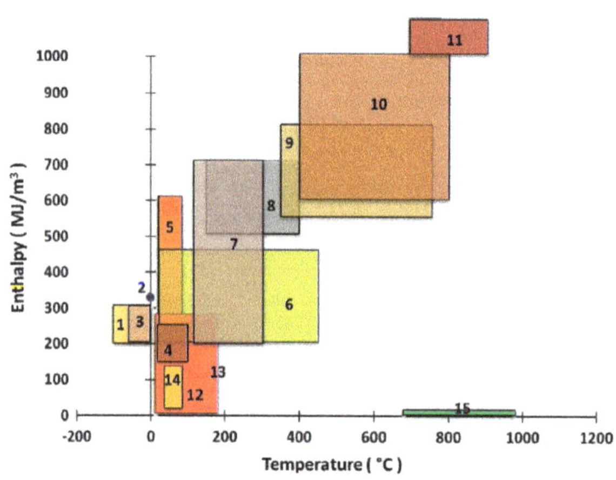

Figure 2. Enthalpy and temperature ranges for SL-PCMs and SS-PCMs; L-PCMs: (1) water–salt solutions; (2) water; (3) clathrates; (4) paraffins; (5) salt hydrates; (6) sugar alcohols; (7) nitrates; (8) hydroxides; (9) chlorides; (10) carbonates; (11) fluorides; (12) polymeric; SS-PCMS: (12) polymeric; (13) organics (polyols); (14) organometallics; (15) inorganics (metallics) [14].

Guldentops et al. [18] proposed a building enclosure system employing SSPCMs to passively regulate the temperature of a south-facing building in a typical central Massachusetts four-season climate, considering both summer and winter conditions. They conducted an analysis of the energy performance of the system by creating a finite element model. The study identified different optimized systems beneficial for summer and winter seasons separately. However, it was noted that further refinement of extinction coefficients and transition temperatures of the SSPCM is necessary for the system to effectively function in both summer and winter conditions. In a related study, Gao et al. [15] integrated a thin layer of SSPCM into the interior side of the inner glass of a south-facing double-glazing window (DGW) of an office room. They performed a numerical analysis to assess the annual energy savings achieved by the system. Utilizing the EnergyPlus for their numerical investigation, the authors encountered a limitation as the software lacked the capability to simulate latent energy storage materials undergoing phase transitions. The authors also mentioned that EnergyPlus cannot simulate transparent PCMs. Consequently, they developed an equivalent model to overcome this limitation. The study demonstrated that the implementation of a 3 mm SSPCM led to energy saving improvements in warm, mixed, and cold climates. Furthermore, the authors argued that the energy savings achieved with the DGW equipped with SSPCM (DGW–SSPCM) outweighed those of low-emissivity windows. Additionally, Ma et al. [19] investigated the integration of both silica aerogel and SSPCM in a glazing system to assess its daylighting and energy performance in a severe cold region of China. For the energy performance analysis, EnergyPlus [20] was employed, while radiance software [21] was utilized for daylighting analysis. Similarly to Gao et al. [15], the authors employed an equivalent model for SSPCM modeling due to EnergyPlus limitations. Through sensitivity analysis, they identified transient temperature, latent heat, absorption coefficient, and refractive index as the most influential parameters. Recommending a 10 mm thickness for silica aerogel in their glazing system, the authors aimed to achieve maximum energy savings while meeting daylighting design standards in China. Their findings suggested the viability of employing DGW–SSPCM in severe cold regions. Moreover, Zhang et al. [22] investigated a similar glazing system containing silica aerogel within the outer gap space and SSPCM within the inner gap space, emphasizing the effectiveness of SSPCMs over SLPCMs. Through simulations conducted for a 24 h real-time period in very cold climates in Anda City, China, they developed a model to predict the heat transfer of the glazing system. Their results highlighted the significant impact of the SSPCM transition temperature and latent heat on heat transfer, while the absorption coefficient and refractive index had subtler effects. Wang et al. [16] developed an inverse approach, based on a hybrid model, to represent useful expressions for the extinction coefficient and refractive index of SSPCMs as a function of temperature for the translucent phase and constant values for opaque and transparent phases. Their results for the optical properties of SSPCMs have been implemented in the current study. Recently, Zhang et al. [22] implemented a two-dimensional numerical parametric study for the influence of optical and thermal properties of a triple-glazed window containing SSPCM within the inner air gap and silica aerogel within the outer air gap using the finite volume method. The study was performed to simulate 24 h of severe cold weather conditions in a city in China. Their sensitivity analysis showed that the thermal efficiency of the glazed window is notably influenced by the melting temperature and the latent heat of PCM, whereas the absorption coefficient and refractive index only have minor effects. In that study, the optimum melting temperature of the PCM was reported as 18 °C, resulting in a 15.4% energy saving rate.

Based on the literature review provided above as well our recent review [12], it is evident that there are few studies investigating the energy performance of SSPCMs in glazing systems using 3D modeling. The existing numerical models use EnergyPlus, which has some limitations in capturing the phase transition phenomenon within the SSPCM, and hence using an equivalent model has been inevitable, or 2D models that assess the SSPCM behavior when fully filling the air gap space of a triple-glazed window. Therefore, to bridge this gap, the main objective of this study is to develop a 3D model and use it to assess the energy performance of a DGW incorporating SSPCMs. The idea of placing the SSPCM onto the interior pane between the air gap has come up to keep the material's temperature as high as possible to yield an almost fully transparent window throughout the whole year. Following model validation, a parametric study was conducted by varying the transient temperature and transient temperature range of a south-facing DGW–SSPCM over a 24 h real-time period. The simulations were performed for sunny and cloudy days during the coldest and hottest days of the year 2022 in Miami.

2. Methodology

2.1. Geometrical Details

The present study comprises two configurations: a double-glazing window (DGW) as a reference called "DGW–REF" and another DGW equipped with an SSPCM applied to the interior pane within the air gap, called "DGW–SSPCM". Both DGW–REF and DGW–SSPCM consist of two panes with dimensions of 20 cm long, 20 cm wide, and 4 mm thick. The DGW–REF features a 1.6 cm air gap, as depicted in Figure 3a, while the DGW–SSPCM includes a 2 mm thick SSPCM, with a mass of 844 g, applied to the interior pane between the panes. This has resulted in a 1.4 cm air gap, as illustrated in Figure 3b.

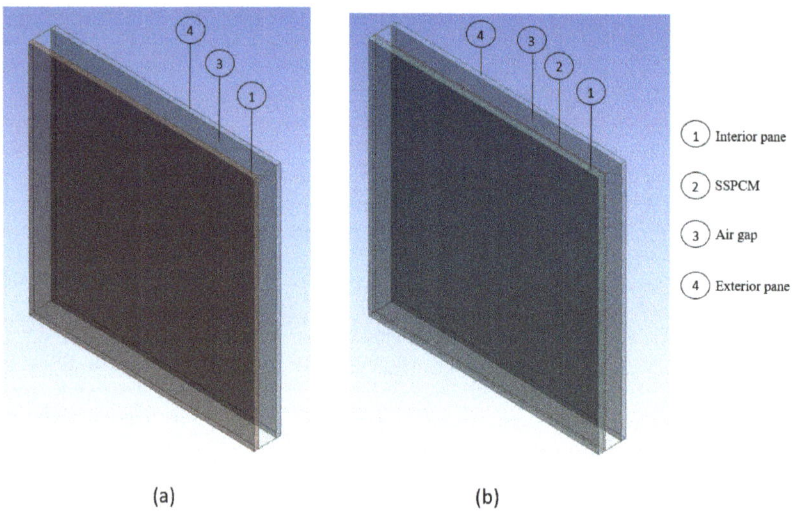

Figure 3. The 3D geometry of the (**a**) DGW–REF and (**b**) DGW–SSPCM.

2.2. Material Perspective

In this study, a typical 4 mm clear glass with an emissivity of 0.9 [23] was chosen, and its thermophysical and optical properties were obtained from [24]. The SSPCM was selected from [25]. It was concluded that in cases involving typical glazing or PCM-glazed setups, where the refractive indices of all materials are relatively low (<1.5), the influence of internal reflections within the glazed systems is insignificant [25]. Additionally, the thermophysical properties of the materials utilized in this study are listed in Table 1. The different phase change temperatures and ranges that have been studied in this paper are provided in Table 2.

Table 1. Materials' thermophysical properties.

Material \ Property	Density (kg/m^3)	Specific Heat (J/kgK)	Thermal Conductivity (W/mK)	Absorption Coefficient (1/m)	Scattering Coefficient (1/m)	Refractive Index	Latent Heat (kJ/kg)
Air *	1.225	1006.43	0.0242	0	0	1	-
Glass	140	840	1.3	19	0	1.5	-
SSPCM	1055	1630	0.36	33.80 (transparent) 25.73 (opaque)	0 (transparent) 119.02 (opaque)	1.11 (transparent) 5.33 (opaque)	110

* Properties at temperature range of 10~30 °C.

Table 2. List of scenarios for the phase change temperature values and ranges.

	Cases	Glazing System		Cases	Glazing System
Summer	0	DGW–REF	Winter	0	DGW–REF
	1	$T_c = 20$ °C, $dT_{pc} = 1$ °C		1	$T_c = 15$ °C, $dT_{pc} = 1$ °C
	2	$T_c = 20$ °C, $dT_{pc} = 3$ °C		2	$T_c = 15$ °C, $dT_{pc} = 3$ °C
	3	$T_c = 20$ °C, $dT_{pc} = 5$ °C		3	$T_c = 15$ °C, $dT_{pc} = 5$ °C
	4	$T_c = 25$ °C, $dT_{pc} = 1$ °C		4	$T_c = 20$ °C, $dT_{pc} = 1$ °C
	5	$T_c = 25$ °C, $dT_{pc} = 3$ °C		5	$T_c = 20$ °C, $dT_{pc} = 3$ °C
	6	$T_c = 25$ °C, $dT_{pc} = 5$ °C		6	$T_c = 20$ °C, $dT_{pc} = 5$ °C
	7	$T_c = 30$ °C, $dT_{pc} = 1$ °C		7	$T_c = 25$ °C, $dT_{pc} = 1$ °C
	8	$T_c = 30$ °C, $dT_{pc} = 3$ °C		8	$T_c = 25$ °C, $dT_{pc} = 3$ °C
	9	$T_c = 30$ °C, $dT_{pc} = 5$ °C		9	$T_c = 25$ °C, $dT_{pc} = 5$ °C

The extinction coefficient is commonly described by the optical thickness (d) of a substance, as shown in Equation (1) [26]. In this equation, s represents the actual thickness of the sample, which is set to 2 mm in this study for the SSPCM–DGW cases.

$$d = (\sigma_{a+}\sigma_s)s \tag{1}$$

In Equation (1), σ_a and σ_s represent the absorption and scattering coefficients, respectively. In this study, the scattering coefficient is considered isotropic. By using Beer–Lambert's law for non-gaseous materials, the transmittance of the PCM can be calculated using Equation (4), as [18]

$$\tau_{PCM} = 10^{-d} \tag{2}$$

where τ_{PCM} represents the transmittance of the PCM. Also, the absorption coefficient can be calculated using Equation (3) [24] and then used to determine the scattering coefficient of Equation (1).

$$\sigma_a = \sigma_s \left[\frac{\tau_{PCM,tr} - \tau_{PCM,op}}{1 - \tau_{PCM,op}} \beta + \frac{1 - \tau_{PCM,tr}}{1 - \tau_{PCM,op}} \right] \tag{3}$$

In Equations (1)–(3), τ and β represent the transmittance and transparency fraction of the material, respectively. The subscripts PCM, tr, and op represent the phase change material, transparent and opaque, respectively. The refractive index and extinction coefficient for the transparent phase are equal to 1.11 and 25.73 m^{-1}, respectively [16], and for the opaque phase, they are equal to 5.33 and 152.82 m^{-1}, respectively [16], of the SSPCM [15].

For the translucent phase, Equations (4) and (5) [16] are used to provide the average optical property as a function of the transparency fraction. This term is used to replace the term "liquid fraction" as there is no liquid phase in the SSPCM that refers to the fraction of the material within the transparent phase. The value of β is equal to 0 when the SSPCM is totally opaque and 1 when the SSPCM is totally transparent. When the transparency fraction is 0, it means the SSPCM temperature is less than or equal to its lower limit of the transient temperature range (opaqueus temperature) and it is in the opaque phase. Conversely, when it is 1, it means that the SSPCM temperature is greater than or equal to its upper limit of the transient temperature range (transparentus temperature) and it is in the transparent phase. However, when the transparency fraction value is between 0 and 1, it indicates that the SSPCM is in the translucent phase, akin to the mushy zone in SLPCMs. Note that at the border between the opaque phase and translucent phase, the temperature is called "saturated-opaque temperature, T_{opaqueus}". Meanwhile, at border between the translucent phase and the transparent phase, the temperature is called "saturated-transparent temperature, $T_{\text{transparentus}}$".

$$\sigma_{a,cell} = 33.8\beta + 25.73(1 - \beta) \tag{4}$$

$$\sigma_{s,cell} = 119.02(1 - \beta) \quad (5)$$

2.3. Governing Equations

To model the SSPCM, the enthalpy–porosity model in FLUENT has been employed utilizing a very high viscosity value (equal to 250 Pa∗s, which is much higher than honey's viscosity range: 6.54–7.01 Pa∗s [27]) for the SSPCM, resulting in nearly zero velocity within it. In this study, the results are represented by neglecting natural convection effects within the air gap of the DGW for all cases (i.e., DGW–SSPCM and DGW–REF), aiming to analyze the applicability and concept of using SSPCM in glazing systems. Utilizing the Miami climate as an illustrative example under natural climatic conditions, this approach seeks to investigate whether energy savings can be achieved in a glazing system equipped with SSPCM compared to the same system without SSPCM. The governing equations of the current study, considering the Discrete Ordinates (DOs) model for radiation and a solidification/melting model for SSPCM, are as follows.

- Mass conservation equation [26]:

$$\frac{\partial \rho}{\partial t} + \nabla.\left(\rho \vec{v}\right) = 0 \quad (6)$$

- Momentum conservation equation [26]:

$$\frac{\partial}{\partial t}\left(\rho \vec{v}\right) + \nabla.\left(\rho \vec{v} \vec{v}\right) = -\nabla \mathrm{p} + \nabla.\left(\mu \nabla \vec{v}\right) + S_m \vec{v} \quad (7)$$

In Equations (6) and (7), ρ, t, $\vec{v}$, p, and μ are density, time, velocity vector, pressure, and dynamic viscosity, respectively. It is noteworthy to mention that the source term "$S_m \vec{v}$" should be added to the right-hand side of Equation (7) according to the solidification/melting modeling. S_m is defined as the negative of the porosity function ($A_m(\beta)$), as described by Brent et al. [28]. The porosity function is formulated in order to make the momentum equations behave according to the Carman–Kozeny equations [29], which are used to describe fluid flow in porous media. However, this study employs an SSPCM, which lacks internal fluid flow, rendering the source term ineffective in the momentum equations.

- Energy equation [26]:

$$\frac{\partial}{\partial t}(\rho H) + \nabla.\left(\rho \vec{v} H\right) = \nabla.(k \nabla \mathrm{T}) + S_h \quad (8)$$

In Equation (11), the enthalpy of the *PCM*, *H*, is calculated as the sum of the sensible enthalpy, h_s, and latent heat, ΔH, as

$$H = h_s + \Delta H, \quad (9)$$

where,

$$h_s = h_{s,\,ref} + \int_{T_{ref}}^{T} c_p dT \quad (10)$$

In Equation (9), the fractional latent heat of the PCM, ΔH, can be expressed in terms of the PCM's latent heat of fusion, L. Note that ΔH can vary between 0 (opaque phase), L (transparent phase), and values between 0 and L when $T_{opaqueus} < T < T_{transparentus}$ (translucent phase).

$$\Delta H = \beta L \quad (11)$$

The transparency fraction, β, can be defined as

$$\beta = \begin{cases} 0 & if\ T \leq T_{opaqueus} \\ 1 & if\ T \geq T_{transparentus} \\ \frac{T - T_{opaqueus}}{T_{transparentus} - T_{opaqueus}} & if\ T_{opaqueus} < T < T_{transparentus} \end{cases} \quad (12)$$

The term S_h in Equation (8) denotes the volumetric heat source/sink related to phase change, which is defined as

$$S_h = -\frac{\partial(\rho\Delta H)}{\partial t} \tag{13}$$

- Radiation equation:

In this research, the DO model is utilized to simulate the radiation effects as it is the most detailed model for radiation in ANSYS FLUENT [26]. This model enables the simulation of radiation scattering and absorption across various optical thicknesses. The DO model addresses radiation heat transfer by converting the equation into a transport equation for radiation intensity, solving it for a finite number of discrete solid angles [26]. Consequently, although the simulation speed is diminished due to the additional equations of the DO model, precision is enhanced.

The radiative transfer equation for an absorbing, emitting, and scattering medium at position $\vec{r}$ in the direction $\vec{s}$ is [26]

$$\frac{dI\left(\vec{r},\vec{s}\right)}{ds} + (\sigma_a+\sigma_s)I\left(\vec{r},\vec{s}\right) = an^2\frac{\sigma T^4}{\pi} + \frac{\sigma_s}{4\pi}\int_0^{4\pi} I\left(\vec{r},\vec{s}'\right)\varnothing\left(\vec{s}.\vec{s}'\right)d\,\Omega' \tag{14}$$

The DO model considers the radiative transfer equation in the direction $\vec{s}$ as a field equation and is written as [26]

$$\nabla.\left(I\left(\vec{r},\vec{s}\right)\vec{s}\right) + (\sigma_a+\sigma_s)I\left(\vec{r},\vec{s}\right) = an^2\frac{\sigma T^4}{\pi} + \frac{\sigma_s}{4\pi}\int_0^{4\pi} I\left(\vec{r},\vec{s}'\right)\varnothing\left(\vec{s}.\vec{s}'\right)d\,\Omega' \tag{15}$$

In Equations (14) and (15), I, n, $\vec{r}$, $\vec{s}$, $\vec{s}'$. $\varnothing$, Ω', and σ represent radiation intensity, refractive index, position vector, direction vector, scattering direction vector, phase function, solid angle, and the Stefan–Boltzmann constant (5.67×10^{-8} W/m^{2}K^4), respectively.

2.4. Weather Conditions

The glazing system in this study represents a south-facing window situated in Miami, with a longitude and latitude of −84.63 (west) and 13.65 (north) degrees, respectively. The time zone in Miami is defined as −5 during Eastern Standard Time (EST) and −4 during Eastern Daylight Time (EDT). Four scenarios for ambient weather conditions have been selected for the coldest (occurred on 30 January) and hottest (occurred on the 18th of August) days of the year in 2022, encompassing both sunny and cloudy conditions. The ambient weather conditions include hourly wind speed [30], hourly ambient temperature [30], hourly solar direct irradiation [26], hourly solar diffuse irradiation [26], and hourly x, y, and z beam direction vectors [26] (equal to the negative value of the sun direction vector) for the coldest and hottest days, as illustrated in Figures 4–6. The geometry's orientation is such that the +z axis points to the north and the −x axis points to the east, as depicted in Figure 7.

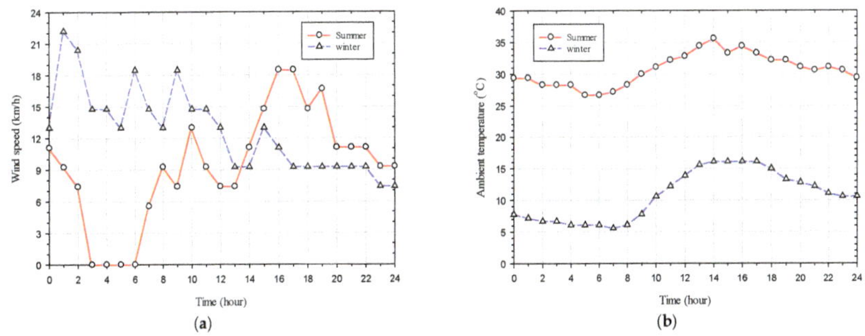

Figure 4. (**a**) Wind speed and (**b**) ambient temperature on 30 January and 18 August 2022 in Miami [30].

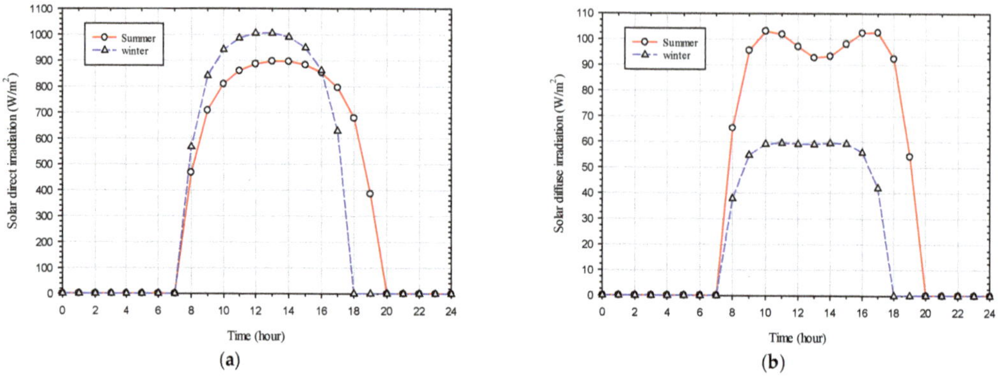

Figure 5. (**a**) Solar direct irradiation and (**b**) solar diffuse irradiation on the 30 January and the 18 August 2022 in Miami [26].

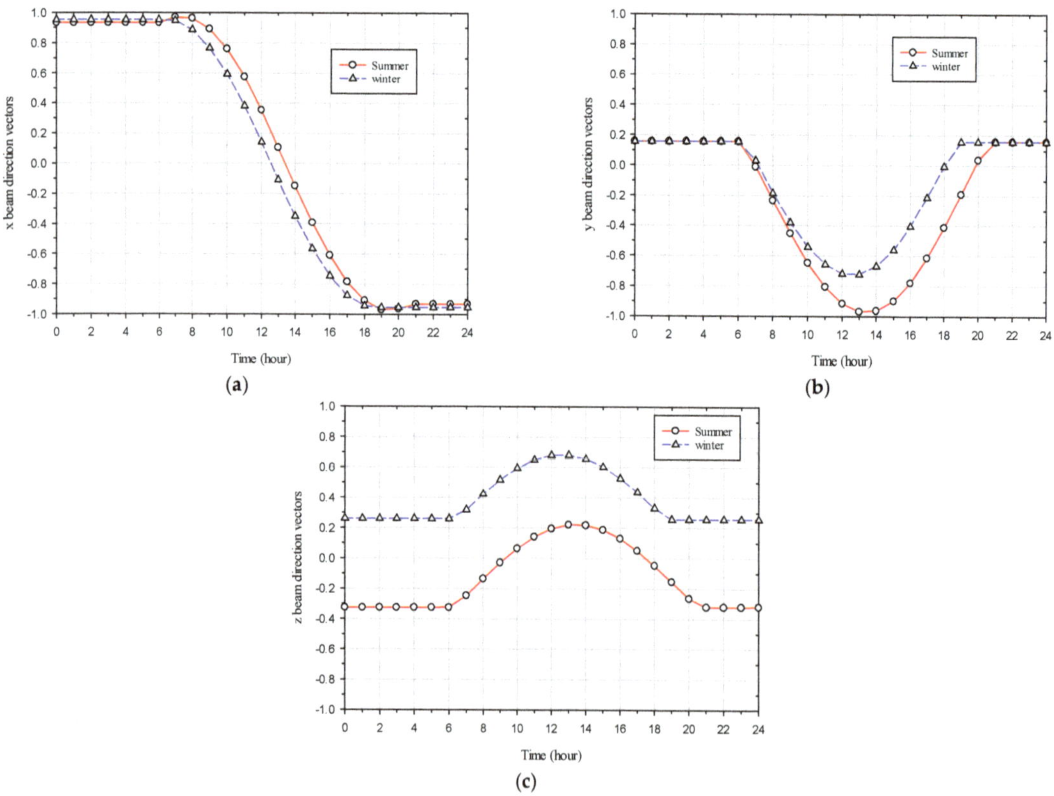

Figure 6. (**a**) x, (**b**) y, and (**c**) z beam direction vectors on the 30 January and the 18 August 2022 in Miami [26].

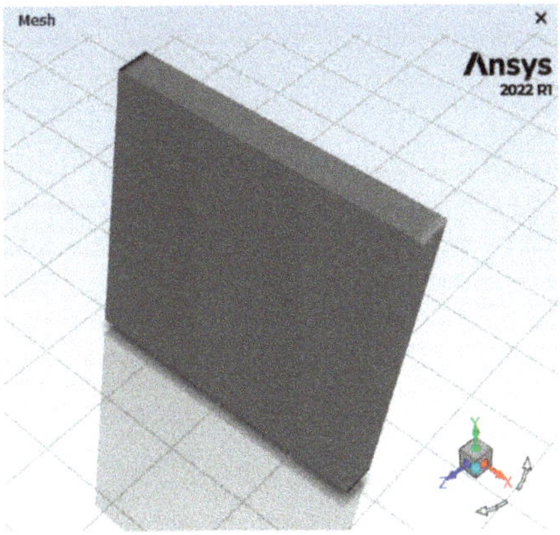

Figure 7. South-facing-orientating DGW–SSPCM.

2.5. Initial and Boundary Conditions

The initial conditions for all parts of the DGW–REF and DGW–SSPCM are set to 26 °C and 24 °C during summer and winter, respectively.

The wall boundary conditions for the side (exterior top, bottom, front, and back) surfaces of the window are set as thermally insulated or adiabatic. The mixed thermal boundary condition including both convection and radiation are set for the indoor and outdoor surfaces of the window. Hence, the parameters, including heat transfer coefficient, free stream temperature, external emissivity, and external radiation temperature, have to be determined for the indoor and outdoor surfaces. Moreover, the semi-transparent boundary condition is set for the radiation boundary condition of the indoor and outdoor surfaces of the window to account for the solar irradiation through the numerical domain. Therefore, the parameters, including direct solar irradiation, diffuse solar irradiation, and x, y, and z beam direction vectors, have to be determined for the indoor and outdoor surfaces. The external emissivity for the typical clear glasses is set to 0.89 for all boundaries. To represent a full sunny day, the sunshine factor is set to 1, while for a cloudy day, this value is set to 0, resulting in a direct solar irradiation value of 0.

Regarding the window's indoor surface thermal and radiation boundary conditions, the heat transfer coefficient is set to 8.7 W/m^{2}K [31]. Both free stream temperature and external radiation temperature are set to 26 °C and 24 °C during summer and winter, respectively. No solar irradiation is set for the indoor radiation boundary condition to obviously not model a direct or diffuse solar irradiation to the indoor surface.

Regarding the window's outdoor surface thermal and radiation boundary conditions, except for the emissivity, all other boundary conditions are imported into ANSYS FLUENT via User-Defined Functions (UDFs) written in C-programming compatible with the FLUENT library to simulate a full day (24 h). The UDFs are written in piecewise linear functions for the available hourly weather conditions. The hourly ambient temperature is set for the free stream temperature in the thermal boundary condition. In addition, the hourly direct and diffuse solar irradiations as well as hourly x, y, and z beam direction vectors are set in the radiation boundary conditions. The hourly heat transfer coefficient as a function of wind speed and hourly external radiation temperature or sky temperature as a function of ambient temperature have been calculated using Equations (16) and (17) [32].

$$h_a = 5.62 + 3.9\ v_{wind} \tag{16}$$

$$T_{sky} = 0.0552\ T_{air,\ o}^{1.5} \tag{17}$$

3. Numerical Procedure

3.1. Model Description

To generate the geometry and grid in this study, both Design Modeler and Ansys Meshing, respectively, were used. The commercial CFD code in ANSYS FLUENT (version 2022 R1 [33]), which is based on the finite volume method, is employed for the 3D numerical simulations in this study. The finite volume method implemented in ANSYS FLUENT discretizes the computational domain into control volumes and applies conservation laws for the mass (Equation (6)), momentum (Equation (7)), and energy (Equation (8)) to these volumes. The resulting algebraic equations are solved iteratively within each control volume, considering boundary conditions and turbulence models as needed. This approach enables the simulation of fluid flow and heat transfer. In this study, the semi-implicit method for pressure-linked equations (SIMPLE) algorithm is used for the velocity–pressure coupling. Additionally, the second-order upwind scheme is employed for the discretization of pressure, momentum, and energy, while the first-order upwind scheme is used for Discrete Ordinates (DOs) and first-order implicit scheme is used for transient formulation. In this study, the DO model is used with one iteration for energy per radiation iteration. The theta and phi divisions are set to 2, while the theta and phi pixels are set to 1. Additionally, the solar calculator is utilized to derive the hourly direct and diffuse solar irradiations for various climatic conditions. These values are then incorporated into a UDF file. Thereafter, the UDF file is imported into the software. The solidification and melting model is also employed to model the solid–solid phase change material (SSPCM). The material properties added to this model, namely the absorption and scattering coefficients, are imported into the software using a UDF file. Additionally, the energy model is used to yield the temperature profile in the numerical domain. Finally, the desired outcome parameters are defined in the report definitions to save the data at each iteration for post-processing purposes. In this study, the convergence criteria were set to be less than 10^{-6}, 10^{-6}, 10^{-6}, 10^{-6}, 10^{-9}, and 10^{-9} for mass conservation, x-velocity, y-velocity, z-velocity, energy, and DO-intensity, respectively. Before using the model to assess the energy performance of the glazing systems considered in this study, it was validated as described next.

3.2. Model Validation

In this study, the same model used for SLPCMs has been applied to SSPCMs using the solidification/melting model. The key distinction lies in the absence of natural convection effects in SSPCMs, as the liquid phase of SLPCM melts. Neglecting the gravity effect and due to the quite high viscosity for the liquid phase of SLPCM allow it to behave similarly to SSPCM.

To validate the numerical model comprising the Discrete Ordinates (DOs) model and the solidification/melting model in glazing systems, transient numerical results obtained over a simulation period of 12000 s have been compared with experimental data reported by Gowreesunkera et al. [24] for the transmittance of PCM-filled glazing units over time. In that study, the researchers devised an experimental setup in order to quantify the changes in radiation effects within the mushy phase, a task that cannot be achieved using a spectrophotometer alone, and to depict the radiation behavior of the PCM within an actual PCM-glazed system, providing a more realistic understanding of its performance. They positioned the entire setup within an environmental chamber where they could control the air temperature. The light source utilized was a 150 W metal halide lamp, emitting a diffuse light with a neutral white color. The regular double-glazing had dimensions of 20 cm × 20 cm, with a total thickness of 24 mm, comprising sequentially a 4 mm glass, a 16 mm air cavity, and another 4 mm glass layer. In contrast, for the PCM-filled glazing configuration, the air within the cavity was substituted with RT27. The irradiation level and air/initial PCM temperature were set to 950 W/m^2 and 13 °C, respectively.

Figure 8 illustrates that the transmittance values derived from the simulation closely match the experimental results, with differences falling within the margin of error. In fact, the overall transmittance was defined by the ratio of radiation flux between the front and back faces. Furthermore, the observed transmittance trends exhibit similarity between the simulation results and the experimental data. Because the numerical results are in good agreement with the experimental data within acceptable margins of error, the model is valid and can be used with confidence in this study.

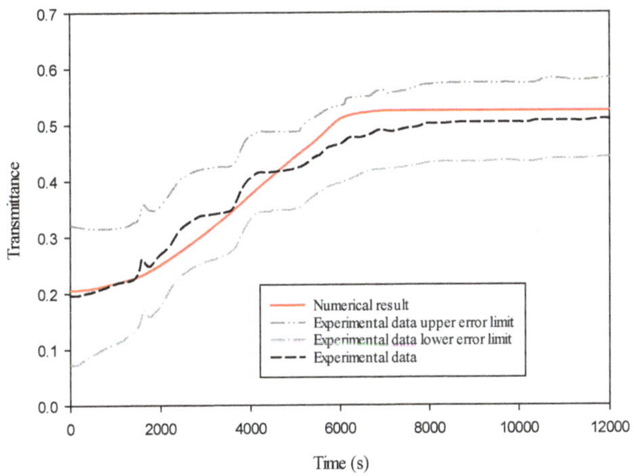

Figure 8. Current numerical model validation study against the experimental data of Gowreesunkera et al. [24].

3.3. Grid Sensitivity

To find the optimum grid size that leads to mesh independent solution, various grid sizes ranging from finer to coarser have been generated over the numerical domain of a DGW–SSPCM. The average total heat flux over 24 h, as defined in Equation (18), on the interior surface of the inner glass pane has been selected as the criterion to assess grid independence. The obtained results are shown in Table 3. As shown in this table, the number of elements of 82,369, shown in the third row, yields to approximately an optimum grid size as its relative error is less than 1%. Thus, the number of elements of 82,369 was used in this study.

$$\overline{q''} = \frac{1}{t}\int_0^t q''\,dt \tag{18}$$

Table 3. Grid sensitivity analysis.

Case	Number of Elements	$\overline{q''}$	Error (%)
1	288,923	15.47104	
2	158,661	15.5293	0.376595
3	82,369	15.58556	0.740245
4	27,440	15.71381	1.569181
5	11,025	15.8746	2.608495
6	4050	16.1296	4.25671

3.4. Time Step Sensitivity

The goal of the time step sensitivity analysis is to find the highest time step size that results in an accurate result. The same procedure as in the previous section was used, where the average total heat flux over 24 h on the interior surface of the inner glass pane has been selected to find the optimum time step size (i.e., the largest time step that would be used to minimize the CPU time). Table 4 shows the results for different time step sizes, and it can be inferred that the time step size of 5 min, shown in the fifth row, would result in a relative error less than 1% and is suitable to be used for the simulations.

Table 4. Time step sensitivity analysis.

Case	Time Step Size	$\overline{q''}$	Error (%)
1	10 s	15.73079	
2	30 s	15.72085	0.06315
3	1 min	15.70591	0.15812
4	2.5 min	15.66088	0.44443
5	5 min	15.58556	0.92319
6	10 min	15.43214	1.89849
7	20 min	15.11929	3.88729
8	30 min	14.82675	5.74694

4. Results and Discussion

Numerical simulations are conducted for a DGW incorporating a 2 mm layer of SSPCM adhered to the inner pane within the gap space. The simulations are carried out over a period of 24 h in real time, encompassing both sunny and cloudy days during the coldest and hottest days of the year 2022 in Miami.

As shown in Table 2, numerical simulations were performed for transient temperatures of 10, 15, and 20 °C, along with transient temperature ranges of 1, 3, and 5 °C in winter and transient temperatures of 20, 25, and 30 °C, and transient temperature ranges of 1, 3, and 5 °C in summer. It is noteworthy to say that the transient temperature represents the central temperature within the temperature range. For example, with a transient temperature of 20 °C and a transient temperature range of 5 °C, the opaqueus (equivalent to the term solidus in SLPCM) and transparentus (equivalent to the term liquidus in SLPCM) temperatures are 17.5 °C and 22.5 °C, respectively. The study analyzes the effects of these parameters on the transparency fraction, interior surface temperature, and energy savings.

As provided next, the numerical results are represented in four sections, each of which contains the parametric study results for one parameter, including transparency fraction, interior surface temperature, total heat flux on the interior surface, and energy savings.

4.1. Transparency Fraction

In this section, the variations of the transparency fraction of the SSPCM during summer and winter for the hottest sunny and cloudy days, as well as the coldest sunny and cloudy days, are analyzed.

Figure 9 shows the dependence of the transparency fraction on time for the studied scenarios during summertime on sunny and cloudy days. In Figure 9a, it is observed that when the SSPCM transient temperature is 20 °C, it remains opaque throughout the day; however, it partially transitions to a translucent state between 12 pm and 4 pm. Conversely, at a transient temperature of 25 °C, a complete phase transition cycle occurs, highlighting the significant influence of transient temperature on the phase transition behavior of the SSPCM and, consequently, the thermal behavior of the glazing system. With the initial temperature of the glazing system set equal to the indoor temperature (26 °C), the transparency fraction of SSPCM decreases from values near 1 during the night as solar radiation is absent at low outdoor temperatures. As shown in Figure 9a, the gradient of the transparency fraction increases with lower transient temperature ranges, with the transparency fraction transitioning from 1 to 0 in about 3.5 h for a transient temperature range of 1 °C and nearly 4.25 h for a transient temperature range of 3 °C. However, for a transient temperature range of 5 °C, the SSPCM does not transition to the opaque phase, and its transparency fraction reduction gradient is lower than that for the other two conditions above. The gradients are consistent across the three different transient temperature ranges. Around 8:00, as solar radiation starts to take place, the temperature of the glazing system starts to rise, leading to an increase in the transparency fraction. All three SSPCMs with different transient temperature ranges reach the transparent phase, remain transparent for a couple of hours, and then start to transition back to a translucent state due to the temperature and solar irradiation reductions. For SSPCMs with a transient temperature of 30 °C, the transparency fraction never decreases, indicating that the SSPCM remains transparent throughout the day and its temperature never exceeds the opaqueus temperature.

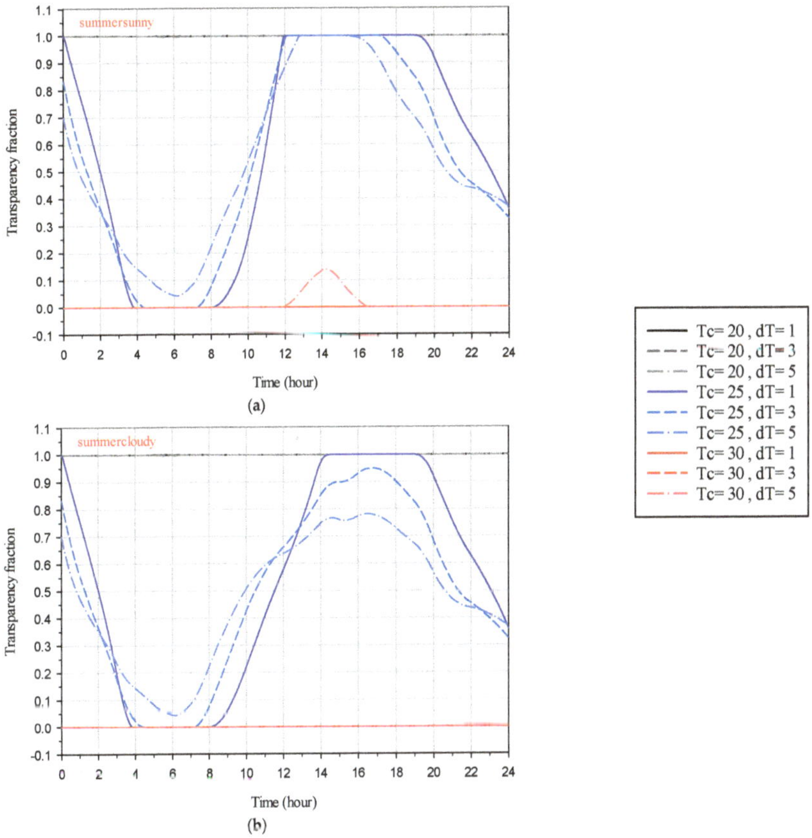

Figure 9. Variations of transparency fraction over time in the cases of (**a**) summer sunny and (**b**) summer cloudy.

Figure 9b provides the results for the summertime on a cloudy day. The transparency fraction variations before sunrise (around 8:00) are identical to those in Figure 9a, as expected. The lack of direct solar irradiation at the presence of diffuse solar irradiation causes less overall solar irradiation, resulting in a lower temperature rise of the system and hence a lower transparency fraction increment gradient during the daytime. It is also evident from this figure that for all transient temperature ranges, the transparency fraction remains at 0 and 1 for transient temperatures of 20 °C and 30 °C, respectively. As shown in Figure 9b, although the SSPCM with a transient temperature range of 1 °C goes through a full phase transition cycle and remains transparent for a couple of hours, it is not the case for the SSPCM with a transient temperature range of 3 °C and 5 °C, as they do not reach the transparent phase. Overall, lower transient temperature ranges have facilitated full phase transition cycles of the SSPCM, allowing the system to fully utilize its latent heat energy storage capacity. Furthermore, the longer the SSPCM remains in its transparent phase, the longer it provides the best visual view for the occupants.

Figure 10 depicts the variations in the SSPCM transparency fraction for the studied scenarios during winter, considering both sunny and cloudy days. It is evident from this figure that the presence of direct solar irradiation on a sunny day greatly affects both the SSPCM temperature and its transparency fraction. Specifically, when the SSPCM transient temperature is 15 °C, on both sunny and cloudy days, the SSPCM remains transparent throughout the entire day. At a transient temperature of 25 °C, the glazing system can utilize the latent heat storage of the SSPCM during sunny days, as the SSPCM undergoes partial phase changes, whereas this is not observed during cloudy days, where the SSPCM remains opaque throughout the day. When the SSPCM transient temperature is set to 20 °C, it undergoes a full phase transition cycle on sunny days, but only a partial phase transition on cloudy days, failing to reach its transparent phase. Figure 10 also shows that that the effect of transient temperature range on the SSPCM transparency fraction is lower during winter sunny days compared to winter cloudy days. Regarding the provision of a visual view for the occupants, during sunny days in winter, a visual view can be provided during the daytime, whereas on cloudy days, the view would mostly be translucent during the daytime.

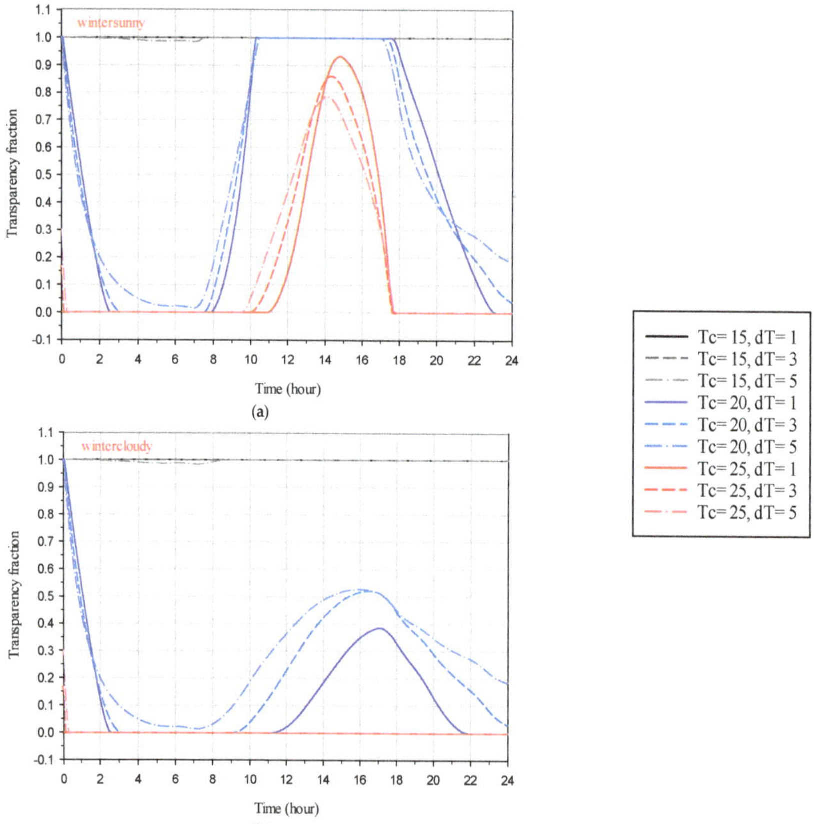

Figure 10. Variations in the transparency fraction over time in the cases of (**a**) winter sunny and (**b**) winter cloudy.

4.2. Interior Surface Temperature

In this section, the simulation results related to the variations in the interior surface temperature of the glazing system during the hottest summer and coldest winter days in sunny and cloudy sky conditions are discussed.

Figure 11 depicts the interior surface temperature distribution during the summer sunny and cloudy days for the studied scenarios of the DGW–SSPCM as well as the DGW–REF (Table 2). During the summertime, this figure shows that the lower the interior surface temperature, the better the thermal performance of the system. Also, the interior surface temperature for the cases with transient temperature of 20 °C and 30 °C are approximately the same, and the trend is the same for both sunny and cloudy days. This is because in these transient temperatures the SSPCM has not gone through any phase transition, and thus the glazing system cannot benefit from its latent heat storage advantage. In addition, the case with a 20 °C transient temperature results in higher inner temperature as it stays in the transparent phase as opposed to the case with a 30 °C transient temperature. Moreover, in these two cases, the transient temperature range has a subtle impact on the interior temperature due to the lack of SSPCM phase change. Regarding these two cases (i.e., transient temperatures of 20 °C and 30 °C), the thermal performance of the system is different on sunny and cloudy days. On sunny days, even when the SSPCM phase transition does not occur, the thermal performance of the DGW–SSPCM (compared to the DGW–REF) has been increased during the daytime, and it has been decreased during the nighttime due to the blockage of the direct solar irradiation.

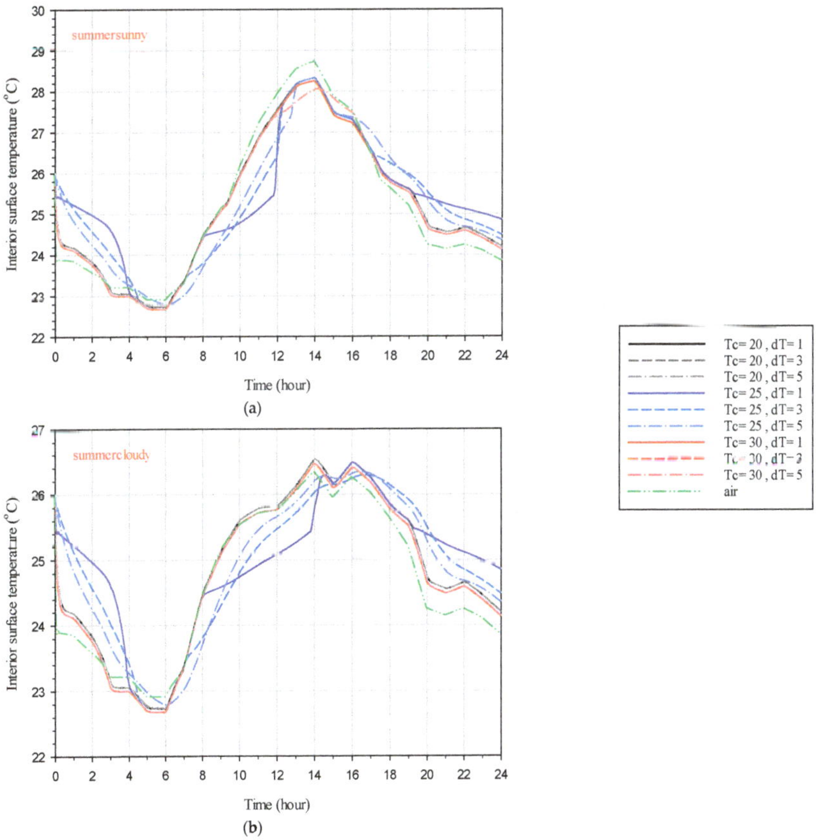

Figure 11. Variations in the interior surface temperature over time in the cases of (**a**) summer sunny and (**b**) summer cloudy.

On cloudy days, Figure 11 shows that the thermal performance of the DGW–SSPCM is lower than the DGW–REF during most of the hours of the day. On the other extreme of the rope, the SSPCM with a transient temperature of 25 °C yields different temperature distributions as the phase transition occurs. It is visible that during the daytime, the interior surface of the DGW–SSPCM is lower than the DGW–REF, and at nights it is reversed. During the day, as the solar irradiation comes to light, the SSPCM absorbs heat using its high latent heat advantage and behaves as a promising insulation material by preventing solar irradiation to heat up the interior surface. At night, however, this absorbed heat is released as the ambient temperature drops, resulting in a higher interior temperature at nights. At this transient temperature (i.e., 25 °C), the influence of the transient temperature range becomes substantial. Finally, on both sunny and cloudy days, the transient temperature range effects on the interior surface temperature are visible, as the lower this value is the greater the temperature gradients are. Also, a lower transient temperature range causes lower interior temperature during the daytime and higher interior temperature at night.

Figure 12 displays the variations in interior surface temperature for the DGW–SSPCM and DGW–REF throughout the day in winter for both sunny and cloudy conditions. A higher interior surface temperature during winter indicates better thermal performance of the system. Referring to Figure 10a, in sunny conditions depicted in Figure 12a, the SSPCM undergoes phase transition at transient temperatures of 20 °C and 25 °C, resulting in differences in the interior surface temperature between the DGW–SSPCM and the DGW–REF. The presence of the SSPCM leads to lower interior temperatures during the day due to its latent absorption and solar direct irradiation blockage, which is released at night when ambient temperatures drop. However, when the SSPCM does not complete a full phase transition cycle, as seen with the transient temperature of 20 °C, the released heat during the night contributes to the low sensible heat storage capacity of the material. In contrast, at a transient temperature of 25 °C, where the full phase transition cycle occurs, higher interior temperatures at night indicate improved thermal performance of the system using the higher latent heat storage capacity of the material.

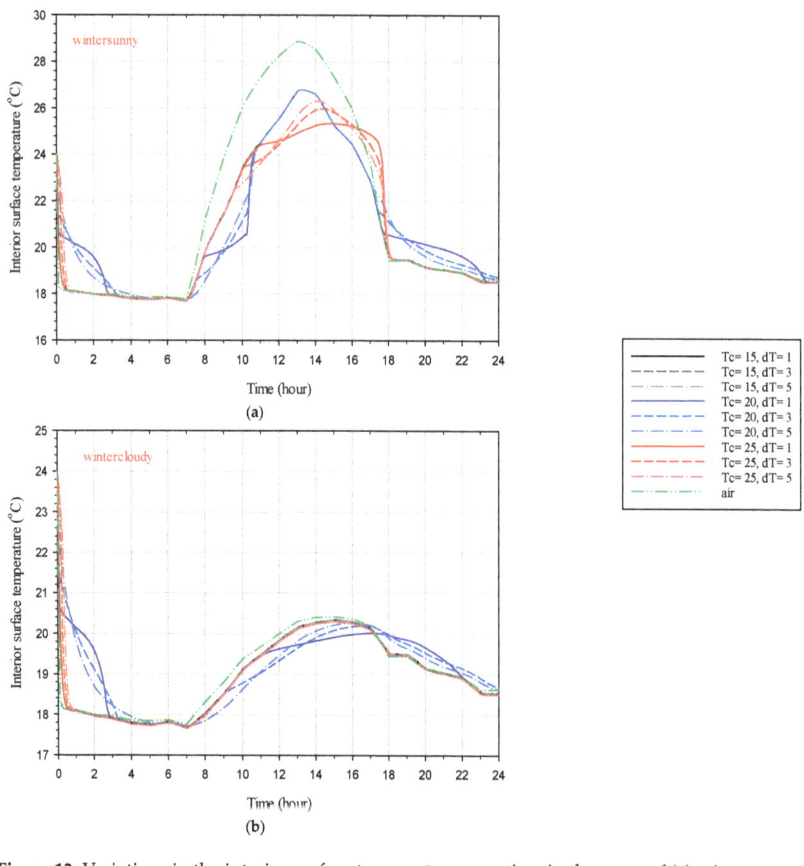

Figure 12. Variations in the interior surface temperature over time in the cases of (**a**) winter sunny and (**b**) winter cloudy.

In cloudy conditions, demonstrated in Figure 12b, only the transient temperature of 25 °C results in beneficial temperature distributions, providing higher interior temperatures during the night. Overall, the transient temperature is a critical parameter for optimizing the thermal performance and energy saving potential of the DGW–SSPCM system compared to the DGW–REF. Incorrect selection of the transient temperature can lead to reduced thermal performance and energy efficiency of the system.

4.3. Total Heat Flux on the Interior Surface

The numerical results for the dependence of the interior surface heat flux on time for the glazing system during the hottest summer and coldest winter days in sunny and cloudy sky conditions are discussed. In Figures 13–15, the negative values for the total heat flux represent the heat flux direction from outside to the inside that contributes to the cooling energy loads.

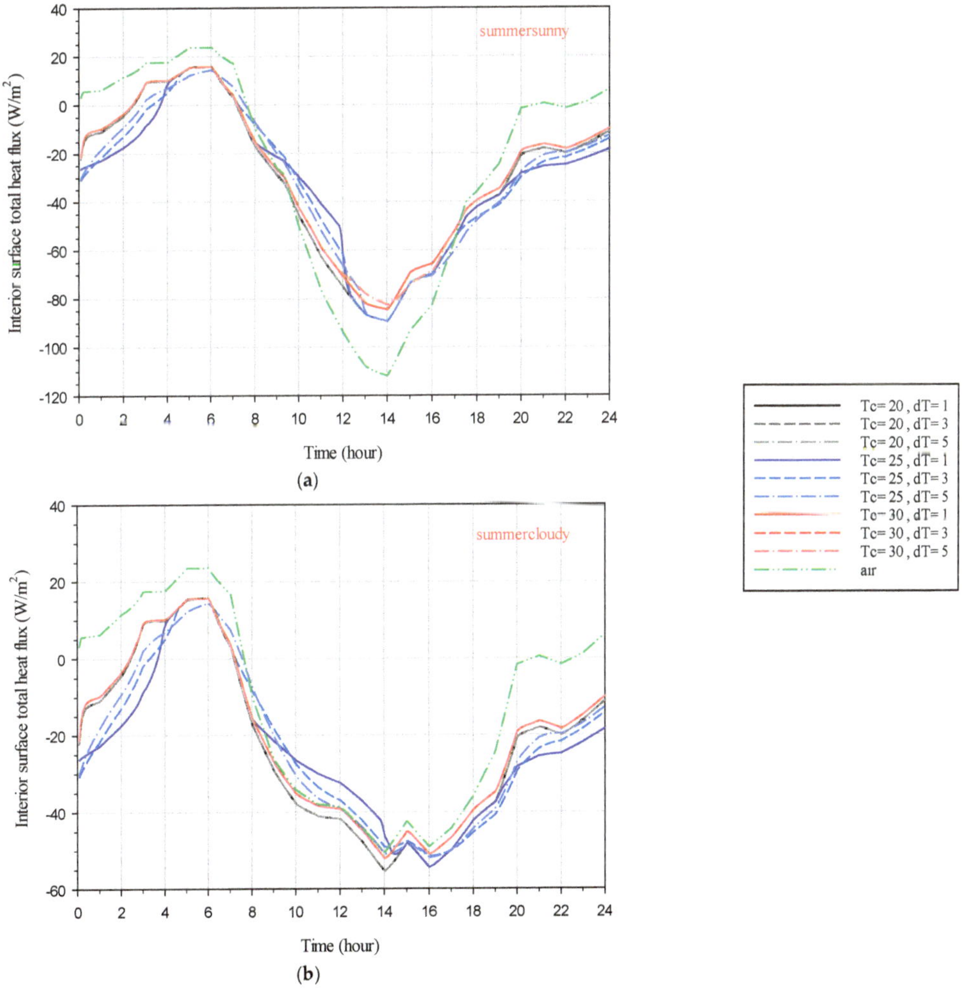

Figure 13. Variations in the interior surface total heat flux over time in the cases of (**a**) summer sunny and (**b**) summer cloudy.

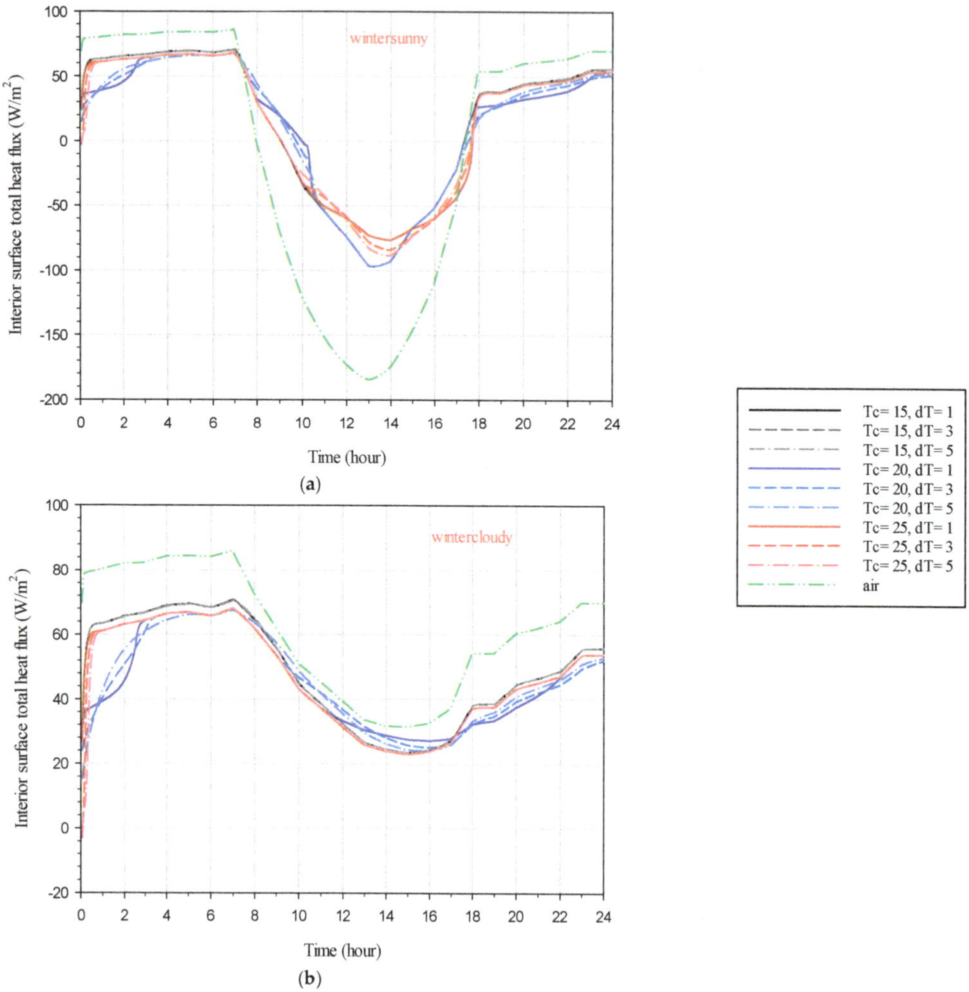

Figure 14. Variations in the interior surface total heat flux over time in the cases of (**a**) winter sunny and (**b**) winter cloudy.

On the contrary, the positive values for the total heat flux represent the heat flux direction from inside to the outside that contribute to heating loads.

In Figure 13, for the DGW–REF case, the heat flux is observed to flow from outside to inside during the night, while during the day, it reverses direction, flowing from inside to outside. On sunny days, using SSPCM results in reduced heat flux entering the interior during the day due to the blocking of solar direct irradiation. However, during the night, heat flux continues to enter the interior as a result of heat release from the SSPCM, which was absorbed during the day. Even on cloudy days, heat release from the SSPCM to the interior is evident during the night, but the heat entering the interior during the day is approximately similar for both DGW–SSPCM and DGW–REF due to the absence of direct solar irradiation. Interestingly, SSPCMs that have not gone through a phase transition still exhibit heat release to the interior during the night, attributed to sensible heat energy storage. However, SSPCMs with a transient temperature of 25 °C, which experience a phase transition during the day, show higher heat flux entering the interior during the night. Comparing different transient temperature ranges reveals that for cases with a transient temperature of 25 °C, lower transient temperature ranges correspond to higher heat flux entering the interior during the night and lower heat flux exiting during the day.

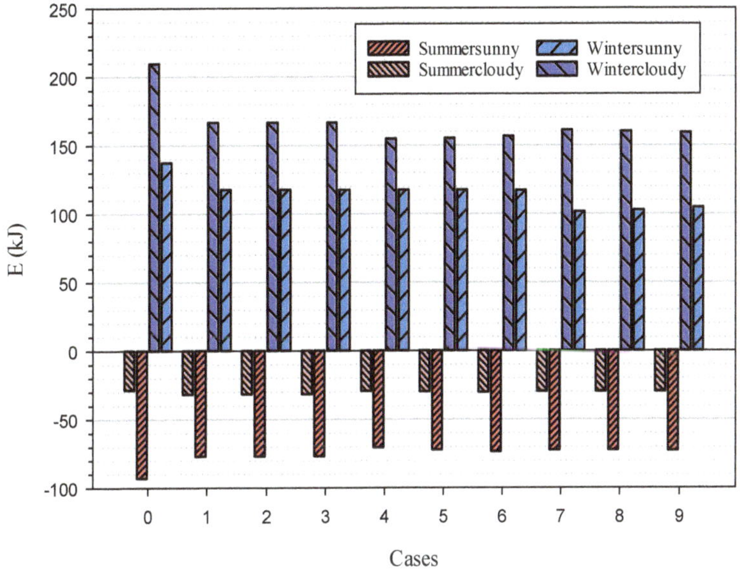

Figure 15. Heat energy vibration for all of the scenarios in the DGW–SSPCM and DGW–REF systems.

In Figure 14, the interior surface total heat flux distribution during the day is depicted for both the DGW–SSPCM and DGW–REF. During winter, the SSPCM can influence the heat flux dynamics, resulting in less overall heat flux being released to the outside during the day, contributing to energy savings. However, it also reduces the amount of heat entering the interior from direct solar irradiation. Figure 14a shows that less heat enters the interior in the DGW–SSPCM compared to the DGW–REF during the day, while the amount of heat flux escaping from the interior in the DGW–SSPCM is also reduced compared to the DGW–REF. Due to the absence of direct solar irradiation, the heat flux throughout the day is directed outside, and the SSPCM contributes positively by reducing the heat flux escaping from the interior to the outside, leading to energy savings (see Figure 14b).

4.4. Energy Savings

To quantify the energy savings with the DGW–SSPCM in relation to DGW–REF, the definition of the energy savings given by Equation (20) is used and depicted in Figure 15 for different scenarios investigated in this paper. Figure 15 shows that the energy savings during cooling and heating degree hours (i.e., the hours during which both cooling and heating are needed) defined as the interior surface temperature of the inner glass pane is above 26 °C in summer and below 24 °C in winter in the DGW system compared to the reference system (i.e., the DGW–SSPCM system).

$$E = \overline{q''} \times \mathrm{t} \times \mathrm{A} \tag{19}$$

$$\Delta E = E_{final} - E_{ref} \tag{20}$$

Figure 15 displays the heat energy of the studied cases (as determined in Table 2) over the peak load hours defined above. Furthermore, Table 5 is provided to summarize the energy savings of all studied cases for various climatic conditions.

Table 5. Energy savings of all studied cases for various climatic conditions.

Climatic Condition	Case 1 (%)	Case 2 (%)	Case 3 (%)	Case 4 (%)	Case 5 (%)	Case 6 (%)	Case 7 (%)	Case 8 (%)	Case 9 (%)
Summer sunny	17.4	17.4	17.4	24.2	22.5	21.1	22.1	22.1	21.9
Summer cloudy	10.9	10.9	10.9	3.1	1.4	6.3	4.1	4.1	4.1
Winter sunny	14.6	14.6	14.6	14.6	14.6	14.6	26.3	25.4	24.0
Winter cloudy	20.4	20.4	20.4	26.2	26.0	25.3	23.1	23.6	24.0

During summer, the SSPCM absorbs solar heat radiation, leveraging its high latent heat energy storage capacity during heating and cooling degree hours. For the summer sunny day, over the cooling degree hours occurring between 9:55 and 17:20, the DGW–SSPCM system exhibits lower heat energy entering the building compared to DGW–REF due to the blockage of direct solar irradiation by the SSPCM. The energy savings of cases 1 to 9 (defined in Table 2) vs. the DGW–REF case on summer sunny days are 17.4, 17.4, 17.4, 24.2, 22.5, 21.1, 22.1, 22.1, and 21.9 percent, respectively. Hence, the highest energy savings of the DGW–SSPCM system vs. the DGW–REF in summer sunny weather conditions occur in case 4, where the transient temperature value and range are 25 °C and 1 °C, resulting in a 24.2 percent energy savings. However, for the summer cloudy day (over the cooling degree hours occurring between 12:55 and 17:05), the DGW–SSPCM system experiences energy loss compared to the DGW–REF system. The energy loss of cases 1 to 9 (defined in Table 2) vs. the DGW–REF case on summer cloudy days are 10.9, 10.9, 10.9, 3.1, 1.4, 6.3, 4.1, 4.1, and 4.1 percent, respectively. Hence, the lowest energy loss of the DGW–SSPCM system vs. DGW–REF in summer cloudy weather conditions occurs in case 4, where the transient temperature value and range are 25 °C and 1 °C, resulting in a 3.1 percent energy loss.

During winter, the SSPCM retains warm air inside of the building and prevents it from escaping. For the winter sunny day, over the heating degree hours occurring between 16:55 and 9:00, the DGW–SSPCM system demonstrates higher energy savings compared to the DGW–REF system. The energy savings of cases 1 to 9 (defined in Table 2) vs. the DGW–REF case on winter sunny days are 14.6, 14.6, 14.6, 14.6, 14.6, 14.6, 26.3, 25.4, and 24.0 percent, respectively. Case 7, with a transient temperature value and range of 25 °C and 1 °C, exhibits the best energy savings among other cases, at 26.3 percent. Conversely, for the winter cloudy day, where the temperature remains below 24 °C throughout the day, the entire day is considered as heating degree hours. The energy savings of cases 1 to 9 (defined in Table 2) vs. the DGW–REF case on winter cloudy days are 20.4, 20.4, 20.4, 26.2, 26.0, 25.3, 23.1, 23.6, and 24.0 percent, respectively. Case 4, with a transient temperature value and range of 20 °C and 1 °C, demonstrates the most energy-efficient system, achieving a 26.2 percent energy savings. While case 7 shows slightly lower energy savings (23.1 percent), it still performs significantly better compared to the DGW–REF system.

Overall, the best SSPCM for the studied south-facing glazing system in Miami is one with a transient temperature value and range of 25 °C and 1 °C, respectively. Choosing an SSPCM with appropriate thermophysical properties and a small thickness applied to the inner glass pane of a DGW within the air gap, resulting in full phase transition cycles along with a lower temperature transient range, could lead to a more energy-efficient glazing system. Despite slightly less energy efficiency during summer cloudy days, the overall energy performance enhancement on other day types during winter and summer compensates for this loss.

5. Recommendations

Future studies are recommended to consider real-world application challenges and to provide comprehensive data for SSPCMs. These studies would include the following:

- Obtaining precise experimental data for the thermophysical and optical properties of SSPCMs. The use of these properties in the numerical simulations would lead to better predictions of the SSPCMs' performance in real-world applications.
- As the application of SSPCMs grows, more manufacturers would enter the PCM market. The increased competition and innovation from multiple manufacturers would drive down costs and improve the quality and variety of SSPCM products.
- There is a notable lack of experimental studies on the application of SSPCMs in the literature. Thus, future studies are needed to understand the real-time challenges of using SSPCMs, including installation, performance under varying environmental conditions, and interaction of SSPCMs with other building components.
- More comprehensive future studies on the cost, manufacturing feasibility, long-term durability, and maintenance challenges are essential to understand the practical implications of SSPCMs in glazing systems. In addition, these studies should include life cycle cost analysis and long-term performance assessments to ensure that SSPCMs are a viable solution for energy efficiency in buildings.
- Research should also explore the integration of SSPCMs with advanced control systems to enhance their efficiency and responsiveness to dynamic environmental changes. This can involve developing smart glazing systems that adjust their properties based on real-time data to optimize thermal comfort and energy savings.

6. Conclusions

This study primarily focuses on assessing the energy performance of the glazing systems subjected to natural weather conditions. The study includes modeling and analysis of a proposed double-glazing configuration that incorporates a 2 mm layer of solid–solid phase change materials (SSPCMs) applied to the inner pane of glass within the air gap. The solid–solid phase change material is chosen due to its advantages over the commonly used liquid–solid phase change materials for glazing system applications. A model was developed using the finite volume method in the ANSYS FLUENT to perform numerical simulations. The model was validated against experimental data, where its predictions were in good agreement with the test results. With the model, different transient temperature values and ranges are investigated in the new proposed glazing system to assess the energy performance of the system during the coldest and hottest days of the year on cloudy and sunny days in Miami, FL weather conditions. A thickness of 2 mm was used for the solid–solid phase change material with transient temperatures ranging from 15 °C to 30 °C and transient temperature ranges between 1 °C and 5 °C. The obtained results showed energy performance enhancement of the glazing system with solid–solid phase change materials as opposed to traditional double-glazing windows. The results showed that the obtained highest energy savings of the system during heating and cooling degree hours are 24%, 26%, and 23% for the case with the transient temperature magnitude and a range of 25 °C and 1 °C in summer sunny, winter sunny, and winter cloudy days, while an energy loss of 3% is observed in this case for the summer cloudy days. Overall, a more energy-efficient glazing system could be achieved by selecting the solid–solid phase change material with a small thickness to be applied to the inner glass pane of a DGW within the air gap with appropriate thermophysical properties leading to full phase transition cycles along with the lowest possible phase transition temperature range.

Author Contributions: Conceptualization, H.A., W.M. and H.H.S.; Methodology, H.A.; Software, H.A.; Validation, H.A.; Formal analysis, H.A. and H.H.S.; Investigation, H.A.; Writing—review & editing, W.M. and H.H.S.; Supervision, W.M. and H.H.S.; Project administration, W.M. All authors have read and agreed to the published version of the manuscript.

Funding: This research received no external funding.

Data Availability Statement: The original contributions presented in the study are included in the article, further inquiries can be directed to the corresponding author.

Conflicts of Interest: The authors declare no conflict of interest.

Nomenclature

A	Area (m^2)	Greek symbols	
c_p	Specific heat (J/kg∗K)	β	Transparency fraction
d	Optical thickness (m)	μ	Dynamic viscosity (Pa∗s)
h	Heat transfer coefficient (W/m^2∗K)	ρ	Density (kg/m^3)
h_s	Sensible enthalpy (J/kg)	σ_a	Absorption coefficient (1/m)
H	Enthalpy (J/kg)	σ_s	Scattering coefficient (1/m)
ΔH	Latent heat (J/kg)	τ	Transmittance
I	Radiation intensity	$\varnothing$	Phase function
L	Latent heat of fusion (kJ/kg)	Ω'	Solid angle
n	Refractive index	Subscripts	
p	Pressure	op	Opaque
q″	Total heat flux (W/m^2)	PCM	Phase change material
$\vec{r}$	Position vector	ref	Reference
s	Sample thickness (path length) (m)	tr	Transparent
$\vec{s}$	Direction vector	Acronyms	
$\vec{s}'$	Scattering direction vector	DGW	Double-glazing window
t	Time (s)	PCM	Phase change material
T	Temperature (°C)	SLPCM	Solid–liquid phase change material
v	Velocity (m/s)	SSPCM	Solid–solid phase change material

References

1. IEA. Buildings—A Source of Enormous Untapped Efficiency Potential. 2024. Available online: https://www.iea.org/topics/buildings (accessed on 12 March 2024).
2. Wang, S.; Jiang, T.; Meng, Y.; Yang, R.; Tan, G.; Long, Y. Scalable thermochromic smart windows with passive radiative cooling regulation. *Science* **2021**, *374*, 1501–1504. [CrossRef] [PubMed]
3. Araújo, G.; Teixeira, H.; Gomes, M.G.; Rodrigues, A.M. Multi-objective optimization of thermochromic glazing properties to enhance building energy performance. *Sol. Energy* **2023**, *249*, 446–456. [CrossRef]
4. Sun, Y.; Liu, X.; Ming, Y.; Liu, X.; Mahon, D.; Wilson, R.; Liu, H.; Eames, P.; Wu, Y. Energy and daylight performance of a smart window: Window integrated with thermotropic parallel slat-transparent insulation material. *Appl. Energy* **2021**, *293*, 116826. [CrossRef]
5. Radwan, A.; Katsura, T.; Memon, S.; Serageldin, A.A.; Nakamura, M.; Nagano, K. Thermal and electrical performances of semi-transparent photovoltaic glazing integrated with translucent vacuum insulation panel and vacuum glazing. *Energy Convers. Manag.* **2020**, *215*, 112920. [CrossRef]
6. Ke, W.; Ji, J.; Wang, C.; Zhang, C.; Xie, H.; Tang, Y.; Lin, Y. Comparative analysis on the electrical and thermal performance of two CdTe multi-layer ventilated windows with and without a middle PCM layer: A preliminary numerical study. *Renew. Energy* **2022**, *189*, 1306–1323. [CrossRef]
7. Sorooshnia, E.; Rashidi, M.; Rahnamayiezekavat, P.; Mahmoudkelayeh, S.; Pourvaziri, M.; Kamranfar, S.; Gheibi, M.; Samali, B.; Moezzi, R. A novel approach for optimized design of low-E windows and visual comfort for residential spaces. *Energy Built Environ.* **2023**, *in press*. [CrossRef]
8. Maduru, V.R.; Shaik, S. Laminated glazing for buildings: Energy saving, natural daylighting, and CO_2 emission mitigation prospective. *Environ. Sci. Pollut. Res.* **2022**, *29*, 14299–14315. [CrossRef] [PubMed]
9. Taleb, H.M.; Antony, A.G. Assessing different glazing to achieve better lighting performance of office buildings in the United Arab Emirates (UAE). *J. Build. Eng.* **2020**, *28*, 101034. [CrossRef]
10. Mohamed, A.F.; Gomaa, M.M.; Amir, A.A.; Ragab, A. Energy, Thermal, and Economic Benefits of Aerogel Glazing Systems for Educational Buildings in Hot Arid Climates. *Sustainability* **2023**, *15*, 6332. [CrossRef]
11. Soares, I.T.; de Araújo, J.R.; Monteiro, S.N.; Marques, M.d.F.V. Novel system using hydrogel with reduced graphite oxide particles as active layer for potential application in smart window. *J. Mater. Res. Technol.* **2023**, *22*, 1924–1934. [CrossRef]
12. Arasteh, H.; Maref, W.; Saber, H.H. Energy and thermal performance analysis of PCM-Incorporated glazing units combined with passive and active techniques: A review study. *Energies* **2023**, *16*, 1058. [CrossRef]
13. Fasano, M.; Borri, D.; Cardellini, A.; Alberghini, M.; Morciano, M.; Chiavazzo, E.; Asinari, P. Multiscale simulation approach to heat and mass transfer properties of nanostructured materials for sorption heat storage. *Energy Procedia* **2017**, *126*, 509–516. [CrossRef]
14. Fallahi, A.; Guldentops, G.; Tao, M.; Granados-Focil, S.; Van Dessel, S. Review on solid-solid phase change materials for thermal energy storage: Molecular structure and thermal properties. *Appl. Therm. Eng.* **2017**, *127*, 1427–1441. [CrossRef]

15. Gao, Y.; Zheng, Q.; Jonsson, J.C.; Lubner, S.; Curcija, C.; Fernandes, L.; Kaur, S.; Kohler, C. Parametric study of solid-solid translucent phase change materials in building windows. *Appl. Energy* **2021**, *301*, 117467. [CrossRef]
16. Wang, P.; Liu, Z.; Zhang, L.; Wang, Z.; Fan, J. Inversion of extinction coefficient and refractive index of variable transparency solid–solid phase change material based on a hybrid model under real climatic conditions. *Appl. Energy* **2023**, *341*, 121098. [CrossRef]
17. Raj, C.R.; Suresh, S.; Bhavsar, R.R.; Singh, V.K. Recent developments in thermo-physical property enhancement and applications of solid solid phase change materials: A review. *J. Therm. Anal. Calorim.* **2020**, *139*, 3023–3049. [CrossRef]
18. Guldentops, G.; Ardito, G.; Tao, M.; Granados-Focil, S.; Van Dessel, S. A numerical study of adaptive building enclosure systems using solid–solid phase change materials with variable transparency. *Energy Build.* **2018**, *167*, 240–252. [CrossRef]
19. Ma, Y.; Li, D.; Yang, R.; Zhang, S.; Arıcı, M.; Liu, C.; Zhang, C. Energy and daylighting performance of a building containing an innovative glazing window with solid-solid phase change material and silica aerogel integration. *Energy Convers. Manag.* **2022**, *271*, 116341. [CrossRef]
20. Tennakoon, T.; Chan, Y.-H.; Chan, K.-C.; Wu, C.; Chao, C.Y.-H.; Fu, S.-C. Energy Performance and Comfort Analysis of Three Glazing Materials with Distinct Thermochromic Responses as Roller Shade Alternative in Cooling- and Heating-Dominated Climates. *Buildings* **2024**, *14*, 1157. [CrossRef]
21. Nazari, M.; Matusiak, B. Daylighting simulation and visualisation: Navigating challenges in accuracy and validation. *Energy Build.* **2024**, *312*, 114188. [CrossRef]
22. Zhang, C.; Yang, R.; Lu, Y.; Arıcı, M.; Ma, Y.; Yang, X.; Qi, Z.; Li, D. Parametric research on thermal and optical properties of solid-solid phase change material packaged in glazing windows. *J. Energy Storage* **2024**, *83*, 110562. [CrossRef]
23. ASHRAE. *ASHRAE Handbook of Fundamental, Effective Thermal Resistance of Plane Air Spaces*; Chapter 26, Table 3; ASHRAE: Peachtree Corners, Georgia, 2021; pp. 26.14–26.15.
24. Gowreesunker, B.; Stankovic, S.; Tassou, S.; Kyriacou, P. Experimental and numerical investigations of the optical and thermal aspects of a PCM-glazed unit. *Energy Build.* **2013**, *61*, 239–249. [CrossRef]
25. PCM Products. Available online: https://www.pcmproducts.net/Phase_Change_Material_Products.htm (accessed on 12 March 2024).
26. Ansys. *ANSYS FLUENT Theory Guide*; (November 2021) Revision: 22.1.0 for ANSYS 2022 R1; ANSYS FLUENT: Canonsburg, PA, USA, 2022.
27. Bambang, N.; Ikhsan, M.; Tensiska; Sukri, N. Mahani Rheological Properties of Honey and its Application on Honey Flow Simulation through Vertical Tube. *IOP Conf. Series Earth Environ. Sci.* **2019**, *334*, 012041. [CrossRef]
28. Brent, A.D.; Voller, V.R.; Reid, K.T.J. Enthalpy-porosity technique for modeling convection-diffusion phase change: Application to the melting of a pure metal. *Numer. Heat. Transf. A Appl.* **1988**, *13*, 297–318.
29. Heijs, A.W.J.; Lowe, C.P. Numerical evaluation of the permeability and the Kozeny constant for two types of porous media. *Phys. Rev. E* **1995**, *51*, 4346–4352. [CrossRef]
30. Available online: https://weatherspark.com/h/d/18622/2022/1/30/Historical-Weather-on-Sunday-January-30-2022-in-Miami-Florida-United-States#Figures-Temperature (accessed on 4 December 2023).
31. Ministry of Housing and Urban-Rural Development of the People's Republic of China. *Code for Thermal Design of Civil Building*; China Architecture & Building Press: Beijing, China, 2016.
32. Goia, F.; Perino, M.; Haase, M. A numerical model to evaluate the thermal behaviour of PCM glazing system configurations. *Energy Build.* **2012**, *54*, 141–153. [CrossRef]
33. Ansys. *ANSYS Fluent*; Ver. 2022 R1; ANSYS FLUENT: Canonsburg, PA, USA, 2022.

energies

MDPI

Article

Experimental Study on Phase Change Energy Storage Flooring for Low-Carbon Energy Systems in Grassland Pastoral

Kai Wang [1], Guoqiang Xu [2,3], Xiaochen Zhao [2], Guo Li [2,3,*] and Lisi Mai [2]

1 School of Architecture and Engineering, Hohhot Vocational College, Hohhot 010051, China; 200900027@hhvc.edu.cn
2 School of Architecture, Inner Mongolia University of Technology, Hohhot 010051, China; xgq5@imut.edu.cn (G.X.); ayzhaoxiaochen@126.com (X.Z.); ngdmls@imut.edu.cn (L.M.)
3 Key Laboratory of Green Building at Universities of Inner Mongolia Autonomous Region, Hohhot 010051, China
* Correspondence: 20211100430@imut.edu.cn

Abstract: Phase change energy storage technology enhances the integration of renewable resources into low-carbon energy systems for grassland pastoral settlements, further addressing the balance between energy needs and environmental sustainability. This study examines a heating system using an experimental platform in an environmental chamber, where the thermal storage and release processes of phase change energy storage flooring were monitored. The results revealed that phase change energy storage flooring exhibits higher heat transfer efficiency and faster heating rates. Under 40 °C heating conditions, the heating rate of the thermal storage layer increased by 12.5% within 1 h. The flooring also demonstrated superior heat release performance, with the peak heat flux of the thermal storage layer delayed by 15 min. Higher heating temperatures shortened the heating time and extended the heat release duration of the phase change energy storage flooring. Under 45 °C heating conditions, the heat transfer efficiency of the surface temperature of the thermal storage layer increased by 38% within 1 hour and by 24.7% over 4 h. In addition, energy consumption in different tests was analyzed, and thermal conductivity was discussed according to the heat transfer model. Phase change energy storage flooring, when coupled with the abundant solar energy resources available in grassland pastoral areas, presents a viable option for the construction of low-carbon energy systems in grassland pastoral settlements.

Keywords: energy storage flooring; phase change materials; thermal performance; grassland pastoral settlements

1. Introduction

Human activities have always been closely connected to their environment. The Inner Mongolian grasslands, one of China's largest grasslands and natural pastures, are located on the Inner Mongolian Plateau. Most of this region lies within areas rich or relatively rich in solar energy resources. According to the "Inner Mongolia Renewable Energy Development Report 2023", the theoretical solar power potential in the Inner Mongolia Autonomous Region is estimated to be 94 billion kW, accounting for approximately 21% of the national total, indicating significant potential for solar energy utilization [1].

The climate of the Inner Mongolia Autonomous Region is classified as severely cold, with year-round cold temperatures and long winter heating periods. In 2022, the average rural household in the pastoral areas of the Inner Mongolia Autonomous Region consumed 1233 kg of loose coal per capita, significantly higher than the 724 kg average in neighboring provinces [2]. The high construction and maintenance costs required for centralized heating systems in grassland pastoral settlements make it essential to integrate renewable energy sources, especially in decentralized systems [3–5].

The abundant solar energy resources in grassland pastoral areas provide favorable conditions for renewable energy-based heating solutions. Considering the unique climate and lifestyle characteristics of grassland pastoral areas, the research team proposed the application of a solar water-based concrete thermal storage flooring system to improve the indoor thermal environment of grassland pastoral settlements [6]. This system includes a solar collector system, an auxiliary heating system, and a radiant floor system.

The use of a solar water-based concrete thermal storage flooring system can significantly reduce the consumption of commercial energy. With the continuous promotion of clean heating in northern China over the past few years, the energy consumption structure in rural areas has undergone significant changes. In 2022, the total energy consumption in the rural pastoral areas of the Inner Mongolia Autonomous Region was equivalent to 9.83 million tons of standard coal, with commercial energy accounting for over 90% of this total [2]. However, the application of solar water heating systems faces challenges, such as instability and complex maintenance [6].

Enhancing the thermal storage capacity of building envelopes and effectively utilizing renewable energy sources, such as solar energy, are key strategies for addressing the current energy consumption issues in grassland pastoral settlements. Integrating existing energy collection technologies with building structures is essential for improving energy efficiency. The use of thermal storage technologies within buildings can address the cyclical instability associated with solar and wind energy utilization [7,8]. Moreover, it can significantly improve the indoor thermal environment, making it a crucial approach to enhancing the energy efficiency of buildings in grassland pastoral areas and optimizing their energy consumption structure.

Advanced material technologies have demonstrated that high-energy-density phase change materials (PCMs) can provide substantial thermal storage within a narrow temperature range [9–12]. By increasing the thermal inertia of building materials and offering favorable thermal regulation effects, these materials can smooth out fluctuations in energy demand to some extent [13,14]. Mohamed Lachheb et al. [15] reviewed the impact of specific factors, such as the type, location, and number of PCMs, on improving building energy efficiency and indoor thermal comfort and showed that the use of PCMs in brick walls is a promising technology that can reduce indoor temperature fluctuations and save energy. Bence Nemeth et al. [16] conducted a study on the thermal performance and electricity demand of two container model houses and found that PCMs plasters can reduce electricity demand by 10.3% and 7.0%, respectively, over the entire winter compared to reference houses using electric radiators or heat pumps. Yuan Zhang et al. [17] designed a building with Trombe walls and PCMs, and the indoor discomfort degree-hour (IDH) and integrated indoor discomfort duration (ID) were analyzed. The results showed that ventilation in Trombe walls using PCMs is the most effective way to reduce IDH. In order to reduce the mismatch between indoor daytime and nighttime temperatures in Pakistani homes during winter, Aurang Zaib et al. [18] investigated a heat exchanger using phase change materials coupled with solar energy, and the results showed that the device could minimize the household heat load while maintaining indoor thermal comfort. Using Energy plus software, Asasei Unarine Casey Ndanduleni et al. [19] simulated temperature fluctuations on residential roofs in South Africa and showed that a 30 mm thick PCMs layer could significantly reduce indoor temperature fluctuations by reducing the temperature to between 14 and 16 °C. Oumayma Babaharra et al. [20] conducted a numerical evaluation of a heated floor with PCMs, showing that the best thermal performance improvement can be achieved by placing PCMs above the heating tubes. Ibtissam Afaynou et al. [21] studied the influence of composite PCMs type, porosity, and other parameters on the thermal properties of the material. The results show that PCMs composites with high porosity have higher heat transfer rates and more uniform melting processes. This contributes to improving overall energy utilization efficiency and reducing reliance on fossil fuels.

Incorporating PCMs into cement mortar results in thermal storage materials with excellent thermal performance, which can be further applied in the construction of solar water-based phase change energy storage flooring [7,22–24]. As shown in Figure 1, slag cement mortar combined with microencapsulated phase change materials (mPCM) is used as the thermal storage layer in the radiant floor system. This advanced energy storage technology addresses the cyclical variations of renewable resources in grassland pastoral areas and aligns with existing peak–valley electricity pricing policies [8], thereby balancing the energy needs of grassland pastoral dwellings with environmental considerations.

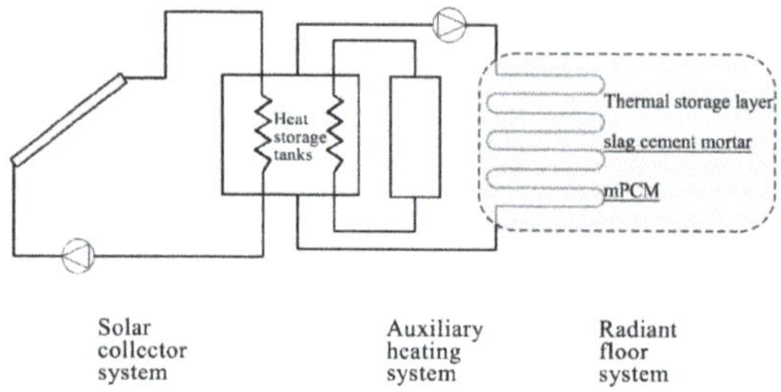

Figure 1. Solar water-based phase change energy storage flooring.

The abundant renewable resources in the Inner Mongolian Plateau offer significant potential for developing low-carbon energy systems in grassland pastoral settlements, examining the heating system using an experimental platform in an environmental chamber, where the thermal storage and release processes of phase change energy storage flooring were monitored. The energy consumption in different tests is analyzed, and the thermal conductivity is discussed according to the heat transfer model. The emphasis is on optimizing the heating terminal while also considering future scenarios where it could be integrated with the supply side of renewable energy sources. This approach holds substantial potential for the development and utilization of renewable energy, contributing to the creation of comfortable thermal environments in grassland pastoral dwellings and advancing green energy-efficient development.

2. Materials and Methods

A field survey was first conducted in a grassland pastoral area in central Inner Mongolia to determine the indoor temperature conditions during winter, both with and without heating, and to understand the actual heating behavior patterns. Figure 2 shows an aerial photograph of a typical grassland pastoral settlement.

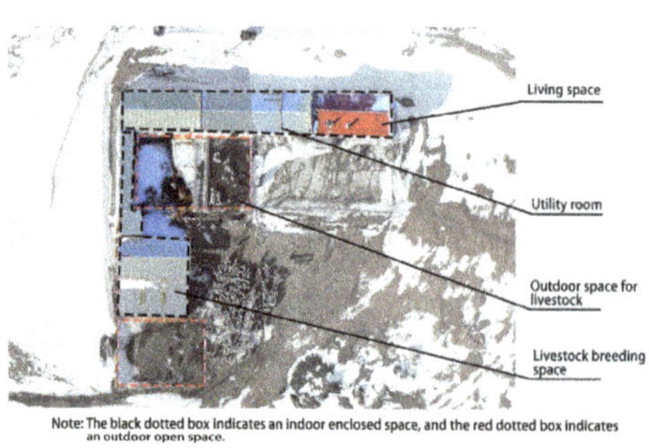

Figure 2. Aerial photograph of a typical grassland pastoral settlement.

The field survey was conducted from 17 February to 24 February 2024. During this period, a temperature data logger (RC-4HC, Jingchuang Electric Co., Ltd., Jiangsu, China) was used to monitor the temperatures both indoors and outdoors at a height of 1.5 m. As shown in Figure 3, the average outdoor temperature at the grassland dwelling was −6 °C, with the lowest nighttime temperature reaching −17 °C. The average daily temperature in the heated room was 15 °C; in contrast, the average daily temperature in the unheated room was 3 °C, with the lowest nighttime temperature dropping to −3 °C.

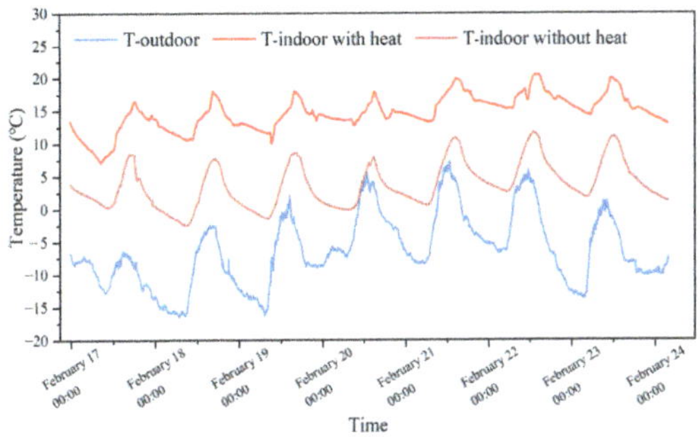

Figure 3. Indoor and outdoor temperatures of grassland dwellings in winter.

The heated room in this building utilizes a decentralized electric heating water tank and radiant flooring as the heating system. Typically, users activate the heating equipment between 16:00 and 20:00. It is important to note that this region implements a peak–valley electricity pricing policy, with a low valley rate of 0.165 CNY/kWh from 18:00 to next day 08:00, and a high peak rate of 0.465 CNY/kWh from 08:00 to 18:00. Additionally, the heat source input is unstable. Advanced thermal storage materials hold promise for leveraging renewable energy to address heating energy consumption during non-valley hours and potentially extending heating duration.

2.1. Fabrication of the Radiant Floor Module

The phase change energy storage flooring was fabricated by incorporating mPCM (Thermalcare 32, Shanghai Xinya New Materials Technology Co., Ltd., Shanghai, China) into the thermal storage layer shown in Figure 1. This PCM consists of paraffin encapsulated in a polymer resin shell, which transitions between solid and liquid phases. The base thermal storage layer utilized slag cement (PSA32.5, Zhucheng 97 Building Materials Co., Ltd., Shandong, China) with good thermal performance [25], standard sand (ISO middle grade, Xiamen ISO Standard Sand Co., Ltd., Fujian, China), and water. Table 1 shows the material quantities used.

Table 1. The material quantities used of the thermal storage layer of the radiant floor module.

Sample	Slag Cement (kg)	Sand (kg)	mPCM (kg)	Water (kg)	W/C Ratio
SSC0	6.48	19.44	0	3.24	0.5
SSC5	6.48	18.468	0.972	3.888	0.6

The phase change energy storage flooring was fabricated by replacing an equal mass of fine aggregate with mPCM [11,26]. The mPCM were introduced early in the raw material mixing process [27], following these steps: (1) The slag cement and sand were weighed according to the specified proportions and mixed evenly to form a dry mix; (2) tap water was then weighed according to the specified proportions and added to the dry mix; and (3) the mortar was mixed at low speed for 60 s to form the mortar mixture.

As shown in Figure 4, the casting of the phase change energy storage flooring involved assembling all pre-prepared materials and using a manual pressurization device to inject water and apply pressure to the module. Four sets of K-type thermocouples were pre-embedded in the thermal storage layer, with each set containing seven thermocouples that recorded temperature changes at different depths within the thermal storage layer.

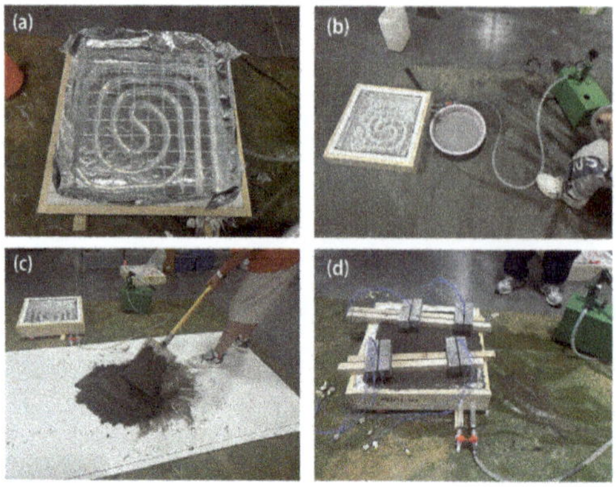

Figure 4. Field installation of radiant floor module: (**a**) radiant floor module production; (**b**) manual water injection and pressure; (**c**) thermal storage mortar preparation; and (**d**) fixation and maintenance.

The construction and casting of the different radiant floor modules were completed within two hours for each module. Afterward, they were placed in a laboratory for natural curing, with water sprayed twice daily. The curing period for the thermal storage layer exceeded 30 days. Before testing, four heat flux sensors were attached near the temperature measurement points using thermal grease.

2.2. Testing Method for the Radiant Floor Module

The use of PCMs in flooring requires consideration of many influencing factors, including the content of phase change materials, heating temperature, and other conditions. In this study, the heat storage and release process of the floor using phase change materials is tested by controlling a single variable in a controlled thermal environment. Secondly, the heating process of grassland pastoral areas also faces the problem of unstable heat sources; in this study, adjustable heat sources are used to achieve different input heating temperatures. To evaluate the thermal performance of the phase change energy storage flooring during heating, cooling, and under different water supply temperatures, an artificial environmental chamber was used as a testing environment that closely simulates the ideal thermal conditions of a grassland pastoral settlement. The environment chamber can adjust the ambient temperature between −30 °C and 30 °C according to the requirements. According to the actual measurement shown in Figure 3, the indoor thermal environment of the residence in the grassland pastoral area can be replicated under the condition of no heating. As shown in Figure 5a, the experimental system primarily consists of a controlled electric heating water tank, a circulation pump, supply and return pipelines, the radiant floor module, a cooling air unit, and a testing system. The arrangement of the temperature measuring points and heat flow measuring points of the phase change energy storage floor is shown in Figure 5b. In the study, the phase change temperature of the thermal management material is 32 °C; thus, the thermocouple is used to monitor the temperature change during the heating and cooling process of the phase change energy storage floor. In addition, a heat flux meter is used to detect the heat release of the phase change energy storage floor.

The controlled electric heating water tank is equipped with an electric heater, allowing for precise adjustment of the water supply temperature. This feature enables the differential adjustment of the heat input to the radiant floor module, simulating the variable heating conditions typical of grassland pastoral dwellings. The parameters of the testing instruments and equipment are listed in Table 2.

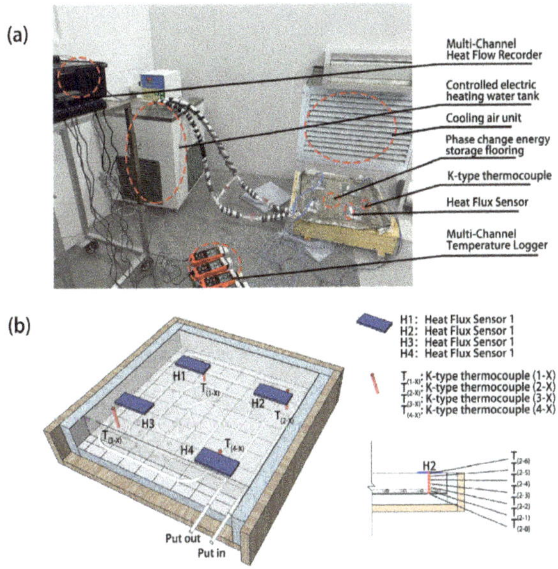

Figure 5. The test system of phase change energy storage flooring: (**a**) components of the test system; (**b**) placement of sensors in the floor.

Table 2. Test equipment and parameters.

Equipment Type	Equipment Parameters
1. Controlled electric heating water tank SN-DHC-3010 (Shangpu Instrument Equipment Co., Ltd., Shanghai, China)	Machine power 1.8 kW, temperature control range −30~100 °C, temperature fluctuation ±0.1 °C, pump flow 4 L/min, maximum head 5 m.
2. Multi-Channel Temperature Logger TA612C (TASI Electronics Co., Ltd., Suzhou, China)	Measurement channel 4, resolution 0.1 °C, ADJ compensation, sampling rate 1 time/s.
3. Multi-Channel Heat Flow Recorder JT2210 (Century JianTong Technology Co., Ltd., Beijing, China)	Number of 3.5 mm coaxial cable connectors 6 pcs.
4. K-type thermocouple TD602-2 (TASI Electronics Co., Ltd., Suzhou, China)	Length 2 m, material rubber (corrosion-resistant, can withstand 200 °C high temperature), measurement range −50~220 °C, accuracy ±2%
5. Heat Flux Sensor JTC08 (Century JianTong Technology Co., Ltd., Beijing, China)	Measurement range −2000~2000 W/m^2, precision ±5%

As shown in Table 3, to test the thermal performance of the radiant floor module during heating, cooling, and under different water supply temperatures, the experiment was divided into four distinct tests. According to the results of the preliminary field investigation, the water temperature of the electric heating water tank used in the grassland pastoral area is generally 40 °C in the operation stage; in the heating phase of this experiment, a controlled electric heating water tank was used to supply hot water, with the supply temperatures set to 35 °C, 40 °C, and 45 °C. During the cooling phase, the cooling air unit of the artificial environmental chamber was used to lower the temperature, with the cooling unit set to 0 °C.

Table 3. Experimental working condition settings.

Sample	Floor Module Number	Temperature of the Heat Source
Test 1	SSC0	40 °C
Test 2	SSC5	35 °C
Test 3	SSC5	40 °C
Test 4	SSC5	45 °C

3. Results and Analysis

3.1. Thermal Storage Characteristics of the Radiant Floor

Figure 6 shows the changes in the temperature and surface heat flux across the thermal storage layer of the radiant floor module during the heating phase in Tests 1 and 3. Figure 6a presents the results for the radiant floor module without added mPCM. As shown in the figure, the temperature trends at various measurement points in the thermal storage layer during the heating phase are generally consistent: within the first hour, due to the temperature difference between the heat source and the thermal storage layer, the average temperature of the thermal storage layer increased from 15 °C to 27 °C, with the corresponding surface heat flux rapidly increasing from 5 W/m^2 to 20 W/m^2.

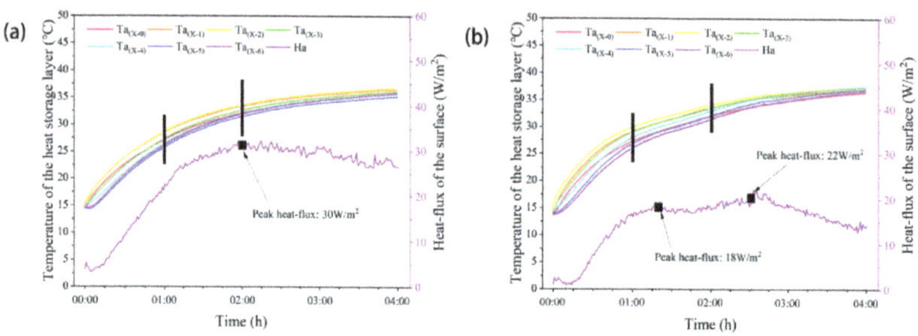

Figure 6. Average temperature and surface heat flux of the thermal storage layer during the heating process: (**a**) Test 1; (**b**) Test 3. The short black line divides the experimental phase.

During the 1 to 2 h heating period, the rate of the temperature increases in the thermal storage layer slowed, with the average temperature rising from 27 °C to 32.5 °C, and the highest temperature reaching 33.6 °C. The corresponding surface heat flux's rate of increase also slowed, peaking at 30 W/m^2 at the 2 h mark. After 2 h of heating, as the temperature difference between the heat source and the thermal storage layer further narrowed, the rate of temperature increases in the thermal storage layer further decreased. The average temperature rose from 32.5 °C to 36 °C, and the temperature variation curves at each measurement point began to level off. During this phase, the surface heat flux of the thermal storage layer remained steady at 27 W/m^2.

Figure 6a also illustrates the temperature differences across various locations within the thermal storage layer. The heat source in the radiant floor module is concentrated about 1~2 cm from the aluminum foil reflective layer, with only the upper surface able to transfer heat to the air. The highest temperature distribution within the thermal storage layer is centered around positions (X-2) and (X-1), gradually decreasing outward. The temperature distribution throughout the heating phase aligns with theoretical expectations and conforms to the basic principles of steady-state heat transfer.

Figure 6b presents the test results for the radiant floor module with added mPCM. As shown in the figure, the temperature increases at various measurement points in the thermal storage layer during the initial heating stage vary significantly, but they become more consistent over time. During the first hour, due to the temperature difference between the heat source and the thermal storage layer, the average temperature of the thermal storage layer increased from 14.5 °C to 28 °C, with the corresponding surface heat flux rapidly increasing from 5 W/m^2 to 17 W/m^2.

In the 1 to 2 h heating period, the rate of temperature increases in the thermal storage layer slowed, with the average temperature rising slowly from 28 °C to 31.7 °C at the 1 h 45 min mark, then gradually to 32.7 °C. The corresponding surface heat flux exhibited a first peak and a trough at 17 W/m^2 at the 1 h 45 min mark. After 2 h of heating, the rate of temperature increases in the thermal storage layer slightly accelerated until 2 h 30 min, after which it slowed down again due to the further narrowing of the temperature difference between the heat source and the thermal storage layer. During this phase, the average temperature rose from 32.7 °C to 37 °C, with the surface heat flux peaking at 22 W/m^2 at 2 h 30 min before decreasing.

Figure 6b also shows the temperature differences at various locations within the thermal storage layer: During the initial heating period, there were significant temperature differences within the thermal storage layer, with positions (X-1) and (X-2), which are closer to the heat source, consistently having the highest temperatures. In contrast, positions (X-5) and (X-6), which are closer to the upper surface, had the lowest temperatures within the first 2 h of heating—a phenomenon that persisted until after 2 h. Additionally, it is noteworthy that positions (X-5) and (X-6) near the upper surface of the radiant floor did not exhibit a reduction in the temperature increase rate due to phase change thermal storage.

From the perspective of surface heat flux in the thermal storage layer, the average heat flux throughout the heating process in Figure 6b was 15.1 W/m^2, lower than the 24.6 W/m^2 observed in Figure 6a. This difference is attributed to the lower thermophysical properties of mPCM, including lower thermal conductivity and thermal diffusivity, as previously demonstrated by earlier studies [7]. Notably, unlike the conventional thermal storage floor, the phase change thermal storage floor exhibited two peak heat fluxes during the heating process. The changes in average surface heat flux reflect the variation in heat transfer from the heat source. The slight decrease in average heat flux between 1 h 15 min and 2 h 30 min corresponds to the phase change thermal storage process, during which the rate of temperature increases in the thermal storage layer slowed.

3.2. Thermal Release Characteristics of the Radiant Floor

Figure 7 shows the changes in temperature and surface heat flux across the thermal storage layer of the radiant floor module during the cooling phase in Tests 1 and 3. Figure 7a presents the results for the radiant floor module without added mPCM. As shown in the figure, the temperature decline trends at various measurement points in the thermal storage layer during the cooling phase are generally consistent: during the first 30 min, the heat source continued to supply heat, causing the temperature of the thermal storage layer to rise further. The highest temperature at (X-1) increased to around 38 °C, while (X-6), due to direct heat exchange with the air, remained around 35.5 °C. After this period, the thermal storage layer maintained a relatively low cooling rate, with the corresponding surface heat flux rapidly increasing and reaching a peak of 186.3 W/m^2 at the 1 h mark.

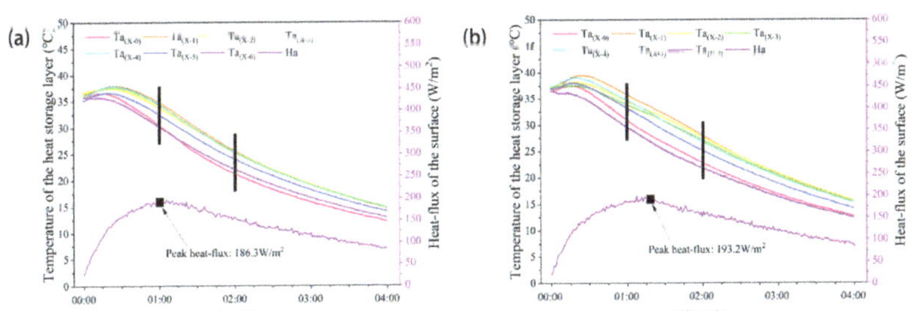

Figure 7. Average temperature and surface heat flux of the thermal storage layer during the cooling process: (**a**) Test 1; (**b**) Test 3. The short black line divides the experimental phase.

During the 1 to 2 h cooling period, the cooling rate of the thermal storage layer increased, with the average temperature of the thermal storage layer decreasing from 32.9 °C to 25.6 °C. From the 2 h mark until the end of the experiment, the temperature difference between the thermal storage layer and the environment diminished, leading to a further reduction in the cooling rate, with the highest temperature dropping from 25.6 °C to 15 °C. Correspondingly, the surface heat flux after the 2 h mark showed a steady decline.

Figure 7a also illustrates the temperature differences at various locations within the thermal storage layer: During the initial cooling stage, the heat source in the radiant floor module is concentrated about 1–2 cm from the aluminum foil reflective layer, with the highest temperature distribution centered around positions (X-1) and (X-2), gradually decreasing outward. The cooling process began first at (X-6), which was closest to the cooling load. It is noteworthy that (X-0) had the lowest temperature at the 1 h cooling mark, and this remained the case until the end of the experiment. After 1 h of cooling, the temperature field distribution within the thermal storage layer stabilized, with the highest temperatures centered around positions (X-2) and (X-3), gradually decreasing outward.

Figure 7b presents the test results for the radiant floor module with added mPCM. As shown in the figure, during the initial cooling stage, the temperature variations at various measurement points in the thermal storage layer differed significantly. During the first 30 min, the heat source continued to supply heat, causing the average temperature of the thermal storage layer to continue rising, with the highest temperature at (X-1) reaching 39.5 °C. Subsequently, due to the high temperature within the thermal storage layer, it maintained a relatively fast cooling rate, with the corresponding surface heat flux rapidly increasing.

During the 1 to 2 h cooling period, the cooling rate of the thermal storage layer decreased, with the average temperature dropping from 33.3 °C to 28.1 °C, and the highest temperature at 1 h 15 min decreasing to 33.2 °C. The corresponding surface heat flux reached a peak of 193.2 W/m^2, after which the cooling rate within the thermal storage layer increased slightly. From the 2 h mark until the end of the experiment, the cooling rate of the thermal storage layer began to decrease again.

Figure 7b also shows the temperature differences at various locations within the thermal storage layer: During the initial cooling stage, the highest temperature distribution within the thermal storage layer was centered around positions (X-1) and (X-2), gradually decreasing outward, with the surface of the thermal storage layer at (X-6) cooling first. During the 1 to 2 h cooling period, the temperature differences within the thermal storage layer were more pronounced, with the highest temperatures centered around positions (X-2) and (X-3), gradually decreasing outward.

From the perspective of surface heat flux in the thermal storage layer, the average heat flux throughout the cooling process in Figure 7b was 139.0 W/m^2, slightly higher than the 137.6 W/m^2 observed in Figure 7a. This increase is attributed to the higher thermophysical properties of the mPCM after melting, as demonstrated by previous studies [7]. Notably, compared to the conventional thermal storage floor, the phase change thermal storage floor reached a peak surface heat flux of 193.2 W/m^2 at 1 h 15 min during the cooling process, with the peak heat release flux delayed by 15 min.

3.3. *Impact of Heating Temperature on Thermal Performance*

Figure 8 shows the changes in surface temperature and surface heat flux of the thermal storage layer during Tests 2, 3, and 4.

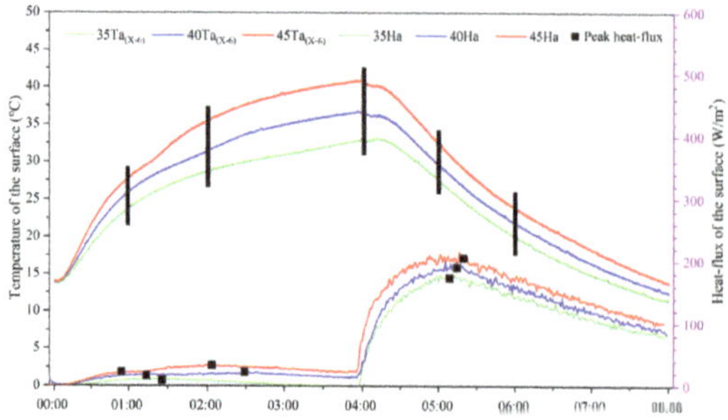

Figure 8. Average surface temperature and surface heat flux under different water supply temperatures. The short black line divides the experimental phase.

As shown in the figure, the surface temperature of the thermal storage layer under different water supply temperatures follows the same trend: Within the first hour of heating, the surface temperature of the thermal storage layer increases rapidly. The rate of temperature rise in Test 4 is greater than in Test 3, which is greater than in Test 2, with temperature increases of 13.8 °C, 12 °C, and 10 °C, respectively. During this stage, sensible heat storage dominates, and the higher the heat source input temperature, the faster the rate of temperature increase. Compared to Test 2, Test 4 with a higher heat source improved the heat transfer efficiency by 38%.

During the 1 to 2 h heating period, the rate of temperature increase in the thermal storage layer decreases in all tests, indicating that the latent heat storage phase requires the absorption and storage of more heat. In Test 4, the phase change heat storage time is shorter, with a noticeable increase in the rate of temperature rise at around 1 h 20 min. In Test 3, the surface temperature starts to increase steadily at 1 h 30 min, while the temperature change in Test 2 during this stage is more gradual. Compared with Test 2, Test 4, with a higher heat source, reduces the time of phase transition by 50%.

After 2 h of heating, the rate of temperature increase in the thermal storage layer surface further decreases. It is noteworthy that in Test 2, the rate of temperature increase remains constant during this phase. At the end of heating, the average surface temperature of the thermal storage layer in Test 4 is 40.9 °C; in Test 3, it is 36.7 °C; and in Test 2, it is 32.8 °C. Compared with Test 2, the heat transfer efficiency of Test 4 with a higher heat source was improved by 24.7%. During the cooling phase, the average surface temperature of the thermal storage layer decreases similarly across all tests. However, due to the higher heat source input in Test 4, more heat is retained, resulting in the highest average surface temperature of approximately 13.8 °C at the end of the cooling phase.

From the perspective of surface heat flux in the thermal storage layer, different water supply temperatures alter the timing of peak heat flux in the phase change thermal storage radiant floor. During the heating phase, the two peak heat fluxes in Test 4 occurred at 50 min and 2 h 5 min; in Test 3, at 1 h 17 min and 2 h 30 min; while in Test 2, only one peak heat flux was observed at around 1 h 27 min. During the cooling phase, the peak heat flux times were 1 h 20 min for Test 4, 1 h 15 min for Test 3, and 1 h 10 min for Test 2.

4. Discussion

This study employs phase change thermal storage technology to enhance the thermal performance of radiant floors. An experimental platform was established in an environmental chamber based on the energy usage patterns of grassland pastoral dwellings. As the first step in constructing the thermal environment of grassland pastoral dwellings, this approach holds significant importance for improving building energy efficiency, optimizing energy consumption structures, and balancing the relationship between energy demand and the environment.

For the heating phase, the low physical properties of mPCM do not impact the final state of the thermal storage floor. Comparing Figure 6a,b, it can be observed that under the same heat source conditions, the addition of mPCM to the radiant floor module alters its heat transfer characteristics: Within the first hour of heating, the phase change energy storage flooring exhibits higher overall heat transfer efficiency and a faster rate of temperature increase, with the rate of temperature rise in the thermal storage layer increasing by 12.5%. During the 1 to 2 h heating period, phase change thermal storage causes the rate of temperature increase in the thermal storage layer to slow down, with the temperature rise decreasing by 14.5%, but the overall average temperature of the thermal storage layer at 2 h increased by 0.2 °C. After 2 h of heating, the phase change energy storage flooring experiences a growth in the rate of temperature increase, with the temperature slowly rising after 2 h 30 min, ultimately leading to an overall average temperature increase of 1 °C. The faster rate of temperature increase helps eliminate the initial uneven temperature distribution in the phase change energy storage flooring and results in a higher final average temperature in the thermal storage layer, which is beneficial for the heat release of the phase change energy storage flooring.

For the cooling phase, the high-energy-density mPCM improves the heat release performance of the thermal storage floor. Comparing Figure 7a,b, it is evident that under the same cooling source conditions, the addition of mPCM in the radiant floor module alters its heat transfer characteristics: The temperature change curves show that during the first hour of the initial cooling phase, the phase change energy storage flooring continues to rise in temperature, with the average temperature of the thermal storage layer increasing by 0.8 °C. During the 1 to 2 h cooling period, the phase change heat release causes the rate of temperature decrease in the thermal storage layer to slow down by 28.8%, delaying the peak heat flux by 15 min, and the overall average temperature of the thermal storage layer at 2 h increased by 2.5 °C. After 2 h of cooling, the cooling rate of the phase change energy storage flooring gradually decreases, with the temperature slowly declining after 2 h 30 min, ultimately leading to an overall average temperature increase of 0.4 °C. The delay in peak heat flux allows the phase change thermal storage floor to maintain a higher temperature and extend the heat release duration.

The power consumption of the controlled electric heating water tank during different tests is shown in Table 4; by comparing Test 1 and Test 4, phase change energy storage flooring can improve the energy efficiency of conventional flooring by 1.7%. A further analysis of the results shown in Figure 8 reveals that a higher-temperature heat source increases the energy consumption of the radiant floor module. The additional energy results in more efficient heat transfer, a faster rate of surface temperature increase, a higher surface temperature at the end of the heating period, and a longer heat release duration. This is particularly suitable for the utilization of solar energy, further coupling renewable resources with the decentralized energy systems of grassland pastoral dwellings.

Table 4. The material quantities used of the thermal storage layer of the radiant floor module.

Sample	Experimental Condition	Initial Power	End Power	Total Power
Test 1	SSC0, 40 °C	7.41 kW·h	12.23 kW·h	4.82 kW·h
Test 2	SSC5, 35 °C	33.98 kW·h	38.69 kW·h	4.71 kW·h
Test 3	SSC5, 40 °C	28.01 kW·h	32.91 kW·h	4.90 kW·h
Test 4	SSC5, 45 °C	40.33 kW·h	45.52 kW·h	5.21 kW·h

The optimal result in this experiment was obtained in Test 4. According to the results shown in Figures 6a and 8, the final surface temperature of the floor in Test 1 after 4 h of heating was 35.7 °C, whereas in Test 4, this temperature was reached in just 2 h of heating. Based on the results shown in Figures 6b and 8, after 2 h of heating, the phase change energy storage flooring had fully undergone the phase change process, with a uniform internal temperature distribution under sensible heat storage. Figure 8 also shows that during the cooling phase, the surface temperature of the floor in Test 4 took approximately 20 min longer to drop to the initial temperature compared to Test 1.

At present, the use of phase change materials in building envelope requires a lot of experimental research in the early stage, and the related research also shows that it has achieved good performance in energy efficiency. In the grassland pastoral area, the winter heating period lasts up to 6 months each year. Currently, grassland pastoral dwellings use decentralized electric heating water tanks for heating between 16:00 and 20:00 daily. If a solar water-based phase change energy storage flooring were used, heating could be provided by solar collector system from 16:00 to 18:00, with an electric auxiliary heating system from 18:00 to 20:00. This system could reduce annual heating costs by 50% per square meter of phase change energy storage floor.

Figure 9 illustrates the variations in temperature across different locations within the thermal storage layer and the surface heat flux during the cooling phase of Test 4. The surface temperature of the flooring decreases to the initial level after 3 h and 56 min of cooling. Due to the low thermal conductivity of the mPCM, the internal temperature of the flooring in Test 4 remains higher, with an average temperature of 16.5 °C. In contrast, in Test 1, as shown in Figure 7a, the surface temperature of the flooring returns to the initial level after 3 h and 32 min of cooling, with an average temperature of only 15.7 °C. Furthermore, at the end of the cooling period, the surface heat flux of Test 4 remains at approximately 100 W/m^2, whereas, for the same cooling duration, the surface heat flux of Test 1 is only 85 W/m^2. The higher internal temperature of the flooring in Test 4 enables a longer heat release duration.

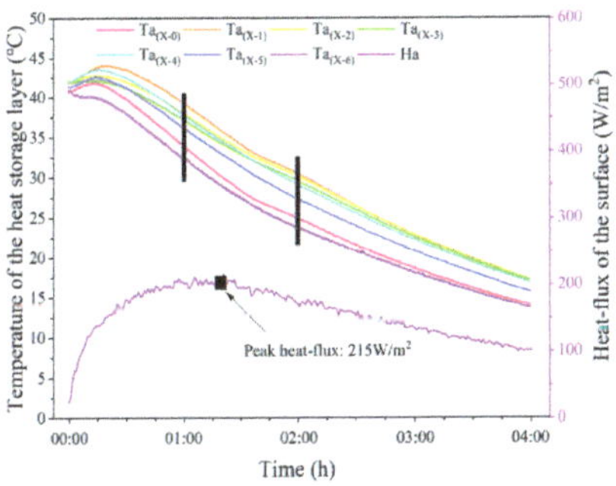

Figure 9. Average temperature of the heat storage layer and surface heat-flux during the cooling process. The short black line divides the experimental phase.

As shown in Figure 5, a large number of thermocouples and heat flow sensors were set up in the phase change energy storage floor in this experiment to form a simple heat conduction test device. For example, Behnam Mobaraki et al. [28,29] used the temperature-based method (TBM) and the heat flux meter (HFM) method to discuss the influence of sensor position on the measurement accuracy of U value. According to Fourier's law (law of heat conduction) formula, the heat flux is proportional to the temperature difference between the two sides of the material, the thermal conductivity of the material, and the thickness of the material is inversely proportional. We extracted the data of surface temperature, bottom temperature and surface heat flux of the phase change storage floor in the experiment and calculated the thermal conductivity of the steady-state heat transfer part.

$$q = \lambda \frac{\Delta\theta}{d} \Rightarrow \lambda = \frac{q \cdot d}{\Delta\theta} \tag{1}$$

where λ is the thermal conductivity of the material (W m^{-1} K^{-1}), q is the heat flux through the material (W), d is the thickness of the material (m), and $\Delta\theta$ is the temperature difference between the two sides of the material (K).

Thermal conductivity reflects the ability of the material to transfer heat. In this experiment, the temperature field is centered on the heat pipe, and the heat can only be transferred towards the upper surface of the phase change energy storage floor. Therefore, Ta(X-2) and Ta(X-6) are selected as reference temperatures, and the thermal conductivity of Test 1 and Test 3 is calculated. As shown in Figure 10, PCMs can significantly reduce the thermal conductivity of the floor. Test 1: In the dynamic heating process, the thermal conductivity is proportional to the heating time, and it increases with the increase in the floor temperature. However, in Test 3, two stages of growth occurred during the dynamic heating process. Figure 6 shows that in Test 3, the temperature rise rate slowed down between 1 and 2 h, and the surface heat flow remained stable, corresponding to the thermal conductivity shown in Figure 10, which remained unchanged at this stage.

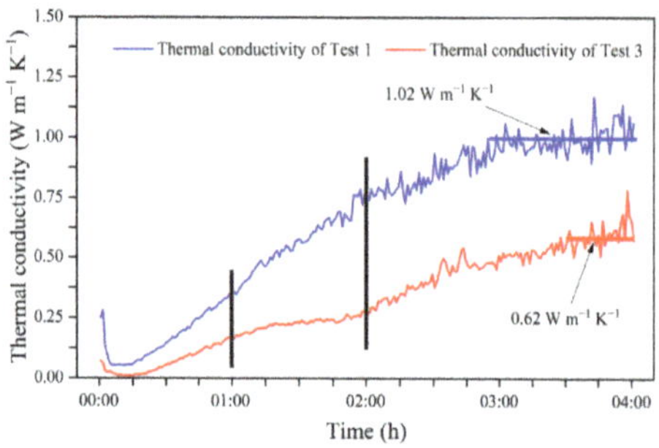

Figure 10. Thermal conductivity calculated according to Fourier's law. The short black line divides the experimental phase.

According to the results shown in Figure 6, the stage with reference temperature difference less than 1 °C is selected as the steady-state heat transfer model. For Test 1, its thermal conductivity is 1.02 W m^{-1} K^{-1}, while that for Test 3 is 0.62 W m^{-1} K^{-1}.

It is important to note that, as shown in Figure 3, the indoor temperature gradient during the night in a grassland pastoral dwelling without heating is relatively small. In this experiment, forced mechanical ventilation was used for cooling in all tests, with indoor temperatures dropping to 0 °C within 1 h, which ultimately delayed the heat release time by approximately 20 min. In the dynamic load environment of an actual grassland pastoral dwelling, the heat release time of the phase change energy storage flooring would be further delayed. Secondly, the cost of phase change energy storage floor is also the difficulty of using PCMs in the current building envelope structure. As shown in Figure 8, the higher heating temperature can improve the thermal performance of the phase change heat storage floor, which is a strong match with high-temperature renewable energy sources, such as solar energy, which will require further in-depth research after coupling with the actual solar heat collection system in the future. The use of phase change energy storage flooring in grassland pastoral areas can be achieved by integrating unstable solar energy into the building; this is very powerful for the development and utilization of renewable energy, reducing commodity energy consumption in heating processes, and reducing carbon emissions from building operation in the grassland pastoral area of Inner Mongolia. In addition, the use of phase change energy storage floor in remote grassland pastoral areas also needs to consider long-term durability issues, which is the next step in our practical experiments relating to grassland pastoral areas.

5. Conclusions

This study focused on a heating terminal system for grassland pastoral dwellings and innovatively uses an environment chamber to study the heat storage and release characteristics of the phase change energy storage floor, and the thermal conductivity was discussed according to the heat transfer model. This research aimed to address the challenges posed by the cyclical variation of renewable resources in grassland pastoral areas and the practical application scenarios of energy supply policies under peak–valley electricity pricing through advanced energy storage technology. The main conclusions are as follows:

(1) Higher Heat Transfer Efficiency and Faster Heating Rate of Phase Change Energy Storage Flooring: Under a 40 °C heating condition, the heating rate of the thermal storage layer increased by 12.5% within 1 h, and the overall average temperature after 4 h of heating increased by 1 °C. The additional stored heat is beneficial for the heat release of the phase change energy storage flooring.

(2) Exhibiting Superior Thermal Release Performance of Phase Change Energy Storage Flooring: In Test 3, the thermal release characteristics of the phase change energy

storage flooring demonstrated significant improvements. The peak heat flux during the heat release process was delayed by 15 min, indicating a more controlled release of stored heat. Additionally, the overall average temperature after 4 h of cooling increased by 0.4 °C. This higher retained temperature contributes to a prolonged heat release period, enhancing the effectiveness of the phase change energy storage flooring in maintaining warmth over extended periods.

(3) Shorter Heating Time and Extended Thermal Release Duration with Higher Heating Temperatures: Under a 45 °C heating condition, the heat transfer efficiency of the surface temperature of the thermal storage layer increased by 38% within 1 h, and the heat transfer efficiency of the final surface temperature after 4 h of heating increased by 24.7%. The overall average temperature after 4 h of cooling increased by 2.4 °C, with the heat release time delayed by 20 min. This suggests significant potential for coupling with high-temperature heat sources such as solar energy, offering great promise for the development and utilization of renewable energy.

To obtain more accurate thermal storage and release patterns of phase change energy storage flooring under real-world conditions, field construction tests should be conducted, and after coupling the phase change energy storage floor with the actual solar heat collection system, further in-depth study was conducted to test its long-term durability. These tests should analyze the dynamic heat transfer characteristics of phase change energy storage flooring in real scenarios and compare them with the thermal comfort of grassland pastoral dwellings to better understand the relationship between economic, energy, and environmental factors.

Author Contributions: Conceptualization, G.X.; Formal analysis, K.W. and G.L.; Investigation, K.W., X.Z. and L.M.; Methodology, K.W., G.X., X.Z. and L.M.; Writing—original draft, K.W.; Visualization: K.W. and X.Z.; Writing—review and editing, G.L.; Funding acquisition, G.X.; Supervision: G.X. and G.L. All authors have read and agreed to the published version of the manuscript.

Funding: This research was funded by the National Natural Science Foundation of China (grant no. 52168006) and Natural Science Foundation of Inner Mongolia Autonomous Region (grant no. 2024MS05023).

Data Availability Statement: The original contributions presented in the study are included in the article; further inquiries can be directed to the corresponding author.

Conflicts of Interest: The authors declare that they have no known competing financial interests or personal relationships that could have appeared to influence the work reported in this paper.

References

1. Energy Administration of Inner Mongolia Autonomous Region, China Renewable Energy Engineering Institute. *Inner Mongolia Renewable Energy Development Report 2023*; China Economic Publishing House: Beijing, China, 2024.
2. Building Energy Conservation Research Center, Tsinghua University. *China Building Energy Efficiency Annual Development Research Report 2024 (Rural Housing Topic)*; China Architecture & Building Press: Beijing, China, 2024.
3. Lu, Y.; Khan, Z.A.; Alvarez-Alvarado, M.S.; Zhang, Y.; Huang, Z.; Imran, M. A Critical Review of Sustainable Energy Policies for the Promotion of Renewable Energy Sources. *Sustainability* **2020**, *12*, 5078. [CrossRef]
4. Nardecchia, F.; Groppi, D.; Lilliu, I.; Astiaso Garcia, D.; De Santoli, L. Increasing energy production of a ducted wind turbine system. *Wind Eng.* **2020**, *44*, 560–576. [CrossRef]
5. Al-Shetwi, A.Q.; Hannan, M.A.; Jern, K.P.; Mansur, M.; Mahlia, T.M.I. Grid-connected renewable energy sources: Review of the recent integration requirements and control methods. *J. Clean Prod.* **2020**, *253*, 119831. [CrossRef]
6. Xu, G.; Wang, K.; Wang, W.; Wang, Z. Application Study of Floor Radiating Heating Technology with Solar Hot Water in Residents in Grasslands and Pastoral Areas. *J. Inn. Mong. Univ. Technol. (Nat. Sci. Ed.)* **2018**, *37*, 142–148. [CrossRef]
7. Li, G.; Xu, G.; Zhang, J. Experimental investigation of thermal and mechanical characteristics of slag cement mortars with PCM for radiant floors. *Case Stud. Constr. Mater.* **2024**, *20*, e02958. [CrossRef]
8. Wang, Z.; Li, Z.; Lu, G.; Gao, Q.; Zhang, R.; Gu, Z. Experimental study on phase change heat storage of valley electricity and economic evaluation of commercial building heating. *Sustain. Cities Soc.* **2022**, *86*, 104098. [CrossRef]
9. Sharshir, S.W.; Joseph, A.; Elsharkawy, M.; Hamada, M.A.; Kandeal, A.W.; Elkadeem, M.R.; Kumar Thakur, A.; Ma, Y.; Eid Moustapha, M.; Rashad, M.; et al. Thermal energy storage using phase change materials in building applications: A review of the recent development. *Energy Build.* **2023**, *285*, 112908. [CrossRef]

10. Zhang, Y.; Zhou, C.; Liu, M.; Li, X.; Liu, T.; Liu, Z. Thermal insulation performance of buildings with phase-change energy-storage wall structures. *J. Clean Prod.* **2024**, *438*, 140749. [CrossRef]
11. Vargas, V.Z.; Claros-Marfil, L.J.; Sandoval, G.F.B.; Rojas, B.H.; Santos, A.G.; González, F.J.N. Experimental assessment of energy storage in microencapsulated paraffin PCM Cement mortars. *Case Stud. Constr. Mater.* **2024**, *20*, e02959. [CrossRef]
12. Refahi, A.; Rostami, A.; Amani, M. Implementation of a double layer of PCM integrated into the building exterior walls for reducing annual energy consumption: Effect of PCM wallboards position. *J. Energy Storage* **2024**, *82*, 110556. [CrossRef]
13. Gencel, O.; Güler, O.; Ustaoğlu, A.; Erdoğmuş, E.; Sarı, A.; Hekimoğlu, G.; Boztoprak, Y.; Subaşı, S. Enhancing sustainability with waste hemp-shive and phase change material: Novel gypsum-based composites with advanced thermal energy storage properties. *J. Clean Prod.* **2024**, *451*, 142000. [CrossRef]
14. Pirasaci, T.; Sunol, A. Potential of phase change materials (PCM) for building thermal performance enhancement: PCM-composite aggregate application throughout Turkey. *Energy* **2024**, *292*, 130589. [CrossRef]
15. Lachheb, M.; Younsi, Z.; Youssef, N.; Bouadila, S. Enhancing building energy efficiency and thermal performance with PCM-Integrated brick walls: A comprehensive review. *Build. Environ.* **2024**, *256*, 111476. [CrossRef]
16. Németh, B.; Ujhidy, A.; Tóth, J.; Ferencz, M.; Kurdi, R.; Gyenis, J.; Feczkó, T. Power consumption of model houses with and without PCM plaster lining using different heating methods. *Energy Build.* **2023**, *284*, 112845. [CrossRef]
17. Zhang, Y.; Zhu, Z.; Peng, Z.; Luo, J.; Sun, X.; Li, J.; Peng, F. The Trombe wall equipped with PCMs for the enhancement of the indoor thermal quality. *Energy Build.* **2023**, *297*, 113407. [CrossRef]
18. Zaib, A.; Mazhar, A.R.; Talha, T.; Inshal, M. Experimental investigation of a solar PCM heat exchanger for indoor temperature stabilization. *Energy Build.* **2023**, *297*, 113478. [CrossRef]
19. Ndanduleni, A.U.C.; Radebe, T.B.; Huan, Z. Reduction of temperature fluctuation in a South African shack house using phase change material insulation. *Build. Environ.* **2023**, *241*, 110376. [CrossRef]
20. Babaharra, O.; Choukairy, K.; Faraji, H.; Hamdaoui, S. Improved heating floor thermal performance by adding PCM microcapsules enhanced by single and hybrid nanoparticles. *Heat Transf.* **2023**, *52*, 3817–3838. [CrossRef]
21. Afaynou, I.; Faraji, H.; Choukairy, K.; Arıcı, M.; Khallaki, K. Heat transfer improvement of phase change materials by metal foams and nanoparticles for efficient electronic thermal management: A comprehensive study. *Int. J. Heat Mass Transf.* **2024**, *227*, 125534. [CrossRef]
22. Moreira, M.; Dias-de-Oliveira, J.; Amaral, C.; Neto, F.; Silva, T. Outline of the incorporation of phase change materials in radiant systems. *J. Energy Storage* **2023**, *57*, 106307. [CrossRef]
23. Liu, Z.; Wei, Z.; Teng, R.; Sun, H.; Qie, Z. Research on performance of radiant floor heating system based on heat storage. *Appl. Therm. Eng.* **2023**, *231*, 120812. [CrossRef]
24. Chen, W.; Liu, Y.; Liang, X.; Luo, F.; Liao, T.; Wang, S.; Gao, X.; Zhang, Z.; Fang, Y. Experimental and numerical investigations on radiant floor heating system integrated with macro-encapsulated phase change material. *Energy* **2023**, *282*, 128375. [CrossRef]
25. Wang, Y.; Xu, G. Numerical Simulation of Thermal Storage Performance of Different Concrete Floors. *Sustainability* **2022**, *14*, 12833. [CrossRef]
26. Shadnia, R.; Zhang, L.; Li, P. Experimental study of geopolymer mortar with incorporated PCM. *Constr. Build. Mater.* **2015**, *84*, 95–102. [CrossRef]
27. Gbekou, F.K.; Benzarti, K.; Boudenne, A.; Eddhahak, A.; Duc, M. Mechanical and thermophysical properties of cement mortars including bio-based microencapsulated phase change materials. *Constr. Build. Mater.* **2022**, *352*, 129056. [CrossRef]
28. Mobaraki, B.; Castilla Pascual, F.J.; Lozano-Galant, F.; Lozano-Galant, J.A.; Porras Soriano, R. In situ U-value measurement of building envelopes through continuous low-cost monitoring. *Case Stud. Therm. Eng.* **2023**, *43*, 102778. [CrossRef]
29. Mobaraki, B.; Castilla Pascual, F.J.; García, A.M.; Mellado Mascaraque, M.Á.; Vázquez, B.F.; Alonso, C. Studying the impacts of test condition and nonoptimal positioning of the sensors on the accuracy of the in-situ U-value measurement. *Heliyon* **2023**, *9*, e17282. [CrossRef]

energies

MDPI

Review

Building the Future: Integrating Phase Change Materials in Network of Nanogrids (NoN)

Ali Kalair [1,*], Elmira Jamei [2], Mehdi Seyedmahmoudian [1], Saad Mekhilef [1] and Naeem Abas [3]

1 Siemens-Swinburne Energy Transition Hub, School of Science, Computing and Engineering Technologies, Swinburne University of Technology, Melbourne, VIC 3122, Australia; mseyedmahmoudian@swin.edu.au (M.S.); smekhilef@swin.edu.au (S.M.)
2 College of Sport, Health and Engineering, Victoria University, Melbourne, VIC 3011, Australia; elmira.jamei@vu.edu.au
3 Department of Electrical Engineering, University of Gujrat, Hafiz Hayat Campus, Jalalpur Jattan Road, Gujrat 50700, Pakistan; naeemkalair@uog.edu.pk
* Correspondence: akalair@swin.edu.au

Abstract: Buildings consume 10% of global energy and 50% of global electricity for heating and cooling. Transitioning to energy-efficient buildings is essential to address the global energy challenge and meet sustainable development goals (SDGs) to limit global temperature rise below 1.5 °C. The shift from traditional to smart grids has led to the development of micro, milli, and nanogrids, which share energy resources symbiotically and balance heating/cooling demands dealing with acute doldrums (dunkelflaute). This scoping review explores the methods by which phase change materials (PCMs) can be used in residential buildings to form a nanogrid. This review examines the components and concepts that promote the seamless integration of PCMs in residential houses. It also discusses the key challenges (e.g., scalability, stability, and economic feasibility in high summer temperatures), proposing the community-scale network of nanogrids (NoN) and the potential of thermochromic and photochromic materials. The findings of this review highlight the importance of latent heat storage methods and ingenious grid architectures such as nanogrids to construct resilient and sustainable houses in the future and thereby offer practical insights for policymakers and industries in the energy sector.

Keywords: phase change materials (PCMs); nanogrid; zero energy buildings (ZEB); nanocomposite; microencapsulation; solar collector; thermochromic; photochromic; sustainable buildings

1. Introduction

Modern societies have increasing energy demands, and the world is heading towards a wilting and warming world. The global average temperature on the planet has remained 5–10 °C higher than the pre-industrial level for the last 500 million years. The global average temperature plunged below the pre-industrial level about 3.5 million years ago, and it is now returning to its historic normal. Changing climate demands extensive research on novel energy storage technologies and management strategies. Climate change acts as a pendulum that causes heat waves in summer and cold waves in winter. Extreme climate events require more energy to cope with heating and cooling. Phase change materials allow us to store summer heat for winter and vice versa. The rapid growth in global energy consumption, alongside rising CO_2 and greenhouse gas emissions, is propelled by population growth, swift economic expansion, and an increasing reliance on energy-intensive devices. This surge in energy consumption poses significant environmental challenges, impacting human health and ecosystems. More fossil fuels inject more greenhouse gases, which mediate positive feedback. Moreover, dwindling oil, coal, and natural gas reserves demand innovative systems for converting and storing sustainable energy. The building sector significantly impacts global energy use and environmental health, chiefly through its high greenhouse gas and CO_2 emissions, which also contribute to ocean acidification. This sector's increased energy usage is evident not only during the production of construction materials and the actual building process but also, more critically, during the operational phase of buildings. Empirical evidence indicates that maintaining indoor comfort accounts for approximately 30% of the sector's total energy demand [1].

Phase change materials represent a viable and potent thermal energy storage solution to reduce energy consumption in the building sector. PCMs can sustain thermal comfort in indoor environments without actively consuming energy. PCMs possess high thermal energy absorption and storage capacity, facilitated by their intrinsic ability to undergo phase transitions. The storage mechanism of these materials (enthalpy-based) is required for phase changes. As the temperature of the PCMs increases, they absorb sensible heat and, upon reaching the transition temperature, store a considerable amount of latent heat of fusion to transition phases. This thermal energy remains sequestered until a temperature decline triggers the reverse phase transition, releasing the stored energy. Phase change materials (PCMs) are grouped based on their phase transition types: solid to liquid, liquid to gas, and solid to solid. Examples of solid–solid PCMs include polyurethanes and cross-linked polyethylene, which exhibit relatively low latent heat in comparison to solid–liquid PCMs, the latter being more suitable for applications in residential buildings. Organic PCMs such as paraffin/fatty acids are especially beneficial for home use. However, their effectiveness is curtailed by their inherently low thermal conductivity. Prevailing methodologies include encapsulation with thermally conductive substrates to mitigate leakage by fabricating form-stable composites incorporating expanded graphite/perlite to improve the limiting thermal performance.

Extensive research has explored integrating PCM in buildings to reduce energy consumption demands [2–5] heating/cooling demands by up to 80% [6], greenhouse gas emissions consumption [7], and indoor thermal comfort [8]. PCMs are embedded inside roofs, walls [9], and ceilings [10], windows [2], outside, inside, and middle of the building envelope [7]. Active uses include heating, ventilation, air conditioning, domestic hot water (DHW), and solar facades. PCMs are integrated into building materials as passive solutions to improve the thermal efficiency of ceilings, walls, and floors, thereby increasing the energy performance of buildings [11] and may result in a 33.02% reduction in heating energy and a 55.48% reduction in cooling energy demands. The literature highlights research gaps in building design with phase change materials (PCMs) and opportunities for future research. The widespread use of PCMs in building envelopes needs further study, especially in high summer temperatures when night-time crystallization hinders PCM efficiency. The whole charging and discharging cycles of PCMs across climate zones require further study to maximize their energy-saving potential. Developing nano- or micro-sized conductive fillers to improve heat conductivity is another priority. Limited study on inorganic and solid–solid PCMs has brought forth an opportunity for thermochromic and photochromic phase change materials.

This scoping review seeks to explore the critical and relevant findings for phase change materials (PCMs) that can be used in the design of nanogrid architecture. A multitude of research has demonstrated and proved the capabilities of PCMs in improving energy efficiency, curtailing local demands, and optimizing thermal regulation in home environments. However, a comprehensive review focusing on what and how to integrate in tandem with PCMs into residential nanogrids was limited to non-existent in the literature, making it much more challenging. This review also addresses this gap by reviewing the literature on different technologies and their possible applications in designing an energy-efficient nanogrid. Moreover, the studies were very diverse, and study designs were equally heterogeneous, making it challenging to converge to commonalities. This review seeks to provide valuable insights for future directions in residential nanogrid systems by synthesizing recent developments and identifying areas requiring further research. The findings of this review will contribute to guiding researchers, policymakers, and industry professionals towards optimizing PCM integration in nanogrid systems, ultimately promoting sustainable and energy-efficient residential solutions. This review encompasses a holistic overview of PCM integration in residential buildings, storage technologies, thermal enhancement methods, and passive design techniques to maintain thermal comfort levels, meet the requirements of a nanogrid, and meet the criteria of net zero/plus energy buildings as summarized in Figure 1.

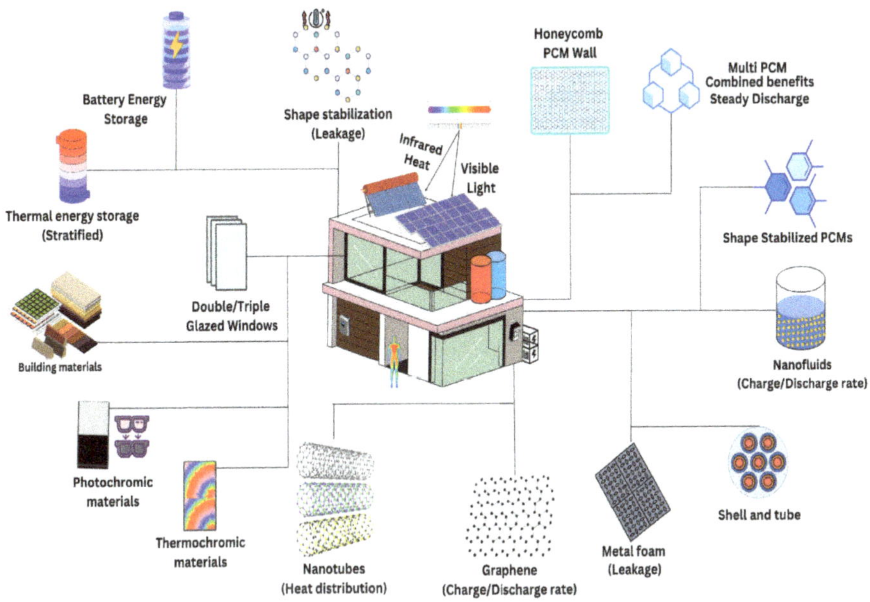

Figure 1. Overview of the components reviewed to be integrated into nanogrid.

2. Method

The paper aims to explore the integration of phase change materials (PCMs) in residential buildings to form nanogrids, enhancing energy efficiency and reducing energy demands. The objective was to investigate the holistic integration of multiple melting-point PCMs in wallboard with solar hot water systems, incorporating advanced additives (e.g., graphite, silicon carbide, multi-walled carbon nanotubes [MWCNT]), innovative coatings (e.g., photochromic cool white, thermochromic vanadium dioxide, titanium oxide, barium sulfate, sodium citrate, polymer paints), and effective heat transfer enhancements (e.g., metal foam, fins) to curtail energy demands and improve energy efficiency indexes while justifying viability in residential nanogrids. It reviews recent literature on the techno-economic feasibility and integration techniques of PCMs, proposes a novel concept of community-scale nanogrids (NoN) for collaborative energy management, highlights software tools for future modeling and evaluation, and emphasizes the importance of sustainable building practices using latent heat storage and innovative grid architectures. These contributions aim to advance PCM applications in residential energy systems and promote sustainable, energy-efficient building practices.

This scoping review was conducted through a structured and systematic approach to justify the relevance and impact of the study. The research articles were sourced from credible science databases including ScienceDirect, IEEE Xplore, and Google Scholar, particularly emphasizing phase change materials (PCMs) and their case studies in residential buildings/nanogrids. These databases were selected based on their comprehensive repositories of peer-reviewed journal articles, conference proceedings, and technical reports pertinent to PCM technology, thermal energy storage systems, and energy-efficient building frameworks. Data are synthesized in tabular form and summarized using a narrative approach that comprehensively reviews the findings of diverse studies and draws meaningful connections between various topics, areas, and disciplines. Themes were framed to identify patterns and extract meaningful insights remaining within the identified scope and boundaries. The data charting process to extract relevant information from the literature involves sorting and articulating data in tabular form over fifteen columns. Data were tabulated using a standardized form, sorting data according to the scope of the research, study design, building size, types of PCMs, integration techniques, performance metrics (energy, exergy, economic), and outcomes. This structured approach ensures that the data charting process is thorough and systematic, providing a solid foundation for analyzing the integration of PCMs in residential nanogrids.

The inclusion criteria for this scoping review encompass various types of screening based on the nature of studies, scope, outcomes, publication dates, and languages to ensure relevance and comprehensiveness. Studies published between 2012 and 2024 were included to capture the most recent advancements in PCM and nanogrid technologies. Eligible studies include experimental research investigating PCM integration in building materials and ways to improve energy efficiency through additives. Only publications from peer-reviewed journals, conference proceedings, and reputable technical reports were considered, focusing on research addressing PCMs in residential energy systems, particularly in the context of nanogrids and the techno-economic feasibility of such implementations. Studies published before 2012 were excluded unless they provided significant context to the development of nanogrid systems, and articles not available in English or focusing on unrelated applications were also excluded. A combination of keywords was used to search the databases, ensuring a comprehensive collection of scientific publications. Keywords included "PCMs", "thermal energy storage (TESS)", "nanogrid", "energy-efficient buildings", "smart grids", "poly-generation", "photochromic PCM", "thermochromic PCM", and "multi-energy systems". Relevant articles were filtered based on their titles, abstracts, scope, and keywords. The selected articles were then thoroughly reviewed. Key information was extracted, focusing on PCM material properties and classifications, phase transition processes, building integration, advancements in nanogrid technologies, and their implementation. The review excluded studies that do not focus on PCM integration in residential buildings and lack empirical evidence or a clear methodological framework. Moreover, any study focusing on commercial buildings or industrial PCM applications was excluded unless a clear advantage was to be highlighted. These criteria ensure that the included studies are directly relevant to the research objectives of understanding and optimizing PCM integration in residential nanogrids.

To guarantee the credibility of the literature, the chosen studies were evaluated by considering the reliability of their publication sources and the number of citations, as well as the studies' methodological rigor and relevance to this review's research questions and objectives. The review process involved initial screening of titles, figures, headings, and abstracts to identify relevant studies. Key concepts and findings were then extracted, synthesized, analyzed, and organized into thematic sections as presented in the literature review. By adopting this structured methodology, the scoping review aimed to provide a thorough and reliable analysis of the potential integration of PCMs into nanogrid systems for residential buildings, highlighting both the advancements and challenges in this field.

3. Literature Review

3.1. Zero Energy Buildings, Plus Energy Buildings, and Nanogrids

A zero-energy building uses salient architectures and integrated solar and wind energy storage systems to regulate temperature. A nanogrid is a decentralized energy system that utilizes local renewable energy generation and storage inside buildings. Thermochromic and photochromic PCMs are integrated inside construction materials to minimize the use of heat and electric energies. PCM-integrated buildings with sun and wind-based systems offer increased energy security, resilience, and efficiency by effectively managing local energy resources. Solar systems go down after sunset, and wind systems may or may not during the day or night. There are times when it is dark doldrums, a situation called dunkelflaute in Germany, which relies more on than any other European country. Contrarily, times come in summer when power prices become negative due to excess production, as has happened several times in Australia. Solar panels increase the urban heat island effect, and wind turbines kill migratory birds. We have to turn to naturally ventilated, sunshine-lit, and geothermal energy-supported systems in addition to thermochromic and photochromic phase change materials for buildings.

Plus energy buildings are prosumer-owned buildings that produce more annual energy than annual energy demands. Enhancing building envelopes and incorporating energy-efficient technologies can reduce energy consumption by 49% [12]. Designing a plus or positive energy building encompasses active and passive design, including building components, insulation materials, advanced storage (nano-enhanced phase change materials, Li-ion batteries, thermally insulated storage tanks), and renewable energy technologies, including solar thermal collectors, solar photovoltaics, biofuels, geothermal heat pumps, air-to-water heat pumps, and photovoltaic systems. Several notable studies have focused on studying plus energy buildings [13]. Retrofit options have also been assessed for old buildings with poor insulation and higher air changes per hour (ACH) with an energy payback period of 2–10 years, depending on the level of retrofitting [14]. The widespread integration of net zero energy buildings poses significant threats to the voltage stability of distribution systems, which can be tackled through nanogrid, microgrid architectures (islanding abilities), or bulk local distributed storage. Phase change materials (PCMs) provide thermal inertia to the residential buildings, enhancing the reliability of net-zero and plus-energy buildings by effectively managing thermal energy.

Nanogrids are residential houses that utilize local renewable energy sources and storage solutions to operate in islanded mode or non-islanded connected to the main grid. Nanogrid houses are gaining traction owing to trigeneration, polygeneration, and multigeneration concepts being investigated at residential scales. There are AC, DC, and hybrid AC/DC nanogrid variations being studied. DC nanogrids are found to be more efficient [15] but are preferred to be implemented in remote communities [16] or new communities. The major challenges that exist in the widespread penetration of nanogrids are the lack of smart grid standards dealing with decentralized architectures such as nanogrids, low inertia, and the requirement for electrical (Li-ion) and thermal energy storage (PCM, insulated tanks) [17]. Other thermal storage options for nanogrids also include Carnot batteries [18] and hydrogen storage. High-energy dense (PCMs) are best suited for residential nanogrids with space limitations because of their energy density capabilities. On a wider scope, nanogrids and microgrids offer ancillary services to the wider grid. Integrating nanogrids with thermal energy storage systems such as PCM and sensible heat storage entails addressing several challenges across technical, economic, social, and regulatory domains. Technical issues include energy management protocols/solutions to optimize use and maintain reliability [19], the intermittency and stability of local intermittent generation [20], and thermal energy storage efficiency [21]. Economic challenges involve high initial investment costs [22]. Regardless of the challenges, nanogrids are now being studied, integrating diverse technologies, including phase change materials within building walls, to decrease energy demands for HVAC applications.

3.2. Phase Change Materials (PCMs)

Phase change materials (PCMs) are energy-absorbing substances that store and release large amounts of latent heat during phase transitions, offering a stable thermal storage solution for residential buildings. PCMs experience phase transitions in (solid–liquid–gas) form with significant volume changes, making them impractical for thermal storage systems despite their high heat of fusion. Solid–solid and solid–liquid phase transitions are majorly used in thermal energy storage applications. Solid–liquid phase change stores significant thermal energy with minimal volumetric expansion, typically 10%. Melting and solidification occur under constant pressure if the container allows for slight volume expansion. During melting, phase change materials (PCMs) absorb thermal energy while maintaining a constant phase change temperature. After melting, additional heat is stored as sensible heat. The energy absorbed during the phase transition, called latent heat, relates to enthalpy change and is not noticeable due to minimal temperature variation. Materials that store thermal energy in this way are known as phase change materials (PCMs) [23,24]. Figure 2 illustrates the thermodynamic characteristics of a single-stage PCMstorage mechanism.

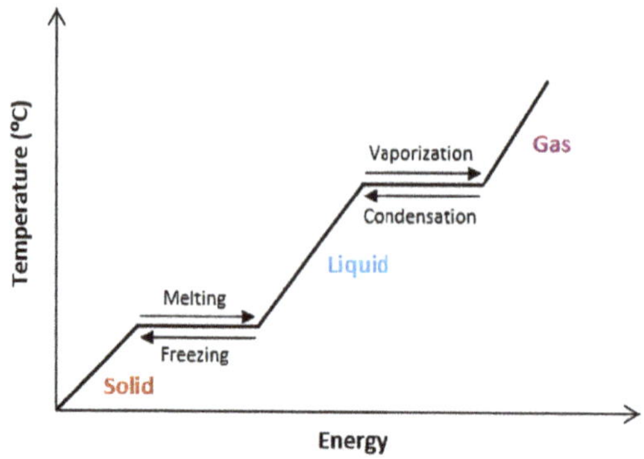

Figure 2. Overview of the components reviewed to be integrated into nanogrid.

Phase change materials (PCMs) were first investigated in the 1940s by researchers Telkes and Raymond. However, their potential was not widely recognized until the energy crisis of the 1970s, which spurred renewed interest in PCMs. By the early 1980s, extensive research had been conducted to explore their applications, with a particular focus on solar energy systems [24–36]. The research into PCMs as thermal energy storage (TES) has since expanded to cover a broad range of applications. Significant contributions have been made in the fields of building air conditioning [37–44], solar heating [45–49], electronic cooling [50–54], waste heat recovery [55–58], textiles [59–61], construction and buildings [62–66], food and drink [67–70], and agriculture [71–76], among others. Holistic use of PCMs integrated with solar thermal storage, hydronic heaters, and other sustainable technologies for residential applications aids in building temperature regulation and thermal energy management. PCMs offer a robust solution for managing surplus solar energy and dealing with the duck curve phenomenon by leveling the load and enhancing solar energy ride-through capability, thereby ensuring a more stable and reliable decentralized energy grid. Advances in nanocomposites, microencapsulated PCMs, and shape-stabilized PCMs are being extensively studied with potential for application in nanogrid architecture, which will be discussed in each subtopic.

PCMs have positive and negative outlooks regarding their applications in polygeneration nanogrid environments. In terms of thermal storage, PCMs can store and release bulk amounts of energy in the form of latent heat in building structures. However, low thermal conductivity results in lower heat transfer rates, response time, and efficiency. Several methods are explored to resolve this issue, as discussed in Section 3.3.5, which includes embedding conductive materials such as graphite, carbon fibers, and metal foams to improve thermal conductivity. Nanogrid houses would require a stable comfort zone temperature range, which PCM can provide, but low conductivity can impact its performance in dealing with sudden fluctuations in thermal loads. A similar approach of adding nano-additives such as carbon nanotubes and metal nanoparticles increases response speed, which is discussed further in Section 3.3.8. Organic biodegradable PCMs such as paraffin and salt hydrates have low environmental impacts but may require encapsulation to prevent leakage, as discussed in Section 3.3.6. PCMs offer passive energy savings by reducing HVAC loads but have a higher price tag. PCMs with low-cost conductive fillers are a good option for providing a reasonable payback period. PCMs must endure long phase change cycles over a period of time to extend their lifespan; this can be improved by using layered, composite, multi-PCMs as discussed in Section 3.3.3. Moreover, PCM thermal response can be improved by integrating heat exchangers and finned structures to improve heat transfer, as discussed in Section 3.2.7. The thermal stability of PCMs is affected by phase segregation and sub-cooling in low-conductivity PCMs. This can be remedied by mixing additives to prevent sub-cooling. A faster response rate limits the application of PCMs in active cooling systems, and hence, hybrid PCM systems are explored. A summary of some of the latest trends is articulated in Table 1.

Table 1. Recent PCM trends and application in nanogrid.

Trends	Key Findings	Residential (Nanogrid) Applications	Challenges	References
Nanocomposites	Enhanced thermal conductivity and energy storage efficiency	Potential for improving energy efficiency in walls and insulation materials	High production costs and potential agglomeration issues	[77–81]
Microencapsulated PCMs	Improved stability and integration with building materials	Suitable for incorporation into drywall, paint, and cement for temperature regulation	Complexity in manufacturing and potential leakage problems	[82–87]
Shape-stabilized PCMs	Better adaptability for residential applications	Effective in floor heating systems and underfloor heating applications	Limited thermal conductivity and scalability concerns	[88–93]
Bio-based PCMs	Increased sustainability and reduced environmental impact	Eco-friendly option for insulation in green buildings and sustainable housing	Variable thermal properties and higher costs	[94–98]
Hybrid PCMs	Combined benefits of multiple PCMs and improved thermal performance	Versatile for use in a variety of integrated thermal storage systems	Complexity in synthesis and phase separation issues	[99–101]
High-temperature PCMs	Industrial applications and higher operational temperature ranges	Less relevant for typical residential use and more applicable in homes using solar thermal systems	Corrosion and material compatibility issues	[102–107]
Thermal Cycling Stability	Enhanced durability and longevity of PCMs through repeated thermal cycles	Ensures long-term reliability of PCM-integrated home heating and cooling systems	Degradation over time and consistency of properties	[108–110]
Additive Manufacturing	Customizable PCM structures for specific applications via 3D printing techniques	Customizable heating and cooling elements that can be designed for specific residential structures	Limited material options and mechanical strength concerns	[111,112]
Phase Transition Temperature Tuning	Tailored thermal properties for specific applications	Allows for precise temperature control in different residential zones or rooms	Complexity in designing and cost of materials	[113,114]
Photochromic and Thermochromic PCMs	Dynamic response to light and temperature changes for improved energy management	Suitable for smart windows and adaptive insulation in residential buildings	Limited availability and higher cost of materials	[115–126]

Latent heat storage (PCM)-based TES systems have high storage capacity. This capacity is 5 to 14 times higher than other heat storage systems [127–129] and compatible with air conditioning applications [130–132], with minimal fluctuations in temperature during the cold storage phase [128,133–135], energy stored/released at a constant temperature [128,134,136], reversible nature, the controllable range of melting point temperatures, minimal volumetric changes [128,134] and low vapor pressures [128]. On the contrary, PCMs pose various challenges, including poor thermal conductivity [136,137], stability issues, and container compatibility issues [133,137,138]. Furthermore, accumulation/deposition of solids on heat exchanger surfaces, density changes, segregation of phases, and high costs are also major challenges [23,133,134,137,139].

3.2.1. Phase Change Materials (Types)

The classification is based on phase transitions. PCMs (solid–solid, solid–liquid phase) retain lower heat during phase transitions but are more relevant to nanogrids because of lower prices [24]. PCMs fall into two primary types: low molecular compounds and polymers. Low molecular compounds are further classified into inorganic (salts, salt hydrates, hydroxides, alloys) and organic (paraffins, alcohols, fatty acids, esters, and others). Polymers include materials such as poly (ethylene glycol). Each PCM presents distinct advantages and limitations as discussed in the literature [25,135,139–141]. While solid–solid phase transitions can address leakage issues associated with solid–liquid PCMs and serve as an alternative energy storage medium, they generally have lower latent heat of phase transition [23,25,140]. Conversely, PCMs with solid–gas and liquid–gas transitions are constrained by specific limitations that hinder their application in thermal energy storage [23,142]. Thermal cycles in PCM's are articulated in Table 2.

3.2.2. Phase Change Materials (Thermal Storage)

Phase change materials (PCMs) offer various applications as reliable thermal energy storage mediums, particularly within residential building-based nanogrids. The effective implementation and integration of PCM thermal energy storage in these systems require specific types of PCMs and techniques [143]. Solar energy, a promising source of heat, can be harnessed for multiple uses, but maintaining the indoor temperature of residential nanogrid buildings is particularly relevant during the winter months [144]. Comparative analysis demonstrated that PCM-fluidized beds outperformed sand-fluidized beds in thermal efficiency and performance [145], highlighting their potential for optimizing PCM-based storage in residential nanogrids. The variations in solar energy output pose many challenges, most significantly the duck curve problem in residential nanogrids, which requires reliable thermal storage and PCM to provide thermal energy ride-through capability during the absence of the sun. Close et al. [146] pioneered solar air heater (SAH) research, resulting in innovative designs such as flat plates, fin plate v-grooves, and Chevron plates. In residential nanogrids, PCM thermal storage enhances air heating efficiency by integrating directly with solar collectors [45], helping to smooth energy supply and address the duck curve issue. Solar dryers are categorized into four types based on their heating mechanisms: (1) direct, (2) indirect, (3) mixed, and (4) hybrid. In residential buildings, the use of solar radiation and air heaters is coupled with thermal energy storage (sensible, latent, chemical), capturing excess energy to level the duck curve issue. Enhancing the efficiency of solar collectors is vital for maximizing the performance of solar heating systems within residential nanogrids. Recent studies, such as those involving the Chevron plate collector [147], have showcased a 20% boost in thermal efficiency, achieving output temperatures that are 10 °C higher at specific mass flow rates.

Table 2. Thermal cycles for commonly used PCMs [138].

Type of PCM	Thermal Cycles	Melting Point (°C)	Latent Heat of Fusion (KJ/Kg)	References
Paraffin (70 wt.%) + Polypropylene (30 wt.%)	3000	44.77	136.16	[148]
Paraffin wax 54	1500	53.32	184.48	[149]
Paraffin wax 58–60	600	58.27	129.8	[149]
Paraffin wax 60–62	600	57.78	129.7	[149]
Acetanilide (C_8H_9NO)	500	113	169.4	[150]
Erythritol	1000	117	339	[149]
Lauric acid ($C_{11}H_{23}COOH$)	1200	42.46	176.6	[151]
Myristic acid ($C_{13}H_{27}COOH$)	450, 1200	50.4, 52.99	189.4, 181	[151,152]
Palmitic acid ($C_{15}H_{31}COOH$)	450, 1200	57.8, 61.31	201.2, 197.9	[152]
Palmitic acid (80 wt.%) + expanded graphite (20 wt.%)	3000	60.88	148.36	[152]
Stearic acid ($C_{17}H_{35}COOH$)	450, 1500	65.2, 63	209.9, 155	[152,153]
Calcium chloride hexahydrate ($CaCl_2.6H_2O$)	1000	29.8, 28, 23.26	190.8, 86, 125.4	[154–156]
Magnesium chloride hexahydrate ($MgCl_2.6H_2O$)	1000	23.26	125.4	[150]
Glauber's salt ($Na_2SO_4.10H_2O$)	500	111.5	155.11	[157]
Sodium acetate trihydrate ($NaCH_3COO.3H_2O$)	500	58	230	[158]
$Na_2SO_4.1/2NaCl.10H_2O$	5550	20	-	[159]
Al–34%Mg–6%Zn alloy	1000	454	314.4	[160]
Ammonium alum ($NH_4(SO_4)_2.12H_2O$)(15%)	1100	53	170	[161]
Capric acid (65 mol%) + lauric acid (35 mol%)	360	19.6	126.5	[161]
Capric acid (73.5 wt.%) + myristic acid (26.5 wt.%)	5000	21.4	152	[162]
Capric acid (83 wt.%) + stearic acid (17 wt.%)	5000	24.68	178.64	[163]
Lauric acid (66 wt.%) + myristic acid (34 wt.%)	1460	34.2	166.8	[164]
Lauric acid (69 wt.%) + palmitic acid (31 wt.%)	1460	35.2	166.3	[164]
Myristic acid (64 wt.%) + stearic acid (36 wt.%)	1460	44.1	182.4	[164]
Myristic acid + glycerol	1000	31.96	154.3	[165]
Palmitic acid + glycerol	1000	58.50	185.9	[165]
Stearic acid + glycerol	1000	63.45	149.4	[165]
$Mg(NO_3)_2.6H_2O$ (93 wt.%) + $MgCl_2.6H_2O$ (7 wt.%)	110	33.8	111.6	[166]

Nanogrid buildings highly depend on the thermal performance of local energy storage systems for optimal energy management. Solar air heaters (double-pass) with sensible heat storage were studied using materials like gravel, limestone, and iron scraps, demonstrating the potential for increasing thermo-hydraulic efficiency in heating systems. Gravel-based TES proved to be 22–27% higher in efficiency, making it suitable for integrating into residential nanogrids to enhance space heating efficiency [167]. Encapsulated TES modules in domestic hot water cylinders, arranged in 57 vertical tubes, enhance storage capacity and reduce operational costs, making them ideal for improving energy efficiency and availability in residential nanogrids during off-peak periods [168]. A recent study [169] combined PCM and solar hot water systems in a heat transfer fluid loop, demonstrating a significant reduction in storage volume, which is one of the main concerns in residential nanogrids that have limited space. TES technologies integrating PCMs may prove to be very critical in residential nanogrids for managing energy storage and dispatch for heating, cooling, and even power generation [170]. Nkwetta et al. [143] reviewed PCM configurations, highlighting key factors influencing PCM selection and application. Although PCM-based TESS systems have higher costs, cascaded TES configurations offer 30% higher energy utilization and 23% improved exergy efficiency, offering a promising solution for residential nanogrids. The integration of transpired solar collectors with PCM storage in residential nanogrids improves energy efficiency, reduces system size and cost, and enhances energy collection, with further optimization and automation recommended for maximizing performance. This study proposes the integration of transpired cascaded solar integrated thermal walls in residential houses with energy-sharing networks to form a near-zero energy network of nanogrids. Investigation by Erro and team [171] encompasses the integration of thermoelectric heat pumps into nanogrids with sensible thermal energy storage systems. They tested charging a sensible thermal energy storage system using air as the heating fluid. The system achieved temperatures up to 139.2 °C and outperformed conventional thermal storage by 30% at a storage temperature of 120 °C.

3.2.3. PCM Thermal Storage (Building)

The development of novel architectures for energy conservation in residential and commercial sectors is crucial, with TES systems offering potential benefits including improved energy and economic efficiency, reduced electrical consumption, and lower environmental pollution and CO_2 emissions. Thermal energy storage (TES) systems in building applications can be categorized into two distinct types: active and passive. Active thermal energy storage (TES) systems, utilizing forced convection and sometimes mass transfer in heat exchangers, aim to regulate indoor conditions and reduce peak energy demand in HVAC systems. The passive TES systems use solar, wind, or natural resources and architectural changes such as insulation of roofs, walls, floors, such as in passive houses, triple-glazed windows, building orientation, thermal mass, green roofs, heat recovery devices, and shading devices to maintain comfortable indoor conditions, aiming to reduce dependence on mechanical heating or cooling. A prototype nano-PCM wallboard (Gypsum and n-heptadecane) with expanded graphite nanosheets was investigated [172] to enhance the thermal conductivity and shape stability of a building. The findings revealed that wallboards significantly reduce heat gains/losses for hot/humid climates. Yuhao Qiao and team [173] employed an artificially controlled environment to simulate and study PCM wallboard for five different climatic conditions. In one climatic condition, the study made a bold claim of reducing temperature accumulation by 88.9%. Another study investigated single- and double-layer PCMs integrated into building envelopes and reiterated the fact that PCM ceilings lower reliance on air conditioning and lower energy consumption. Moreover, a double-layer PCM building envelope can reduce maximum temperature variation by 8.5 °C as compared with a single layer. The study [174] claimed a 6 °C decrease in temperature (sub-tropical climate) when using OM35 (157 J/g) and Eicosane as PCM material.

3.2.4. PCM Thermal Storage (Solar Heating)

An experimental investigation [72] was conducted using indirect solar drying and PCMs for drying applications specifically, but the results are significant for potential applications in microgrid applications. The setup included a pair of flat plate solar air heaters, a drying chamber, a PCM storage unit, and an air blower. Each heater featured a 1 mm thick copper absorber plate covered by two 5 mm thick glass sheets with a 25 mm gap and insulation (40 mm foam). The drying chamber (galvanized iron), measuring 0.6 by 1.2 by 1.7 m^3, was also insulated with 40 mm foam and contained three stainless steel trays supported by aluminum frames. PCM energy storage was provided by two plastic cylinders filled with paraffin wax, insulated with foam and glass, and equipped with 32 copper tubes for heat transfer. The use of PCM maintained a constant drying temperature for 7 h daily, improving air temperature by 3.5–6.5 °C after 2:00 p.m. compared with systems without storage [73].

Another study with similar implications was reported [175,176], in which PCM (calcium chloride hexahydrate) integrated solar dryers with reflecting mirrors for trapping more solar radiation. Another study with H-68 PCM [177,178] also concluded improved performance using PCMs. All the above studies have used mass flow rate and discharge period as the main parameters of interest. Hybrid solar technologies with PCM result in 10–25% savings on total energy consumption [70,179,180], which prompts its use in nanogrids. Whenever we talk about thermal energy storage, it is imperative to discuss heat transfer fluids. Blood in the human body facilitates the distribution of nutrients and oxygen. Similarly, heat transfer fluids in thermal energy systems play an equally critical role in the efficient transfer. The integration of U-pipe evacuated glass tube (EGT) solar collectors into residential nanogrids may enhance energy efficiency by optimizing thermal storage. Experimental investigations were conducted for such a solar collector model. Thermal efficiency mainly depends on the mass flow rate of HTF and not on the length of the tube. A low heat loss coefficient collector was deemed suitable for cold climates. The absorption coefficient, ranging between 0.85 and 0.94, did not significantly impact efficiency under different weather conditions [181]. Double U-tube collectors demonstrated an efficiency of 80% under solar irradiance of 900 W/m^2 and a thermal conductivity of 100 W/mK, outperforming single U-tube collectors by 4% [182].

An absorption refrigerator is a refrigerator that uses a heat source (e.g., steam and hot water) to provide the energy needed to drive the cooling process. However, in adsorption refrigeration, the refrigerant or adsorbate vapor molecules adsorb onto the surface of a solid instead of dissolving into a liquid like LiBr and ammonia. Both absorption and adsorption chillers are well suited for residential applications, but an absorption chiller is proposed for the nanogrid because of the integration of evacuated tube collectors and insulated thermal storage, which use hot water to drive the cooling process with a higher coefficient of performance C.O.P (0.7–1.8) and cooling capacity as compared with an adsorption chiller. Nano-enhanced PCMs (Ne-PCMs) [183] find applications in solar cookers, water heaters, air warmers, and desalination systems, emphasizing their role in enhancing the thermal performance of all nanogrid elements and, most importantly, increasing heat storage capacity and energy efficiency. Table 3 outlines the various operating temperature ranges for different solar thermal technologies.

Table 3. Temperature ranges for different solar thermal technologies.

No.	Component	Melting Temperature (°C)
1.	Air conditioning	<15 [4]
2.	Absorption refrigeration	>90 [4]
3.	Solar heating	15 to 90 [4]
4.	Residential heating/cooling	0 to 65 [68]
5.	Agricultural drying products	40 to 75 [131]
6.	Solar thermal plant	>500 [111,141]

3.2.5. PCM Thermal Storage (Integrated Hot Water Storage Unit)

In a numerical study conducted with TRNSYS, a domestic hot water tank was coupled with a thermal energy storage (TES) module containing phase change materials (PCM). The results demonstrated that integrating PCM effectively improved the tank's energy storage density. The optimal PCM was identified as a 10% graphite and sodium acetate trihydrate composite due to its superior storage potential. Key considerations included the thermal conductivity and thickness of the PCM module, as well as ensuring safe and durable containment. The study demonstrated that PCM integration not only enhanced energy density but also effectively met energy, load shift, and peak power demands [143]. The integration of PCM in hot water tanks within residential nanogrids enhances energy storage density, enabling effective load shifting and peak demand management. This leads to optimized thermal management and reduced reliance on instantaneous heating. Figure 3 illustrates the charging durations for a variety of PCMs.

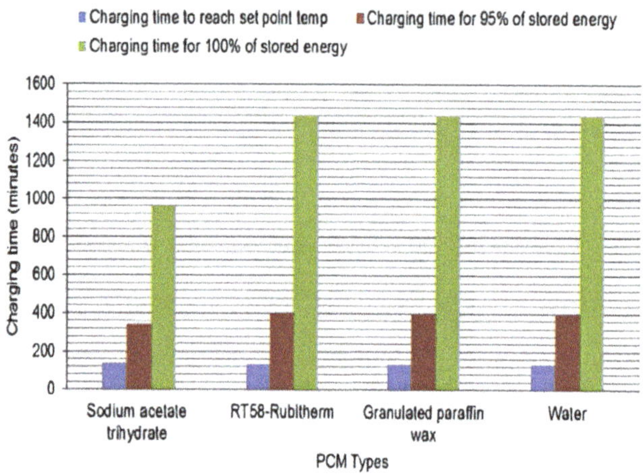

Figure 3. Overview of the components reviewed to be integrated into nanogrid.

3.2.6. PCM Thermal Storage (Feasibility of PCM-TES)

There is a disparity in demand/supply at centralized and decentralized levels in the grid with the penetration of renewable technologies, especially solar and wind. Such disparities can be best addressed by reliable electrical and thermal energy storage, which also improves the energy and exergy performance of intermittent sources. Hybrid energy storage, such as synergetic electro-thermal storage technology, effectively reduces overall energy consumption. TES systems are pivotal in achieving successful thermal systems by significantly reducing thermal energy losses [184]. Sensible heat storage (energy stored through heating/cooling) and PCM thermal energy storage (melting/ vaporization/ solidification/ liquefaction stores the energy) are all feasible options for nanogrid integration. Sensible heat storage, being a mature technology, is integrated in residential buildings due to its cost-effectiveness, despite its large volumetric requirements for small temperature variations [134]. In contrast, latent heat storage offers higher energy density (kWh/m^3) and superior properties overall. Latent heat storage is self-regulating, discharging energy at a constant temperature, and can store five to ten times more energy than sensible heat storage units, making them ideal candidates for small residential nanogrids [133,140]. Based on prior research, latent heat storage systems appear to be a promising and practical solution for integration in micro and nanogrids.

3.2.7. PCM Thermal Storage (Key Challenges)

The effectiveness of PCM-integrated thermal energy storage (TES) systems in nanogrids depends largely on their thermophysical characteristics at melting temperature. Key attributes include heat of fusion, specific heat, thermal conductivity, stability, non-toxicity, inertness, and density. However, significant challenges persist, such as constrained space, elevated capital costs, and the intermittent availability of solar energy. Integrating PCM thermal storage into residential nanogrids can bridge the energy supply–demand gap, enhancing energy conservation and reliability. However, the poor thermal conductivity of paraffin wax limits efficient charging and discharging in TES systems, necessitating performance improvements in connected solar air heaters. Low thermal conductivity is a major hurdle for PCMs; research favors integrating highly conductive additives such as fins and metallic foams to improve overall thermal performance. However, techno-economic case-based studies must be conducted to reach a consensus on its use in nanogrids. To advance PCM thermal energy storage technology, operational parameters must be optimized. Desired traits include a high heat transfer rate, stable inlet and outlet temperatures, consistent phase transition temperatures, uniform heat transfer fluid (HTF) temperatures, rapid charging/discharging, minimal circulation power for HTF, controlled temperature regulation, and uniform temperature distribution. Additional considerations are the compatibility of PCM with the container, minimized heat loss, and efficient solid/liquid motion during freezing and melting processes [94]. Proper structural adjustments can thus lead to a well-designed thermal energy storage system. Proper selection of PCM based on its melting characteristics and recommended working temperature is critical. Additionally, the effectiveness of PCMs depends on the convective heat exchange between the PCM walls and the surrounding room air. The current heat transfer coefficients specified in building codes might not be suitable for PCM applications [185].

3.3. Phase Change Materials (Heat Transfer Enhancement)

Contemporary research on phase change materials (PCMs) focuses on various facets such as the materials themselves, heat transfer fluids, and their properties such as temperature, flow rate, configurations, structural designs, and additives. This exploration spans both theoretical analysis and experimental investigations to uncover their diverse applications [186]. One of the most effective ways of enhancing heat transferability is the use of PCM-packed beds, improving storage density and efficiency. Figure 4 illustrates a spherical capsule-packed bed with multiple PCM arrangements.

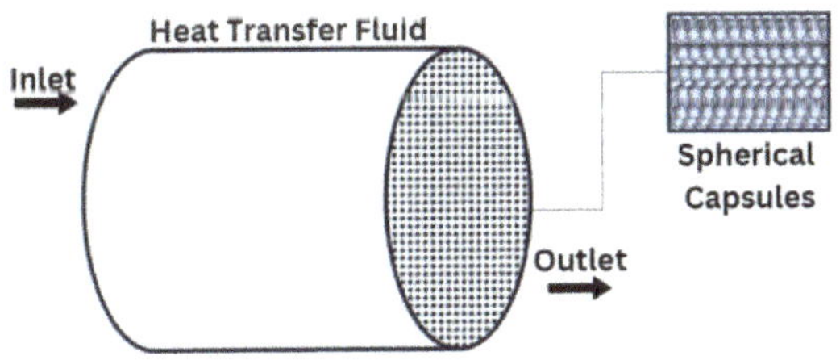

Figure 4. Spherical capsule packed bed storage.

Nanoparticles significantly boost the thermophysical properties of phase change materials (PCMs), making them ideal for nanogrid applications. The key to these enhancements lies in the nanoparticles' high volume-to-surface area ratio, which optimizes interfacial heat transfer, thermal conductivity, and thermal diffusivity [187]. As a result, PCMs with embedded nanoparticles exhibit better energy storage and discharge capabilities. Titanium oxide TiO_2 nanoparticles are claimed to achieve up to a 37% increase in efficiency but highlighted the need for further research on long-term stability and higher nanoparticle concentrations [188]. Increasing nanoparticle volume fraction can reduce energy storage efficiency due to decreased capacity and increased viscosity, while optimized long and narrow triangle fins, especially unequal length ones, enhance heat transfer and outperform rectangle fins in improving PCM melting rates without compromising storage capacity [189].

The findings indicate that nano-enhanced PCMs exhibit 8.3% more heat charged in and 25.1% more heat discharged compared with pure PCMs under winter test conditions [190].

3.3.1. Heat Transfer Fluids and Mass Flow Rates (Thermophysical Property)

A detailed study [191] on the choice of heat transfer fluids considered air, compressed air (10 bar), supercritical CO_2 (100 bar), steam (10 bar), solar salt, and liquid sodium with flat plate PCM as thermal energy storage (140 MW). The study found that liquid sodium reduced the thermal storage charging duration by 25% compared with solar salt and delivered 99.4% of its stored energy during discharge, outperforming other fluids. It concluded that the choice of heat transfer fluid had a minor impact on electricity generation, suggesting possible cost savings with less efficient fluids. The implications of the aforementioned study extend to application in residential nanogrids. Residential buildings with diurnal energy demands require rapid charging cycles to provide reliable energy ride-through capabilities. Using liquid sodium as a heat transfer fluid may enhance the responsiveness and reliability of thermal storage. Additionally, the potential cost–benefit analysis for residential applications needs to be studied.

The predicted outlet air temperature during discharge in a solar air heater (SAH) equipped with a phase change material (PCM) thermal energy storage (TES) unit was assessed across mass flow rates ranging from 0.05 kg/s to 0.19 kg/s. The optimal values were determined by evaluating the outlet air temperature and the freezing time of the PCM. A paraffin wax and aluminum powder composite (0.5% by weight) enhanced thermal conductivity, revealing an inverse relationship between mass flow rate and PCM freezing time [192]. Experimental and numerical studies showed discrepancies during PCM melting and solidification, confirming the TES unit's applicability to other air-based systems [193]. Optimizing mass flow rates and enhancing the thermal conductivity of PCM composites can develop efficient TESS for residential nanogrids, ensuring a stable energy supply and improved sustainability.

3.3.2. Shell and Tube PCM (Arrangement)

Researchers in the study [194] explored a shell and tube thermal energy storage (TES) system for district heating, using paraffin (RT100) as the phase change material (PCM) and water as the heat transfer fluid (HTF). Their findings revealed that a paraffin/graphite composite with 15% graphite content exhibited impressive performance. The configuration included vertically arranged HTF pipes in a 15 × 15 matrix. The addition of expanded graphite improved the thermal conductivity of the storage medium, maintaining the PCM's essential thermophysical properties. The TES system reliably met heating demands ranging from 130 kW to 400 kW, depending on the operating conditions. Figure 5 illustrates a shell-and-tube PCM thermal energy storage unit.

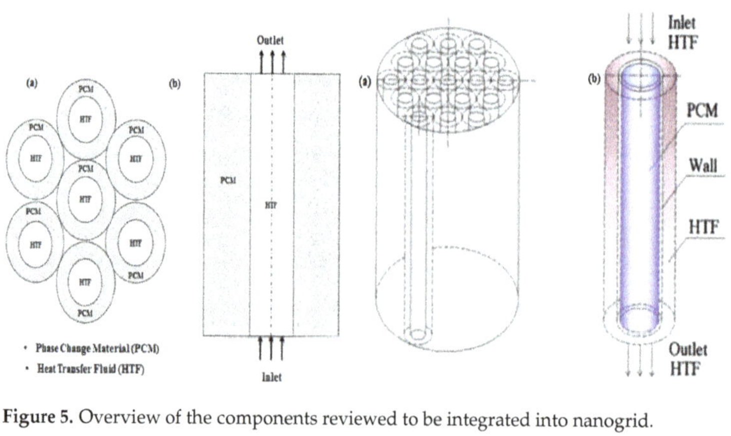

Figure 5. Overview of the components reviewed to be integrated into nanogrid.

The inclusion of nanoparticles such as Al_2O_3 and graphene nanoplatelets (GNPs) significantly enhances the thermal performance of PCMs in horizontal triple-series shell-and-tube heat exchangers by accelerating melting times and increasing effective temperatures [195,196]. These enhancements suggest promising applications for improved energy storage in a nanogrid context.

3.3.3. Multi PCM Systems (Arrangement)

An array of multiple PCMs with multiple melting points can be fused/arranged together vertically or horizontally, as shown in Figure 6, in tanks to improve the overall heat transferability of the system. To improve the heat transfer ability of the systems, multiple phase change materials (PCMs) with varying melting points are linked in horizontal/vertical tanks in ascending/descending order. This combinational approach leads to optimized charging/discharging. Employing packed bed storage with multiple PCMs significantly enhances the heat transfer efficiency of thermal storage systems. In such a unit, the temperature difference between the heat transfer fluid (HTF) and the PCMs is minimized, leading to an improved system coefficient of performance (COP). Multi-PCMs (series/parallel combinations) have significant advantages in building applications, especially in reducing energy consumption. Multiple PCMs transform the flexibility of storage mediums to adjust to changing operational conditions. In a multi-system, each PCM has a different melting point, thus providing a wider temperature control range and faster charging/discharging [197]. In building contexts, systems with multiple PCMs can balance heating and cooling demands by storing and releasing heat at various temperatures, making them ideal for year-round use with reduced energy consumption. A multi-PCM configuration (RT-35-RT27) in a PV-PCM window has been shown to lower peak interior wall temperatures by up to 46% compared with standard glass panes [198]. The use of multiple PCMs in air-phase change material heat exchangers (HEX.) reduced incoming fresh air temperatures before reaching air conditioning systems, lowering cooling requirements by more than 11% [199].

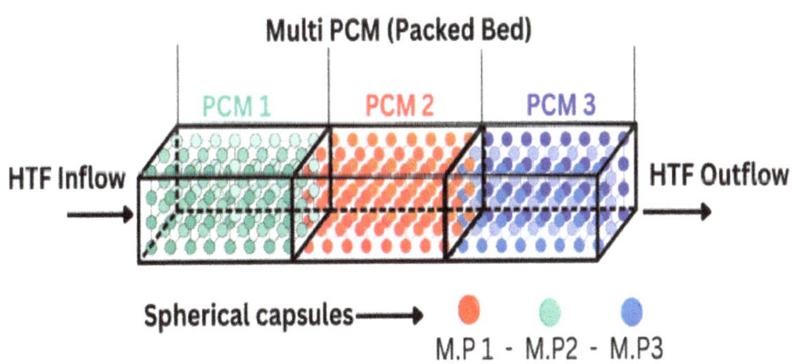

Figure 6. Multiple PCMs spherical capsules packed bed storage unit for solar systems.

3.3.4. Hybrid Technologies (Arrangement + Thermophysical Properties + Structure)

Residential nanogrids may also benefit from combining thermoelectric (TE) modules with evacuated glass tube heat pipe solar collectors [200], enabling simultaneous water heating and electricity generation with validated theoretical and empirical performance models. Despite having a thermal efficiency of 1–2%, slightly lower than the 3–4% efficiency of Rankine Cycle systems, the TE/evacuated glass systems offer benefits such as simple design, no moving parts, and flexible component replacement. These systems can provide both thermal energy for domestic hot water (DHW), space thermal conditioning, and electrical power generation simultaneously, realizing combined heat and power nanogrids. It has a packed bed of spherical PCM capsules that demonstrated a consistent heat output of 200 W/m^2 for 11 h at night during discharge. The net daily energy and exergy efficiencies ranged from 32% to 45% and 13% to 25%, respectively. The ability of such a device to provide consistent heat output during night-time is ideal for nanogrid energy ride-through capabilities, energy sharing, and reliability. The team in [201] emphasized the use of latent heat TES for thermal storage applications. They studied the inclusion of nanoparticle additives and fins to enhance thermal conductivity. Their design is a triple-tube heat exchanger with centrally installed PCM. The team compared cases with/without arc-shaped fins, different shapes of fins (upward, downward), and different angles. Higher angles improve circulation and heat transfer within PCM. The best performance was observed at 90-degree upward fins. Moreover, a higher Reynolds number and inlet temperatures enhance the system's performance. This study has significant implications for nanogrid buildings, which depend on energy efficiency and the ability to store excess energy to provide thermal energy ride-through capabilities. The team in [202] pointed out that the solutions to low thermal conductivity include nanoparticle distribution, high conductivity materials, and composite PCMs. The study's findings indicate that the implementation of a framed structure led to a storage rate of 20.4 watts, demonstrating 115% improvement over the 9.5-watt storage rate observed with an unframed structure. Furthermore, the research highlighted the efficacy of zigzag surface shapes over smooth ones in boosting the thermal energy storage rate through the melting process. Additionally, the study observed that an increase in the number of pitches on the zigzag surfaces correlated with a better thermal energy storage rate. The reverse arc-shaped structure emerged as the most effective, achieving a thermal energy storage rate of approximately 34 watts and completing the melting process in just over an hour (63.15 min). This superior performance was recorded when the structure was heated by water with a Reynolds number of 1000 and an inlet temperature of 50 °C. Lastly, the research revealed that both an increase in the Reynolds number, from 1000 to 2000, and a rise in the heat transfer fluid's (HTF) inlet temperature, from 50 °C to 55 °C, contributed to enhancements in the thermal energy storage rate by around 6 watts and 10 watts, respectively. An innovative solar air heater (SAH) [74] with an illustration of the setup shown in Figure 7.

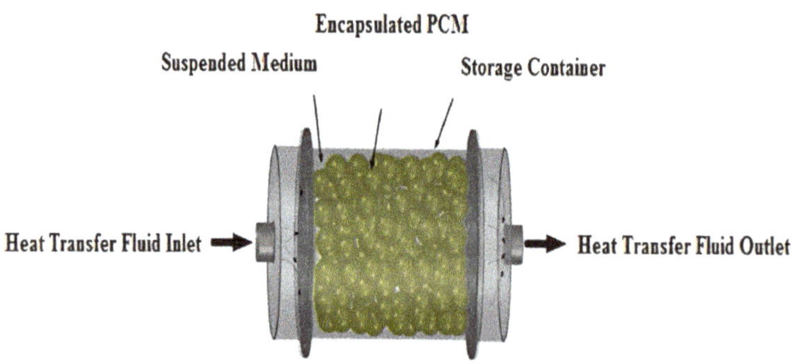

Figure 7. Multiple PCMs spherical capsules packed bed storage unit for solar systems.

Solar air heaters are designed based on several factors. Collector covers are classified into three types: bare plate SAH, single cover SAH, and double cover. The absorber material in SAHs can be metallic, non-metallic, or a matrix structure. The shapes include slats, porous media, or fins to enhance heat absorption efficiency. The absorber flow patterns are another essential aspect that can occur over, under, or on both sides of the surface. Flow directions within SAHs can vary, including single pass, parallel pass, double pass, and triple pass configurations. Hybrid collectors are also noteworthy, combining different mediums such as water, air, or a combination of both (water–air) [203]. Low-temperature solar air heating with PCMs is an ideal candidate for nanogrids designed in cold climates. PCM with encapsulated spherical PCMs was analyzed [42] to identify three main parameters influencing the charging of thermal storage. The parameters include the size of the heat transfer surface area reducing the size of the encapsulated balls, the temperature differential between the phase transition and HTF inlet, and finally the mass flow rate of HTF. An extensive study was conducted on a compact PCM air heat exchanger. The thermal storage system comprised compact storage modules constructed from aluminum containers filled with PCM. These modules were arranged vertically, with air forced to flow vertically from the upper to the lower part of the storage unit. The study [204,205] utilized 135 kg of paraffin (RT27) as the storage medium, taking into account the thermo-physical properties of the phase change material (PCM) and a variety of parameters to refine the model. Key factors such as the air mass flow rate, surface roughness, and the thermal conductivity of the encapsulating material were specifically evaluated in the numerical analysis. Paraffin wax-integrated PCM-TES systems ensure consistent temperature regulation and energy supply during intermittent sunlight conditions. The characteristics of paraffin wax as a phase change material (PCM), as detailed by the supplier Pure Temp, are presented in Table 4. A study [206] investigated such a system with and without PCM storage in clear and semi-cloudy conditions. In contrast to a similar system without storage, the plate temperature increased at 2.5 m from the entrance point, remaining stable over the subsequent 7.6 m of the storage path. During periods of limited or no sunlight, the liquid phase change material (PCM) released its stored heat energy to the circulating water until it fully solidified. The literature on ground-source heat pumps (GSHPs) coupled with TES systems highlighted that the coefficient of performance (COP) for GSHP with TES systems varied from 2.0 to 6.49 for both cooling and heating applications and from 2.3 to 7.95 for standalone setups [207].

Table 4. Thermal characteristics of paraffin PCMs.

PCM	Heat (Latent) (J/g)	Phase Transition Temperature (°C)	Density (g/cm^3)	Solid (J/g °C)	Liquid (J/g °C)
PT-40	198	40	0.85	1.98	2.12
PT-43	180	43	0.88	1.87	1.94
PT-48	245	48	0.82	2.10	2.27
PT-50	200	50	0.86	1.82	1.94
PT-56	237	56	0.81	2.47	2.27
PT-61	199	61	0.84	1.99	2.16
PT-68	198	68	0.87	1.84	1.91

3.3.5. Nanocomposite PCMs (Structure)

A reliable nanogrid requires consistent operation over multiple charge/discharge cycles, efficiency of heat storage, rapid heating/cooling, and maximum discharge efficiency. Through the in situ miniemulsion polymerization process, the team in [208] crafted nano-encapsulated phase change materials (PCMs) featuring n-octadecane (core), encased in a resilient styrene/methyl methacrylate copolymer shell. This robust polymeric armor adeptly safeguarded the n-octadecane, showcasing exceptional thermal stability and enduring up to 360 cycles of charging and discharging. An investigation was conducted on a nanocomposite latent heat storage system, employing stearic acid as the phase change material (PCM) and incorporating multi-walled carbon nanotubes (MWCNTs) as an additive for enhanced performance. Adding MWCNTs resulted in a lower melting point during charging and an increased freezing point during discharge compared with pure stearic acid. While the inclusion of MWCNTs improved the thermal heat transfer of stearic acid, it simultaneously diminished its natural convective heat transfer. Thus, the overall thermal performance of the stearic acid/MWCNT nanocomposite depends on a balance between the benefits and the drawbacks. When catering to consumer energy demands, the discharge rate of latent heat storage is more crucial than the charge rate. Findings revealed that the charge rate only saw improvement when the volumetric concentration of MWCNTs in stearic acid was kept below 5%. To ensure efficient charging and discharging, establishing an effective heat transfer network is pivotal, as it enables rapid heating or cooling of the PCM's inner regions [132]. López et al. [209] performed experimental investigations to scrutinize the performance attributes of an organic PCM storage tank. The analysis further delved into the design parameters for the latent heat storage tank, with a particular emphasis on evaluating paraffin's efficacy in cold storage applications. The experiments revealed that precise control over the target supply temperature could be accomplished by incorporating a sensible heat storage unit downstream of the latent heat storage tank. Measurements of both the PCM and the heat transfer fluid (HTF) indicated pronounced vertical stratification, notably within the liquid phase of the PCM. Within 4 h, the storage tank reached 78% of its maximum capacity, with stored energy showing greater sensitivity to supply temperature variations during melting tests. Despite testing various coil designs, a full phase change was not accomplished during solidification tests at temperatures 6 K below the phase change threshold. Combining latent heat storage with sensible heat storage holds potential to improve the efficiency and dependability of energy systems for homes. Table 5 summarizes the standard pricing of the organic latent heat storage (LHS) tank.

Table 5. Cost of organic PCM latent heat storage tank [209].

Element	Cost (Euro)	Cost (AUD)	% Cost
Wall (tank) + lid + drainage	1700	2788	36.5
Insulation	1500	2460	32
Copper (tubes)	400	656	8.6
Collectors	240	393.6	5.1
PCM (165 kg)	814	1334.96	17.5
Total	4654	7632.56	100

Parameshwaran et al. [210] found that embedding hybrid nanocomposites into phase change materials (PCM) significantly enhanced thermal properties and heat storage capacity, making them suitable for cooling applications in buildings. Similarly, Li et al. [211] studied the effects of various carbon nanofibers (CNFs), carbon nanofillers—short and long multi-walled carbon nanotubes (S-MWCNTs, L-MWCNTs), and graphene nanofillers (GNPs)—on paraffin-based nanocomposite PCMs. They observed a nearly linear decrease in melting/solidification enthalpies with the addition of nanofillers, with minimal dependency on filler size and shape. The thermal conductivity improvement, particularly notable with S-MWCNTs and planar GNPs, was attributed to reduced steric impedance and low thermal interface resistance, respectively. However, the performance of graphene nanofillers was highly dependent on their size and thickness, with larger and thicker layers being more effective. Consequently, graphene nanofillers emerged as optimal candidates for enhancing thermal conductivity in nanocomposite PCMs while only moderately reducing energy storage capacity. These findings have significant implications for residential nanogrids, where optimized nanocomposite PCMs can enhance energy efficiency and reliability, crucial for managing energy demands and cooling needs.

Nanocomposite phase change materials (PCMs) incorporating paraffin infused with hexagonal boron nitride (h-BN) nanosheets have been evaluated for their potential in advancing thermal energy storage solutions. A 10 wt.% concentration of h-BN nanosheets increased thermal conductivity by 60% and decreased latent heat of fusion by 12% compared with pure paraffin wax, while also accelerating solidification and melting rates by 25%. Dimensional attributes of h-BN nanosheets significantly affected thermal conductivity, and the absence of surfactants caused instability due to nanosheet precipitation during cyclical phase changes, necessitating further stabilization efforts [212]. Additionally, Teng et al. [213] investigated modified PCMs with multi-walled carbon nanotubes (MWCNTs) and graphite added to paraffin at various concentrations (1.0, 2.0, and 3.0 wt.%). Their direct-synthesis method and differential scanning calorimetry (DSC) experiments revealed that MWCNTs were more effective than graphite in enhancing the thermal storage performance of paraffin, reducing temperature differentials between PCM and heating fluids, and improving several experimental parameters. The phase change temperature variation was minimal, with a maximum decrease of only 3.69% compared with pure paraffin, indicating the significant potential of MWCNT-enhanced paraffin for future thermal storage applications.

3.3.6. Microencapsulated PCM (Arrangement)

Microencapsulation is a process of enclosing PCM materials within protective coatings, forming microcapsules. Microencapsulated phase change materials (PCMs) range from 1 to 1000 μm, offering enhanced heat transfer, chemical stability, and volumetric control during phase transitions. A high-concentration (45 wt.%) PCM slurry of microencapsulated Rubitherm RT6 has been developed, achieving a heat transfer coefficient up to five times that of water [186,214]. Microencapsulated PCMs, combined with nanoparticle additives, significantly enhance thermal conductivity in storage systems. Additionally, absorption and sorption technologies have been introduced for negative temperature applications. Ma et al. [215] developed a model to study heat transfer in PCM slurries, finding that TBAB CHS showed temperature variations but incomplete melting in microencapsulated PCM. PS/n-heptadecane micro/nano encapsulated PCM, synthesized via miniemulsion polymerization, exhibited good thermal properties and reliability [216]. Encapsulated PCM with tetradecane and donkey-hide gelatin measured distinct melting/freezing points and latent heats [217], while PMMA/docosane microcapsules displayed compact surfaces and high thermal reliability [218]. Polystyrene-shelled PCMs were also explored, with PMMA/octadocosane identified as the most effective PCM [219]. In the context of residential nanogrids, the improved heat transfer rates and thermal conductivity of microencapsulated PCMs, augmented by nanoparticle additives, can substantially enhance energy efficiency and storage capabilities. The application of such advanced materials in residential nanogrids could lead to more efficient thermal management, thereby contributing to the reduction of energy consumption. The characteristics of microencapsulated PCMs are given in Table 4 [220]. Encapsulation techniques at macro, micro, and nanoscales were discussed in [221] focusing on their application in thermal energy storage systems. It highlights the potential of nano-encapsulated PCMs (nano-PCMs) to enhance thermal conductivity and efficiency. Various methods, such as pan coating, micro-fluidic techniques, and interfacial polymerization, are discussed, emphasizing their role in improving the thermal performance of PCMs for nanogrid applications.

3.3.7. Shape-Stabilized PCMs (Structure)

Shape-stabilized PCMs tend to maintain their structural integrity by being embedded in polymer matrices during phase changes, minimizing leakage. Advancements in stabilized, shape-stabilized PCMs significantly impact residential nanogrids by enhancing energy storage and efficiency. Integrating high-capacity, thermally stable composites like paraffin/HDPE and bio-based PCMs with boron nitride addresses issues such as leakage and low thermal conductivity. Additionally, materials like expanded graphite and aluminum nitride improve heat transfer, optimizing overall performance. Encapsulation is expensive; an alternative to enhance thermal conductivity is to use stabilized and shape-stabilized PCMs. Polymer matrix prevents leakage of molten paraffin, resolving issues related to encapsulation [220,222–231]. Trigui et al. [232] conducted a study on PCM composites made of low-density polyethylene and paraffin wax, identifying poor heat transfer capabilities and shape stabilization as obstacles to widespread PCM application. Another study prepared bio-based PCMs with boron nitride using vacuum impregnation, highlighting their high latent heat and thermal stability but noting issues with low thermal conductivity and leakage, which boron nitride addressed effectively [233]. Polyethylene was identified as a supportive material for shape-stabilized PCMs due to its compatibility with paraffins, and the melt mixing method achieved well-dispersed paraffin/HDPE composites [234]. Further studies showed that blending PCMs with PMMA and other polymers could support up to 80 wt.% PCM without leakage [235], and adding aluminum nitride improved heat transfer capability [236]. Other findings included the successful stabilization of paraffin/expanded graphite composites through capillary forces without leakage [237], and thermal characterizations revealed the optimal mass fractions for maximizing latent heat [238]. The long-term stability of Rubitherm 21 (RT21) and a propyl stearate–palmitate ester mixture was assessed by examining changes in their thermophysical properties under sustained heating above their melting points. RT21 exhibited irreversible changes in weight and latent heat of fusion, while the ester mixture remained stable without mass loss or alteration in thermal properties [239]. These results suggest that the stable ester-based PCM is a viable option for enhancing the efficiency and durability of TESS in residential nanogrids.

3.3.8. Nitrides, Additives and Nanofluids (Additives)

A study [240] explored how sodium nitride, potassium nitride, and their combinations with expanded graphite could act as additives to boost thermal conductivities, uncovering potential opportunities for residential nanogrids. At 25 °C, the thermal conductivities of pure sodium nitride and potassium nitride were measured at 0.224 W/mK and 0.156 W/mK, respectively. The addition of graphite significantly improved these conductivities, with mixtures containing 10% graphite showing a 30–40% increase and those with 5% graphite showing a 10–20% increase. Additionally, graphite altered the phase change temperatures and reduced the freezing/melting points of the mixtures by 5–18% compared with pure nitrides. These findings suggest that integrating graphite-enhanced nitrides could improve the efficiency of solar energy storage in residential nanogrids. An experimental study claimed that embedding metal foam in phase change materials (PCM) can boost heat transfer performance by nearly ten times compared with pure PCM. This enhancement is attributed to the metal's superior thermal conductivity merging with the PCM. Moreover, smaller porosity and pore size within the metal foam outperformed larger ones, further elevating heat transfer efficiency [241]. When it comes to residential nanogrids, boosting the thermal conductivity of phase change materials (PCMs) is essential for optimal energy management. A study explored the incorporation of carbon fibers to enhance the thermal conductivity of fiber/paraffin composites. Two different setups were examined: fibers dispersed randomly and a suspended carbon fiber brush. The random arrangement and fiber length showed minimal effect, but the fiber brush markedly improved thermal conductivity [242]. Similarly, paraffin/copper nanofluid composites were examined, revealing that the integration of copper foam substantially boosted both heat transfer rate and thermal conductivity [243]. Another investigation focused on paraffin/graphite composites, which exhibited a 20–30% higher thermal conductivity compared with pure paraffin [244]. Furthermore, tests on paraffin integrated with SiO_2 and expanded graphite, as well as paraffin combined solely with SiO_2, showcased impressive enhancements in thermal conductivity—registering improvements of 94.7% and 28.2%, respectively, compared with pure paraffin [245]. Adding silica-alumina nanoparticles into $NaNO_3$-KNO_3 binary salt is claimed [246] to enhance thermal storage properties. Incorporating 1.0 wt.% of these nanoparticles into the nanofluid enhances its specific heat capacity, boosting it by a remarkable 15% to 57% when in the solid phase and by 1% to 22% in the liquid phase. These improvements could significantly benefit the integration of efficient PCM-TES systems in nanogrids. In a comparative study, the charging times for copper, polyethylene, and polyvinyl chloride (PVC) capsules were found to be 7 h, 8 h, 9 h, and 20 min, respectively. The minor differences in charging times suggest that polyethylene capsules are a cost-effective option for constructing TES storage tanks [247].

3.4. *Photochromic and Thermochromic (Materials)*

Photochromic and thermochromic materials are gaining traction as phase change materials (PCMs), which have the ability to transform their properties in response to changes in light and temperature, respectively. Photochromic materials change their colors when exposed to light (UV). These materials undergo a reversible molecular transformation when exposed to certain wavelengths of light, with a change in their absorption properties and colors. Thermochromic materials change their color in response to temperature changes. These materials can change color at specific temperature thresholds due to physical or chemical changes in their molecular structure. An ideal photochromic or thermochromic material must switch from black color at low temperature to white color at high temperature. Thermochromic material Bis(dimethylammonium) tetrachloronickelate(II) changes color from green (25 °C) to yellow (50 °C) as photochromic glasses change color from transparent (25 °C) to dark (50 °C) in sunshine. Nature is replete with color-changing reptiles (Chameleon), insects (Chrysso), and birds (Surakav). Photochromic and thermochromic materials often change color from one (say green) to another (say yellow), like chameleon with a change in temperature instead of ideal black and white coatings as shown in Figure 8.

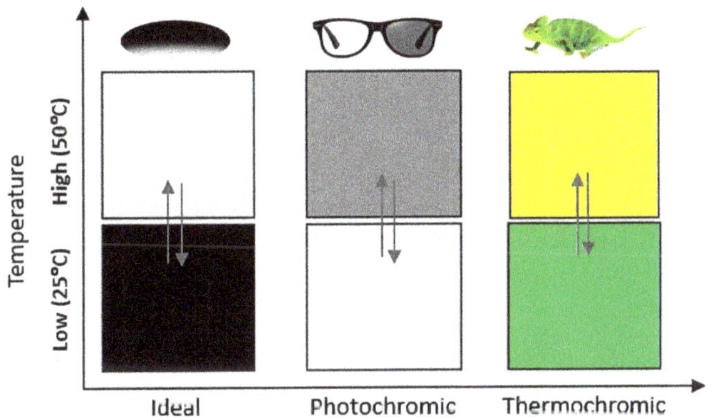

Figure 8. Materials change color with temperature.

Photochromic materials are used for smart glasses, which turn dark in the sun and translucent in the shade. Experimental studies demonstrated that cool coatings can reduce cooling loads by 18–93% and peak cooling demand by 11–27% in air-conditioned buildings [248]. Cool white coatings like titanium oxide and barium sulfate can reflect 98% visible light, whereas sodium citrate can reflect infrared. There are some polymer paints that can reflect UV too. Titanium dioxide-based solar reflective paint cost dollars 8/L compared with dollars 4/L for others. Super therm cool roof paints are good reflectors in the visible range but not very efficient in the IR and UV ranges. Ultra-white paint technologies can replace air conditioners. Measurements show at an ambient temperature of 35 °C the indoor temperature of a white-painted roof building reaches 37 °C compared with 60 °C in an unpainted grey rooftop building, where others claim even less than the ambient temperature inside an ultra-white building. Civil engineers are deploying thermochromic materials for cool surface pavements [249]. Bitumen membranes can achieve higher solar reflectance when coated with white materials, while elastomeric or cementitious cool white coatings typically reflect 0.7 to 0.85 of solar radiation. For thermal energy storage using granular encapsulated PCMs, the PCM serves as the core, encapsulated by a shell made of inorganic metal oxides like zirconium dioxide (ZrO_2). The temperature discoloration function is used as an indicator [250,251]. Cool tiles in residential buildings offer high solar reflectance and may achieve energy savings of up to 15% in hot, dry climates [248].

3.4.1. Photochromic (Materials)

Photochromism refers to a reversible change of color upon exposure to light. This color change is mediated by photoisomerization of chromic material mixed in silicon lenses or coated on plastic lenses. This change is not long-lasting and reverses with an increase in temperature. The photochromic compounds undergo reversible photochemical reactions altering absorption bands in the visible spectrum. Exposure of silver halide to UV breaks the ionic bond, and silver atoms create a blackish shade due to absorption of light. However, the temperature rise may reverse this opaque state back to a transparent condition. We may engineer the thermochromic properties by adding copper atoms. All photochromic molecules, like tenebrescence, back-isomerize at certain rates, which are accelerated by heat. Any material like nitrospiropyran that takes ten minutes to back-isomerize at room temperature is considered photochromic material. Thermally stable photochromic compounds like diarylethenes do not back-isomerize for 90 days even after heating to 80 °C. There is a strong relationship between photochromic and thermochromic compounds. The textile industry uses photoreactive compounds, which undergo color change after exposure to light. Various classes of photochromic materials are triarylmethanes, stilbenes, azastilbenes, nitrones, fulgides, spiropyrans, naphthopyrans, spirooxazines, quinones, etc., as shown in Figure 9.

Smart photochromic compounds and meta surface or semiconductor phase change materials respond to UV, IR, light, heat, stress, humidity, and electromagnetic fields [252,253]. Ge, Sb, and Te (GST) exhibit large change in n and change in k changes, whereas Ge2Sb2Se4Te1 has a broadband response from NIR to FIR parts of the electromagnetic spectrum [254]. Photochromic technology is used in the sunglasses, data storage disks, cosmetics, and clothes industries. Photoisomerization may be used to store solar energy in dihydroazulene (DHA)/ vinylheptafulvene (VHF) photo/thermo-switch material [255,256]. The IEA technology roadmap expediting energy transition recommends using solar thermal systems where photo and thermochromic materials can play a role [257]. Rampant rise in rooftop photovoltaic and solar thermal panels impacts the urban heat island (UHI).

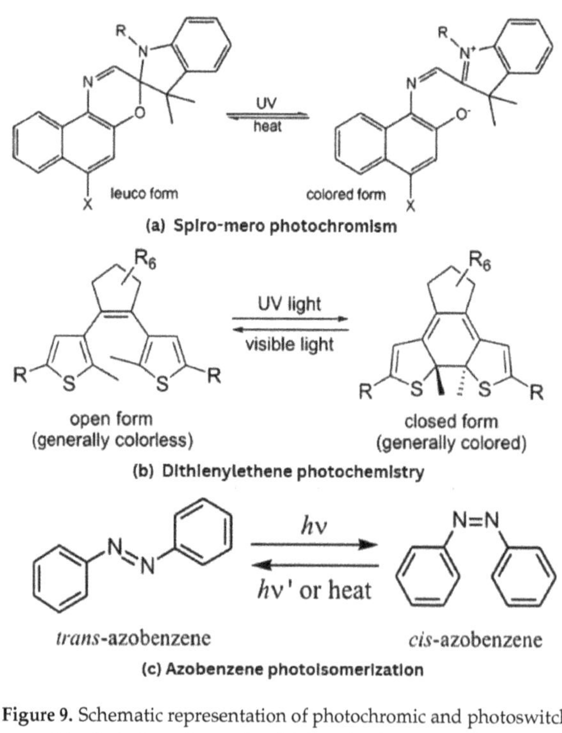

Figure 9. Schematic representation of photochromic and photoswitchable molecular systems. (**a**) A reversible photochromic reaction (**b**) A molecular system that reversibly switches between colorless and colored forms under UV and visible light. (c) The reversible photoisomerization of trans-azobenzene to cis-azobenzene under UV light.

3.4.2. Thermochromic (Materials)

Thermochromism refers to a change in color caused by a change in the substance's temperature. Thermochromism is found in both organic and organic compounds. Thermochromic liquid crystals with a limited color range and leuco dyes with widespread color and temperature ranges are commonly used for precision and rough estimates. The color of liquid crystals depends upon the crystalline structure of the material, which changes from crystal (cool), nematic (warm), and isotropic (hot) phases. However, liquid crystals exhibit thermochromism in the nematic phase, wherein light undergoes Bragg diffraction. Thermochromic material colors switch from black to visible colors. Low temperature results in red–orange, and high temperature results in blue–violet colors. Low (3–5 °C), medium (17–23 °C), and high (37–40 °C) temperatures may be obtained by mixing cholesteryl oleyl carbonate, Cholesteryl nonanoate, and Cholesteryl benzoate in various proportions. A mood ring or baby feeder changes colors depending on temperature. Liquid crystal phase change material is used in thermometers, refrigerators, and aquariums. Thermochromic dyes (leucos) are used with other chemicals to change from no colour to some color; litmus papers are pH indicators that correspond to various colors. Commonly used dyes include spirolactones, flurans, spiropyrans, and fulgides, which with bisphrenol a, parabens, 1,2,3-triazole, and 4-hydroxycoumarin exhibit leuco switching from no to some color states. Liquid crystals have better temperature response than leuco dyes, which are used in thermochromic papers, thermoplastics, polymers, and inks. Most inorganic materials exhibit to some extent thermochromism. Titanium oxide, zinc sulfide, and zinc oxides look white at room temperature and yellow at high temperature. Indium oxide switches color from yellow to yellow–brown on heating. This phenomenon is dramatic in phase transition materials. Cuprous mercury iodide undergoes a phase transition at 67 °C, switching from bright red to dark brown [258]. Silver mercury iodide switches from yellow at room temperature to orange at 47–51 °C. Bis(dimethylammonium) tetrachloronickelate switches from red at low temperature to blue at 110 °C but returns back to red in one week [259]. Chromium and aluminum oxides in a 1:9 ratio are red at room temperature and gray at 400 °C [260]. Chromio–Chromium richsare reddish purplish at low temperatures, which become green at 80 °C. Vanadium dioxide blocks IR from the rooftop to cool the building [261]. DR is the main cause of overheating in summer, so paint windows with vanadium dioxide to keep the heat out. However, copper iodide and manganese violet are irreversible thermochromic materials [262]. Thermochromic materials are applied in the building to keep the heat out to reduce energy consumption on air conditioning by up to 17% in half a million big cities facing UHI problems [263,264]. Buildings consume 30–40% of global energy consumption. Global building energy consumption was 23.7 PWh in 2010, which will increase to 38.4 PWH by 2040 at a growth rate of 490 TWh per year. The average increase of the cooling demand was 23% from 1970 to 2010, while the average reduction of the heating was 19% in the same period, though both increased by 11% in four decades, which is attributed to global warming and the urban heat island effect [265].

4. Proposed Nanogrid Integrated Architecture

4.1. Overview of Studies

The review highlights the importance of PCM technology in improving the techno-economic indicators of nanogrid buildings. Advanced PCM systems, particularly those with enhanced thermal properties and integrated designs, can significantly contribute to energy efficiency and sustainability. The findings from experimental studies on phase change material (PCM)-based thermal energy storage (TES) systems in residential nanogrids reveal several significant insights. The reduction in charging time is notably achieved through the utilization of metallic capsules, smaller capsule sizes, lower Heat Transfer Fluid (HTF) temperatures, and higher HTF flow rates. Conversely, an extension in discharge time is observed with non-metallic capsules, larger capsule sizes, lower HTF temperatures, and lower HTF flow rates. The energy storage capacity benefits from the use of metallic capsules, larger capsule sizes, increased HTF flow rates, and lower HTF temperatures.

Metallic capsules, larger capsule sizes, and lower HTF flow rates also contribute to a superior energy recovery ratio. Despite their lower energy recovery ratios, non-metallic capsules offer a more cost-effective and easily manufacturable alternative. Enhancing the solid mass fraction is achieved with metallic capsules, smaller capsule sizes, higher HTF flow rates, and lower HTF temperatures. Conversely, increasing the melted mass fraction and recovering stored energy is facilitated by metallic capsules, smaller capsule sizes, higher HTF flow rates, and higher HTF temperatures. Interestingly, the thermal conductivity of metallic capsules has a minimal effect on overall charging and discharging times, solid and melted mass fractions, and the energy recovery ratio, suggesting that other factors are more critical in enhancing system performance. When comparing capsule types, metallic capsules demonstrate superior charging characteristics and shorter charging times, whereas non-metallic capsules, being economical and easy to manufacture, offer shorter charging/discharging times with lower energy recovery ratios, making them suitable for specific thermal energy storage applications. Furthermore, innovative heat transfer enhancement techniques, including double-pipe and shell-and-tube designs, have been recognized as highly effective solutions. Integrating PCMs into medium-scale air conditioning systems is advised to lower costs and minimize the size of energy storage systems. These conclusions suggest various strategies and configurations to optimize PCM-based TES systems, balancing cost, efficiency, and application suitability.

4.2. Nanogrid (Architecture)

A proposed nanogrid architecture is introduced based on the information analyzed in this review. This architecture leverages PCM technology to enhance energy efficiency and thermal storage within residential buildings. Polygeneration and multi-energy hubs are gaining attention in the decentralization of smart grids. Concepts of combined cooling, heating, and power (CCHP), combined heating and power (CHP), and trigeneration concepts have been employed in the industries at commercial levels, but only recently has the focus shifted towards residential buildings. Passive house concepts have also been introduced as the main architecture for nanogrids. In the proposed network, each nanogrid house acts as a building block to form a holistic microgrid with energy, data, and resource sharing hubs of multiple nanogrids, as illustrated in the AI-generated community of nanogrids on the left and floor plan on the right (Figure 10).

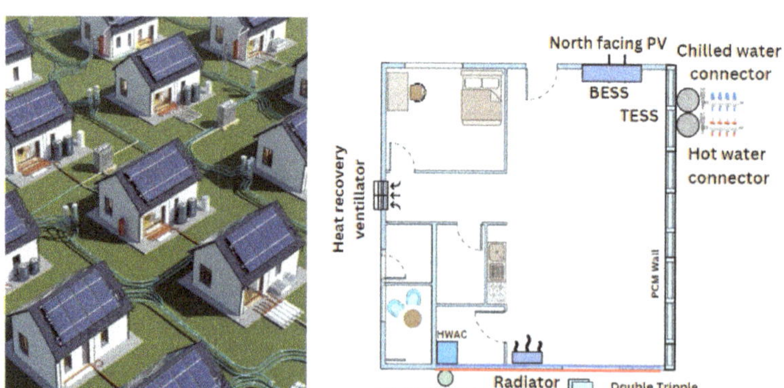

Figure 10. Architecture of network of nanogrids (NoN).

4.3. Nanogrid (Design)

A nanogrid house design should address/include consistent operation over multiple charge/discharge cycles, reliable/efficient thermal energy storage, rapid heating/cooling through any means of hydronic/radiative/convective methods, maximum discharge efficiency, and reliable backup. Reviewing various trends in phase change materials (PCM) technologies for nanogrids, it is evident that each trend offers distinct advantages and poses specific challenges. The selection of PCM should cover heat of fusion, specific heat, thermal conductivity, stability, non-toxicity, inertness, and density. The challenges in integrating PCMs into nanogrids include limited space, higher capital costs, the intermittent nature of solar, and the thermal conductivity of PCM material. The proposed nanogrid features a north-facing (Australia) PCM wall-integrated residential building forming an airtight (0.6 to 2 air changes per hour) chassis encompassing an area of 144–256 m^2. PCM with inorganic composites is installed in either the attic or wall of the house with a thickness of 20 mm [266]. The nanogrid will aim to produce heating, cooling, and power [267] simultaneously, with each node/nanogrid connected to another through pipes and cables to improve the reliability of microgrid architecture. A nanogrid will be assumed to accommodate a family of 2–4 occupants with an average peak energy demand of 1800 kWh (winters) and 1500 kWh (summers). For residential nanogrids, the optimal parameters for a solar water heating system in a single-family house include a 6 m^2 collector area, a 0.42 m^3 tank capacity, and a 1.2 m tank height. Using HFOs R-1234yz and R-1234yf, which have minimal global warming potential, enhances thermal efficiency. CO_2 (0.720289) and R-410a (0.723428) showed the highest annual solar fraction readings under various weather conditions [268]. The PCM may have a height in the range of 2–4 m, a width of 4–12 m, and a thickness of 0.2–0.535 m. The wall includes one glass cover with a glazing emittance of 0.1, an extinction coefficient of 0.0026, and a refractive index of 1.526. The vent outlet area is 1 m^2, with vents spaced 3.5 m apart. The thermal demands of this nanogrid would be around 60%, while other power demands will be 40%. Nanogrid will be integrated with two insulated thermal energy storages with a surface area of 2 m^2 and a tank capacity of 0.5 m^3. A multilayer envelope of 0.535 m made of wood, foam, and brick, designed for low thermal mass on the outside, insulation in the middle, and high thermal mass inside. The goal is to keep air changes per hour below 1, ideally below 0.6. Nanocomposites offer improved thermal conductivity for PCMs, but the payback is higher. Microencapsulated PCMs offer improved stability and seamless integration into nanogrid buildings, resulting in efficient temperature regulation, but do pose leakage risks. Shape-stabilized PCMs may be useful for hydronic underfloor heating systems. Bio-based PCMs are very marketable due to their sustainable nature and lower environmental impact but incur higher costs. Multi-PCMs offer enhance thermal performance for TES applications but face phase separation issues. High-temperature PCMs are best suited for solar thermal applications. Photochromic/thermochromic PCMs, suitable for adaptive insulation, offer dynamic responses to environmental changes but are constrained by limited availability and higher material costs.

Nanogrid design may be validated by comparing energy flows across system boundaries with internal energy changes as shown in Figure 11. In the nanogrid system, QGains represents total energy inputs, including QCol, energy from the evacuated tube collector, and QAux, auxiliary energy from backup systems. QPump refers to energy for pumping heat transfer fluids, while QWall accounts for heat gained through PCM walls. On the losses side, QLosses measures all energy leakage within the system. IEC (Internal Energy Changes) tracks variations in stored energy, and TLoss indicates thermal losses to the environment. PLoss denotes power losses due to resistance and inefficiencies in conversion or storage, with Qconv capturing heat lost through fluid movement. Qeq represents energy losses from operating nanogrid equipment. Together, these terms quantify both energy inputs and losses within the nanogrid. Consistent energy transfer among components indicates proper timestep and convergence tolerances. An energy imbalance suggests improper model development. An annual mean energy balance of 3% is acceptable.

$$Q_{\text{gains}} = Q_{\text{col}} + Q_{\text{aux}} + Q_{\text{pump}} + Q_{\text{wall}} \tag{1}$$

$$Q_{\text{losses}} = \text{IEC} + T_{\text{loss}} + P_{\text{loss}} + Q_{\text{conv}} + Q_{\text{eq}} \tag{2}$$

$$\text{IEC} + Q_{\text{col}} + Q_{\text{aux}} + Q_{\text{pump}} + Q_{\text{wall}} + Q_{\text{load}} - T_{\text{loss}} - P_{\text{loss}} - Q_{\text{conv}} - Q_{\text{eq}} = 0 \tag{3}$$

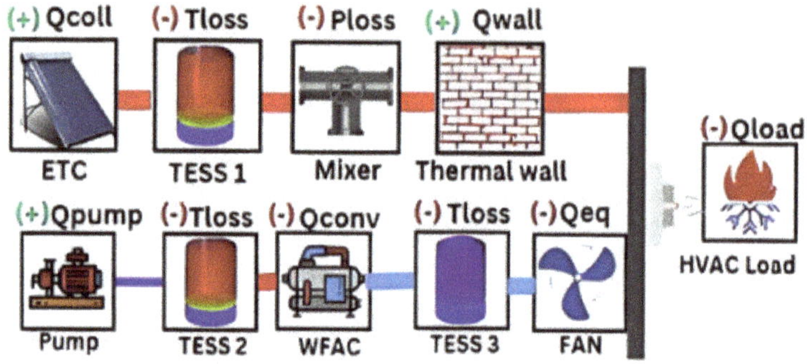

Figure 11. Energy balance boundary conditions (nanogrid).

The payback period for phase change materials typically depends on factors such as installation costs, type of materials, and the specific application in buildings. Conducting a detailed cost analysis considering these factors would be essential in determining the payback period for detailed nanogrid projects. Designing an efficient residential nanogrid requires integrating thermal energy storage (TES) systems and phase change materials (PCMs) to optimize space heating and energy efficiency. The proposed design can incorporate water/gravel-based TES, which uses materials like water, gravel, limestone, and iron scraps, making it ideal for space heating in residential nanogrids. The TES system, with aluminum containers filled with paraffin wax (RT27, RT100), provides a constant thermal output and energy ride-through capability. To enhance storage density, the system includes a packed bed of spherical PCM capsules made from cost-effective polyethylene or metallic materials.

The optimal positioning of PCMs in walls [269] is influenced by several factors, especially the weather conditions that dictate the building's heating/cooling demands. Moreover, the goal of integrating the PCM wall should be clear, such as reducing heating or cooling loads. The melting temperature and heat of fusion of the PCMs determine their effectiveness in different temperature ranges. The quantity of PCMs used affects the total thermal energy storage capacity, which decides the energy ride-through capability of the nanogrid, while the thermal properties of wall materials influence heat transfer rates as per the air conditioning requirement. Integrating PCMs with solar hot water systems is crucial for reducing storage volume, especially in space-constrained residential nanogrids. The PCMs are stored in plastic cylinders insulated with foam and glass containing paraffin wax. Enhanced thermal performance can be achieved by incorporating graphite, silicon carbide, or multi-walled carbon nanotubes (MWCNT) into the PCM. Furthermore, nano-encapsulated PCMs with an n-octadecane core and a styrene/methacrylate copolymer matrix, or microencapsulated Rubitherm RT6, offer improved thermal reliability and heat transfer coefficients.

Various heat transfer fluids (HTFs) are considered to optimize the TES, including air, compressed air (10 bar), supercritical CO_2 (100 bar), steam (10 bar), solar salt, and liquid sodium, with each fluid having specific advantages. Liquid sodium, in particular, offers enhanced responsiveness and reliability, although its economic feasibility for residential applications requires further investigation. The HTFs mass flow rates can be adjusted between 0.05 kg/s and 0.19 kg/s, depending on environmental conditions, to optimize the system's performance. Transpired solar collectors integrated with PCM storage significantly enhance energy efficiency and reduce costs. Double U-tube solar collectors, which offer a 4% higher efficiency with additional capital investment, are suitable for high solar insolation areas, while flat plate collectors and U-pipe evacuated glass tube collectors can be selected based on specific regional solar insolation levels. The walls of the nanogrid incorporate PCM with a paraffin core (melting point: 25–30 °C), featuring additives like graphite and silicon carbide to improve thermal conductivity and stability. Additionally, the structure includes specialized coatings, such as a photochromic cool white coating for overall thermal performance enhancement. Thermochromic vanadium dioxide blocks IR radiation, reducing energy demands and keeping the nanogrid cool. Coatings with titanium oxide and barium sulfate reflect 98% of visible light, sodium citrate reflects infrared light, and specific polymer paints reflect UV light, contributing to indoor thermal comfort. The design emphasizes high thermal conductivity, facilitated by additives like graphite and silicon carbide, which can increase thermal conductivity by up to 94.7% compared with pure paraffin. The size and configuration of heat transfer surfaces are critical in optimizing the TES's efficiency. While carbon nanotubes offer superior performance, their high cost must be justified for residential use. Proper stratification and insulation are essential for maintaining stable inlet/outlet temperatures and efficient energy storage. Overall, this detailed nanogrid design effectively employs advanced materials and technologies to create a sustainable and energy-efficient residential system. Future studies should focus on economic analysis and practical applications to refine and validate the proposed design.

5. Future Works

The integration of phase change materials (PCMs) in nanogrids offers promising benefits, primarily by improving energy efficiency through latent heat storage. Latent heat storage is critical to managing heating and cooling demands in net-zero energy nanogrid buildings. However, challenges persist, including the economic feasibility of niche technologies like liquid sodium, carbon nanotubes, and graphene-based materials for residential applications. While advanced materials like nanocomposites do offer PCM performance, they entail higher costs and complex integration/retrofit in building applications. Current studies frequently lack long-term data, leading to uncertainties about the durability and degradation of PCMs in real-world conditions, and scaling these technologies for widespread residential use presents significant challenges. These gaps highlight the need for comprehensive economic analyses of each individual element discussed in this review, the feasibility of the range of technologies through optimization modeling, and long-term performance studies to evaluate the viability and reliability of PCM systems under varying climatic conditions and through multiple charge/discharge cycles. This review underscores a range of future research opportunities focused on developing PCM-integrated nanogrids. Future studies should utilize software tools such as TRNSYS, EnergyPlus, ASPEN, MATLAB/Simulink, ANSYS/COMSOL, HOMER Energy, Helioscope, Power Factory, and PSSE to conduct detailed analyses of nanogrids connected to main grids.

Future studies could explore the cost-effectiveness of advanced PCM technologies, such as liquid sodium and nanocomposites for residential nanogrids. By evaluating initial investments, operational costs, payback periods, and potential cost savings through tools like TRNSYS and EnergyPlus, researchers can identify economic viability. Further research should assess the long-term stability, durability, and degradation of PCMs under real-world nanogrid conditions to ensure lasting performance. These investigations should compare the economic and energy performance of PCM walls against traditional solar thermal collectors, conduct exergy analyses, and assess the techno-economic performance of various PCM materials integrated into buildings. Additionally, the impact of these integrations on overall energy demand must be evaluated. A comparative analysis of energy efficiency and economic viability between PCM-integrated walls and traditional solar thermal collectors using ANSYS/COMSOL and EnergyPlus could clarify the benefits of each approach. Exergy performance evaluations of different PCM materials for thermal energy storage, facilitated by ASPEN and MATLAB/Simulink, could offer insights into optimal material selection. Additionally, the impact of PCM integration on overall energy demand in residential nanogrids should be studied, with a focus on peak load reduction, energy savings, system efficiency, and energy balance using TRNSYS, Heliscope, and Power Factory. Conducting lifecycle assessments (LCA) to quantify the environmental impacts of PCM use, including global warming potential, embodied energy, and recyclability, using tools like SimaPro, OpenLCA, GaBi, TRNSYS, and EnergyPLAN, could clarify sustainability aspects. For scalability, researchers should investigate large-scale PCM applications in networks of nanogrids, identifying challenges and using TRNSYS, PSSE, Homer, and SketchUp for modeling. An overview of cross-disciplinary approach in PCM integrated nanogrid in summarized in Figure 12.

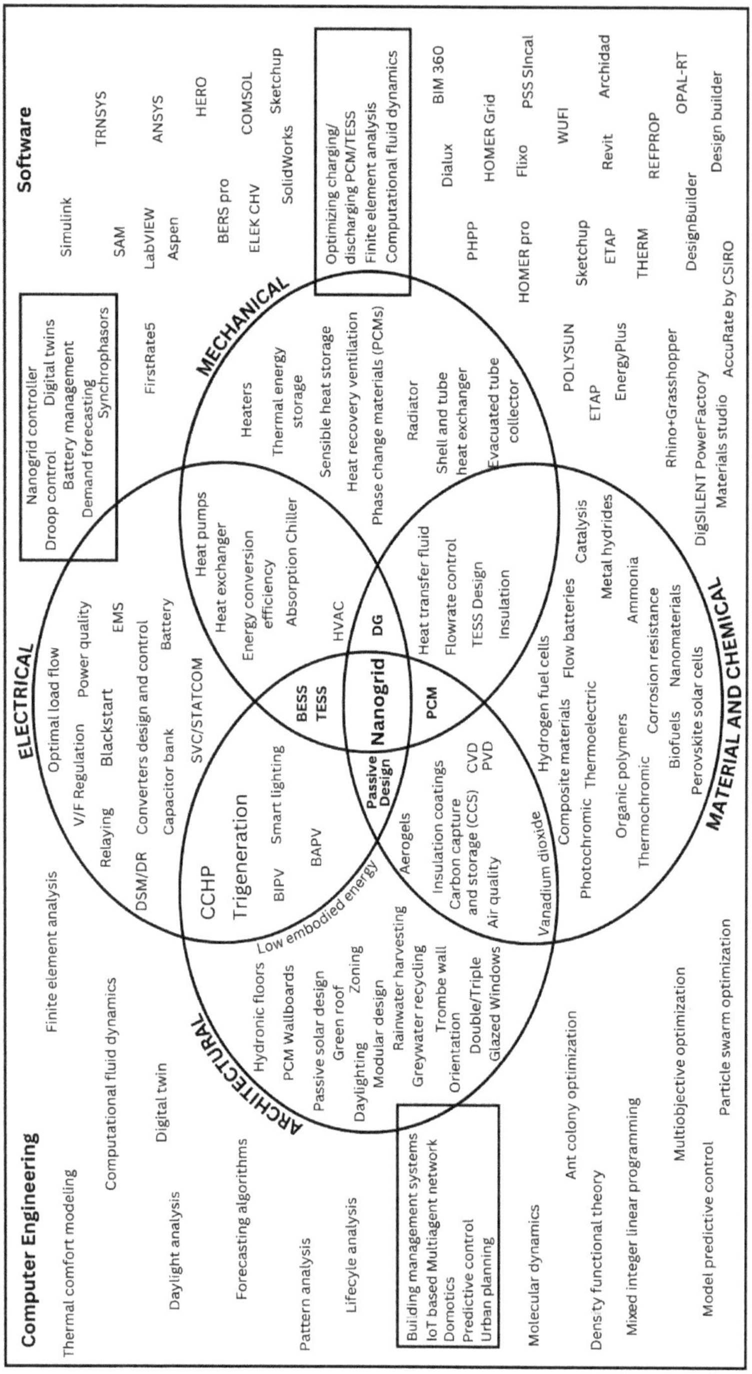

Figure 12. Network of nanogrids technology and discipline overview.

Further studies may compare the thermal performance of PCM-integrated wall systems (e.g., gypsum, concrete, bricks) across various climates via EnergyPlus and TRNSYS, focusing on energy savings and thermal comfort. Optimal PCM placement within walls could be examined using ANSYS/COMSOL to identify configurations that maximize thermal performance and minimize material use. Computational fluid dynamic (CFD) simulations can assess the heat transfer efficiency of hexagonal PCM structures for wall integration, and finite element analysis (FEA) could analyze other geometric PCM structures (e.g., hexagonal, honeycomb, lattice) for their heat storage and transfer capabilities. The stability and dispersion of nanofluids within PCMs should be studied using particle size analysis and thermal cycling tests, evaluating the effectiveness of nanofluid-PCM combinations. Integrating PCMs with advanced insulation materials like aerogels and vacuum insulated panels could further improve thermal management. Research into the effect of phase transition temperature tuning on PCM wall applications using ASPEN and EnergyPlus would also be valuable. Future smart wall systems integrating PCMs with sensors and control mechanisms for real-time thermal management, using IoT multi-agent systems in a nanogrid or microgrid, are worth investigating. Graphene incorporation in PCMs for enhanced heat transfer could be studied using COMSOL, and an in-depth analysis of nanogrids linked to main grids using TRNSYS, EnergyPlus, and MATLAB/Simulink could reveal performance optimization strategies.

Moreover simulations of PCMs with varied metallic and non-metallic capsules can be studies for thermmal performance and energy recovery, weighing cost and efficiency trade-offs. Studies on nanofluid types and concentrations for residential buildings could summarize the comparative performance. Additionally, hybrid PCM walls integrated with heat pumps, solar thermal collectors, and hydrogen storage may offer versatile thermal solutions. An analysis of PCM layer thickness and composition for walls using FEA in ANSYS would help determine optimal thickness for economic and energetic efficiency. Research into photochromic and thermochromic materials for PCM or building integration, focusing on color change efficiency and thermal durability, may offer energy-saving innovations. Integrating these materials in smart windows within passive houses or nanogrids, with an emphasis on exergy analysis and payback, would enhance thermal regulation. Finally, hybrid coatings with photochromic and thermochromic properties, along with long-term thermal cycling tests for PCMs in nanogrids, could reveal sustainable, efficient approaches to residential energy management. Interdisciplinary research teams, including electrical, mechanical, materials, energy, and architectural engineering experts, are crucial for addressing the complexities of PCM-integrated nanogrids. Collaborative efforts will ensure these systems are both economically viable and technically robust, advancing sustainable energy solutions for residential use.

6. Conclusions

The integration of phase change materials (PCMs) in nanogrids offers significant benefits, primarily enhancing energy efficiency through latent heat storage methods, which is vital for managing heating and cooling demands in net-zero energy buildings. Additionally, additives like carbon fibers and copper foam notably improve the thermal conductivity of PCMs, facilitating better energy management in residential nanogrids. However, challenges remain, such as the economic feasibility of advanced technologies like liquid sodium for residential applications and the stability and leakage risks associated with microencapsulated PCMs. Different viewpoints highlight the potential of advanced materials like nanocomposites to boost PCM performance, though these innovations often come with higher costs and complex manufacturing processes. Environmental benefits, such as the use of low global warming potential refrigerants in solar water heating systems, are promising, yet the overall lifecycle impact of PCM materials requires further scrutiny. Existing studies often lack long-term data, leading to uncertainties about the durability and degradation of PCMs in real-world conditions, and scaling these technologies for broad residential use presents significant challenges. Identified gaps include the need for comprehensive economic analyses and long-term performance studies to evaluate PCM systems' viability and reliability under varying climatic conditions and through multiple charge/discharge cycles. By critically analyzing these aspects, this review provides a balanced perspective on PCM integration in nanogrids and underscores areas for future research.

The proposed concept is a holistic approach in designing a nanogrid. Each nanogrid house is designed to operate independently yet contribute to a shared microgrid, allowing water, electricity, data, and resource sharing. These nanogrids, when networked, form a "Network of Nanogrids" (NoN) with improved energy reliability. Each unit supports a family of 2–4, with tailored energy demands for seasonal variations (1800 kWh in winter, 1500 kWh in summer). Energy demands are divided into two main categories 60% thermal demands and 40% electrical demands. Each nanogrid includes a north-facing PCM wall (for optimal solar exposure in Australia), aiming for an airtight envelope with 0.6 to 2 air changes per hour and covering an area between 144–256 m^2. PCM is incorporated into the wall or attic (20 mm thick), enhancing the thermal envelope. A multi-layered wall structure (wood, foam and brick) aims to balance thermal mass, insulation, and low thermal mass for energy retention and efficiency. Each nanogrid is equipped with a range of technologies PV (5–10 kW), Battery (5–15 kWh), Tripple glazed windows (U = 0.4, g = 0.408, glass 2500 mm), hot water fired absorption chillers 2–5 kW with COP 0.8. TESS tanks are aimed to be in range from 200–500 L per insulated tank. The solar hot water system includes a 3–6 m^2 solar collector and TESS with 1.2 m height. The system uses eco-friendly refrigerants (HFOs R-1234yz and R-1234yf) with minimal global warming potential. Different heat transfer fluids (HTFs) enhance TES, including options like compressed air, supercritical CO_2, and solar salt. For PCM storage, transpired and double U-tube solar collectors are recommended for efficient energy capture. The walls integrate PCM with paraffin, enriched with additives like graphite for enhanced conductivity. A variety of coatings (photochromic cool white, thermochromic vanadium dioxide) regulate heat absorption, with reflective layers to maintain thermal comfort. Coatings with titanium oxide and barium sulphate reflect visible light, while sodium citrate and polymer paints address infrared and UV light reflection, respectively.

This review paper paves the way for a multitude of future studies focused on developing phase change material (PCM) integrated nanogrids. Interdisciplinary research teams, including electrical, mechanical, materials, energy, and architectural engineering experts, are essential to address the complexities of PCM-integrated nanogrids. Collaborative efforts will ensure these systems are both economically viable and technically robust, advancing sustainable energy solutions for residential use. As we look towards "Building the future", the focus must shift from merely innovating new PCM technologies to establishing international standards that ensure their effective and widespread adoption.

Author Contributions: A.K.: Conceptualization, Original draft preparation, Formal analysis, Data curation, E.J.: Conceptualization, Review, Supervise, N.A.: Data, Supervise, M.S.: Review, Supervise, S.M.: Review, Supervise. All authors have read and agreed to the published version of the manuscript.

Funding: This research received no external funding.

Data Availability Statement: Data available in a publicly accessible repository.

Conflicts of Interest: The authors declare no conflicts of interest.

Abbreviations

The following abbreviations are used in this manuscript:

CHS	Clathrate Hydrate Slurry
CNF	Carbon Nanofibers
CO_2	Carbon Dioxide
COP	Coefficient of Performance
DHA	Dihydroazulene
DHW	Domestic Hot Water

DR	Demand Response
DSC	Differential Scanning Calorimetry
FIR	Far Infrared
GSHP	Ground Source Heat Pump
GNPs	Graphene Nanofillers
GST	Germanium-Antimony-Telluride
h-BN	Hexagonal Boron Nitride
HDPE	High-Density Polyethylene
HTF	Heat Transfer Fluid
HVAC	Heating, Ventilation, and Air Conditioning
IEA	International Energy Agency
IR	Infrared
LHS	Latent Heat Storage
L-MWCNTs	Long Multi-Walled Carbon Nanotubes
MWCNT	Multi-Walled Carbon Nanotubes
MWCNTs	Multi-Walled Carbon Nanotubes
$NaNO_3$	Sodium Nitrate
Ne-PCMs	Nano-enhanced Phase Change Materials
NIR	Near Infrared
ORC	Organic Rankine Cycle
PC-PCM	Photochromic Phase Change Material
PCMs	Phase Change Materials
PMMA	Poly(methyl methacrylate)
PVC	Polyvinyl Chloride
RT6	Rubitherm Type 6
SAH	Solar Air Heater
S-MWCNTs	Short Multi-Walled Carbon Nanotubes
TBAB	Tetra-n-butylammonium Bromide
TCO	Transparent Conductive Oxide
TC-PCM	Thermochromic Phase Change Material
TES	Thermal Energy Storage
TESS	Thermal Energy Storage Systems
UHI	Urban Heat Island
UV	Ultraviolet
VHF	Vinylheptafulvene
W/mK	Watts per meter Kelvin
wt.%	Weight Percent
ZrO_2	Zirconium Dioxide

References

1. Kalair, A.R.; Seyedmahmoudian, M.; Stojcevski, A.; Mekhilef, S.; Abas, N.; Koh, K. Thermal comfort analysis of a trombe wall integrated multi-energy nanogrid building. *J. Build. Eng.* **2023**, *78*, 107623. [CrossRef]
2. Reddy, V.J.; Ghazali, M.F.; Kumarasamy, S. Advancements in phase change materials for energy-efficient building construction: A comprehensive review. *J. Energy Storage* **2024**, *81*, 110494. [CrossRef]
3. Ahmad, A.; Memon, S.A. A novel method to evaluate phase change materials' impact on buildings' energy, economic, and environmental performance via controlled natural ventilation. *Appl. Energy* **2023**, *353*, 122033. [CrossRef]
4. Jacob, J.; Paul, J.; Selvaraj, J.; Vaka, M. Phase change materials integrated buildings: A short review. In Proceedings of the IOP Conference Series: Earth and Environmental Science, Penang, Malaysia, 7–8 June 2023. [CrossRef]
5. Aghoei, M.M.; Astanbous, A.; Khaksar, R.Y.; Moezzi, R.; Behzadian, K.; Annuk, A.; Gheibi, M. Phase Change Materials (PCM) As A Passive System in The Opaque Building Envelope: A Simulation-Based Analysis. *J. Energy Storage* **2023**, *101*, 113625. [CrossRef]
6. Jahangir, M.H.; Alimohamadi, R. A comparative evaluation on energy consumption of a building using bio-based and paraffin-based phase change materials integrated to external building envelope. *Energy Rep.* **2024**, *11*, 3914–3930. [CrossRef]
7. Hrncic, B.; Vusanovic, I. Analyzing the impact of using phase change materials on energy consumption in buildings: A case study. *J. Phys. Conf. Ser.* **2023**, *2766*, 012226. [CrossRef]
8. Selvasofia, S.D.A.; Shri, S.D.; Kumar, P.M.; B, P.K. Characterization of bio-nano doped phase change material (bio-nano/PCM) for building energy management. *Energy Sources Part A Recovery Util. Environ. Eff.* **2024**, *46*, 3760–3778. [CrossRef]

9. Zhang, L.; Zhang, Q.; Jin, L.; Cui, X.; Yang, X. Energy, economic and environmental (3E) analysis of residential building walls enhanced with phase change materials. *J. Build. Eng.* **2024**, *84*, 108503. [CrossRef]
10. Rashid, F.L.; Dulaimi, A.; Hatem, W.A.; Al-Obaidi, M.A.; Ameen, A.; Eleiwi, M.A.; Jawad, S.A.; Bernardo, L.F.A.; Hu, J.W. Recent Advances and Developments in Phase Change Materials in High-Temperature Building Envelopes: A Review of Solutions and Challenges. *Buildings* **2024**, *14*, 1582. [CrossRef]
11. Asghari, M.; Fereidoni, S.; Fereidooni, L.; Nabisi, M.; Kasaeian, A. Energy efficiency analysis of applying phase change materials and thermal insulation layers in a building. *Energy Build.* **2024**, *312*, 114211. [CrossRef]
12. Hawila, A.; Alsalloum, H.;Merabtine, A.; Fardoun, F. Plus Energy Buildings: A Numerical Case Study. *Fluid Dyn. Mater. Process.* **2023**, *19*, 117–134. [CrossRef]
13. Healey, T.T.; Andresen, I.; Gaitani, N. Energy performance and scenario analyses of a multistorey apartment building in Norway. *E3S Web Conf.* **2022**, *362*, 10004. [CrossRef]
14. Passer, A.; Ouellet-Plamondon, C.; Kenneally, P.; John, V.; Habert, G. The impact of future scenarios on building refurbishment strategies towards plus energy buildings. *Energy Build.* **2016**, *124*, 153–163. [CrossRef]
15. Graillet, O.; Genon-Catalot, D.; Lucas de Peslouan, P.-O.; Bernard, F.; Alicalapa, F.; Lemaitre, L.; Chabriat, J.-P. Optimizing Energy Consumption: A Case Study of LVDC Nanogrid Implementation in Tertiary Buildings on La Réunion Island. *Energies* **2024**, *17*, 1247. [CrossRef]
16. Abbas, A.H.R.; Rajeswari, M.; Sharma, D.; Singh, R.; Jeyakani, P.; Dhabliya, D. Optimization of Nanogrids for Remote Off-Grid Communities. *E3s Web Conf.* **2024**, *540*, 1014. [CrossRef]
17. Santoro, D.; Delmonte, N.; Simonazzi, M.; Toscani, A.; Rocchi, N.; Sozzi, G.; Cova, P.; Menozzi, R. Local Power Distribution—A Review of Nanogrid Architectures, Control Strategies, and Converters. *Sustainability* **2023**, *15*, 2759. [CrossRef]
18. Jacob, R.; Hoffmann, M.; Weinand, J.M.; Linßen, J.; Stolten, D.; Müller, M. The future role of thermal energy storage in 100% renewable electricity systems. *Renew. Sustain. Energy Transit.* **2023**, *4*, 100059. [CrossRef]
19. Bitar, M.; El Tawil, T.; Benbouzid, M.; Dinh, V.B.; Benaouicha, M. On Hybrid Nanogrids Energy Management Systems—An Insight into Embedded Systems. *Appl. Sci.* **2024**, *14*, 1563. [CrossRef]
20. Rakib, M.W.; Munna, A.H.; Farooq, T.; Boker, A.; He, M. Enhancing Grid Stability and Sustainability: Energy-Storage-Based Hybrid Systems for Seamless Renewable Integration. *Eur. J. Electr. Eng. Comput. Sci.* **2024**, *8*, 1–8. [CrossRef]
21. Alemam, A.; Ferber, N.L.; Malm, T.; Eveloy, V.; Calvet, N. Experimental Demonstration of a Solar Powered High Temperature Latent Heat Storage Prototype: Preliminary Results of an Experimental Validation Campaign. In Proceedings of the SolarPACES 2023, 29th International Conference on Concentrating Solar Power, Thermal, and Chemical Energy Systems, Sydney, Australia, 10–13 October 2023; TIB Open Publishing: Hannover, Germany, 2024; Volume 2. [CrossRef]
22. Assis, F.A.; Coelho, F.C.; Castro, J.F.C.; Donadon, A.R.; Roncolatto, R.A.; Rosas, P.A.; Andrade, V.E.; Bento, R.G.; Silva, L.C.; Cypriano, J.G.; et al. Assessment of Regulatory and Market Challenges in the Economic Feasibility of a Nanogrid: A Brazilian Case. *Energies* **2024**, *17*, 341. [CrossRef]
23. Mehling, H.; Cabeza, L. *Heat and Mass Storage with PCM: An Up-to-Date Introduction into Basics and Applications*; Springer: Berlin/Heidelberg, Germany, 2008.
24. Telkes, M.; Raymond, E. Storing solar heat in chemicals—A report on the Dover house. *Heat. Vent.* **1949**, *46*, 80–86.
25. Abhat, A. Low temperature latent heat thermal energy storage: Heat storage materials. *Sol. Energy* **1983**, *30*, 313–332. [CrossRef]
26. Henze, R.H.; Humphrey, J.A.C. Enhanced heat conduction in phase-change thermal energy storage devices. *Int. J. Heat Mass Transf.* **1981**, *24*, 459–474. [CrossRef]
27. Abhat, A. Short term thermal energy storage. *Energy Build.* **1981**, *3*, 49–76. [CrossRef]
28. Bareiss, M.; Beer, H. An analytical solution of the heat transfer process during melting of an unfixed solid phase change material inside a horizontal tube. *Int. J. Heat Mass Transf.* **1984**, *27*, 739–746. [CrossRef]
29. Bushnell, D.L. Performance studies of a solar energy storing heat exchanger. *Sol. Energy* **1988**, *41*, 503–512. [CrossRef]
30. Prakash, J.; Garg, H.P.; Datta, G. A solar water heater with a built-in latent heat storage. *Energy Convers. Manag.* **1985**, *25*, 51–56. [CrossRef]
31. Charach, C. Solidification (Melting) in Finite Extent Pcm Bodies. In *Intersol Eighty Five*; Bilgen, E., Hollands, K.G.T., Eds.; Pergamon: Oxford, UK, 1986; pp. 786–790, ISBN 9780080331775.
32. Sokolov, M.; Keizman, Y. Phase Change Storage in Solar Pipes. In *Advances In Solar Energy Technology*; Bloss, W.H., Pfisterer, F., Eds.; Pergamon: Oxford, UK, 1988; pp. 1217–1221, ISBN 9780080343150.
33. Theunissen, P.-H.; Buchlin, J.-M. Numerical optimization of a solar air heating system based on encapsulated P.C.M. storage. *Sol. Energy* **1983**, *31*, 271–277. [CrossRef]
34. Hasnain, S.M.; Gibbs, B.M. Low Temperature Thermal Energy Storage for Solar Heating Applications Preliminary Investigations. In *Advances In Solar Energy Technology*; Bloss, W.H., Pfisterer, F., Eds.; Pergamon: Oxford, UK, 1988; pp. 1197–1201, ISBN 9780080343150.
35. Brandstetter, A. Phase Change Storage for Greenhouses. In *Advances In Solar Energy Technology*; Bloss, W.H., Pfisterer, F., Eds.; Pergamon: Oxford, UK, 1988; pp. 3353–3357, ISBN 9780080343150.
36. Aboul-Enein, S.; Ramadam, M.R.I. Storage of low temperature heat in salt-hydrate melts for heating applications. *Sol. Wind. Technol.* **1988**, *5*, 441–444. [CrossRef]

37. Parameshwaran, R.; Harikrishnan, S.; Kalaiselvam, S. Energy efficient PCM-based variable air volume air conditioning system for modern buildings. *Energy Build.* **2010**, *42*, 1353–1360. [CrossRef]
38. Gowreesunker, B.L.; Tassou, S.A.; Kolokotroni, M. Coupled TRNSYS-CFD simulations evaluating the performance of PCM plate heat exchangers in an airport terminal building displacement conditioning system. *Build. Environ.* **2013**, *65*, 132–145. [CrossRef]
39. Al-Abidi, A.A.; Mat, S.; Sopian, K.; Sulaiman, M.Y.; Mohammad, A.T. Experimental study of PCM melting in triplex tube thermal energy storage for liquid desiccant air conditioning system. *Energy Build.* **2013**, *60*, 270–279. [CrossRef]
40. Chou, H.-M.; Chen, C.-R.; Nguyen, V.-L. A new design of metal-sheet cool roof using PCM. *Energy Build.* **2013**, *57*, 42–50. [CrossRef]
41. Cerón, I.; Neila, J.; Khayet, M. Experimental tile with phase change materials (PCM) for building use. *Energy Build.* **2011**, *43*, 1869–1874. [CrossRef]
42. Karthikeyan, S.; Solomon, G.R.; Kumaresan, V.; Velraj, R. Parametric studies on packed bed storage unit filled with PCM encapsulated spherical containers for low temperature solar air heating applications. *Energy Convers. Manag.* **2014**, *78*, 74–80. [CrossRef]
43. Sharma, A.; Shukla, A.; Chen, C.R.; Dwivedi, S. Development of phase change materials for building applications. *Energy Build.* **2013**, *64*, 403–407. [CrossRef]
44. Noro, M.; Lazzarin, R.M.; Busato, F. Solar cooling and heating plants: An energy and economic analysis of liquid sensible vs. phase change material (PCM) heat storage. *Int. J. Refrig.* **2014**, *39*, 104–116. [CrossRef]
45. Summers, E.K.; Antar, M.A.; Lienhard, J.H.V. Design and optimization of an air heating solar collector with integrated phase change material energy storage for use in humidification–dehumidification desalination. *Sol. Energy* **2012**, *86*, 3417–3429. [CrossRef]
46. Harikrishnan, S.; Magesh, S.; Kalaiselvam, S. Preparation and thermal energy storage behaviour of stearic acid–TiO_2 nanofluids as a phase change material for solar heating systems. *Thermochim. Acta* **2013**, *565*, 137–145. [CrossRef]
47. Shukla, A.; Buddhi, D.; Sawhney, R.L. Solar water heaters with phase change material thermal energy storage medium: A review. *Renew. Sustain. Energy Rev.* **2009**, *13*, 2119–2125. [CrossRef]
48. Chaabane, M.; Mhiri, H.; Bournot, P. Thermal performance of an integrated collector storage solar water heater (ICSSWH) with phase change materials (PCM). *Energy Convers. Manag.* **2014**, *78*, 897–903. [CrossRef]
49. Islam, M.R.; Sumathy, K.; Khan, S.U. Solar water heating systems and their market trends. *Renew. Sustain. Energy Rev.* **2013**, *17*, 1–25. [CrossRef]
50. Tan, F.L.; Tso, C.P. Cooling of mobile electronic devices using phase change materials. *Appl. Therm. Eng.* **2004**, *24*, 159–169. [CrossRef]
51. Weng, Y.-C.; Cho, H.-P.; Chang, C.-C.; Chen, S.-L. Heat pipe with PCM for electronic cooling. *Appl. Energy* **2011**, *88*, 1825–1833. [CrossRef]
52. Baby, R.; Balaji, C. Experimental investigations on phase change material based finned heat sinks for electronic equipment cooling. *Int. J. Heat Mass Transf.* **2012**, *55*, 1642–1649. [CrossRef]
53. Lohse, E.; Schmitz, G. Performance assessment of regularly structured Composite Latent Heat Storages for temporary cooling of electronic. *Int. J. Refrig.* **2012**, *35*, 1145–1155. [CrossRef]
54. Jaworski, M. Thermal performance of heat spreader for electronics cooling with incorporated phase change material. *Appl. Therm. Eng.* **2012**, *35*, 212–219. [CrossRef]
55. Zhang, H.; Wang, H.; Zhu, X.; Qiu, Y.J.; Li, K.; Chen, R.; Liao, Q. A review of waste heat recovery technologies towards molten slag in steel industry. *Appl. Energy* **2013**, *112*, 956–966. [CrossRef]
56. Shon, J.; Kim, H.; Lee, K. Improved heat storage rate for an automobile coolant waste heat recovery system using phase-change material in a fin–tube heat exchanger. *Appl. Energy* **2014**, *113*, 680–689. [CrossRef]
57. Ebrahimi, K.; Jones, G.F.; Fleischer, A.S. A review of data center cooling technology, operating conditions and the corresponding low-grade waste heat recovery opportunities. *Renew. Sustain. Energy Rev.* **2014**, *31*, 622–638. [CrossRef]
58. Pitié, F.; Zhao, C.Y.; Baeyens, J.; Degrève, J.; Zhang, H.L. Circulating fluidized bed heat recovery/storage and its potential to use coated phase-change-material (PCM) particles. *Appl. Energy* **2013**, *109*, 505–513. [CrossRef]
59. Sarier, N.; Onder, E. Organic phase change materials and their textile applications: An overview. *Thermochim. Acta* **2012**, *540*, 7–60. [CrossRef]
60. Nejman, A.; Goetzendorf-Grabowska, B. Heat balance of textile materials modified with the mixtures of PCM microcapsules. *Thermochim. Acta* **2013**, *569*, 144–150. [CrossRef]
61. Sánchez, P.; Sánchez-Fernandez, M.V.; Romero, A.; Rodríguez, J.F.; Sánchez-Silva, L. Development of thermo-regulating textiles using paraffin wax microcapsules. *Thermochim. Acta* **2010**, *498*, 16–21. [CrossRef]
62. Sun, X.; Zhang, Q.; Medina, M.A.; Lee, K.O. Energy and economic analysis of a building enclosure outfitted with a phase change material board (PCMB). *Energy Convers. Manag.* **2014**, *83*, 73–78. [CrossRef]
63. Cabeza, L.F.; Castell, A.; Pérez, G. Life cycle assessment (LCA) of phase change materials (PCMs) used in buildings. *Eco-Efficient Construction and Building Materials*; Pacheco-Torgal, F., Cabeza, L.F., Labrincha, J., de Magalhães, A., Eds.; Woodhead Publishing: Sawston, UK, 2014; pp. 287–310, ISBN 9780857097675
64. Memon, S.A. Phase change materials integrated in building walls: A state of the art review. *Renew. Sustain. Energy Rev.* **2014**, *31*, 870–906. [CrossRef]

65. Waqas, A.; Din, Z.U. Phase change material (PCM) storage for free cooling of buildings—A review. *Renew. Sustain. Energy Rev.* **2013**, *18*, 607–625. [CrossRef]
66. AL-Saadi, S.N.; Zhai, Z. Modeling phase change materials embedded in building enclosure: A review. *Renew. Sustain. Energy Rev.* **2013**, *21*, 659–673. [CrossRef]
67. Zsembinszki, G.; Solé, C.; Castell, A.; Pérez, G.; Cabeza, L.F. The use of phase change materials in fish farms: A general analysis. *Appl. Energy* **2013**, *109*, 488–496. [CrossRef]
68. Oró, E.; Cabeza, L.F.; Farid, M.M. Experimental and numerical analysis of a chilly bin incorporating phase change material. *Appl. Therm. Eng.* **2013**, *58*, 61–67. [CrossRef]
69. Lu, W.; Tassou, S.A. Characterization and experimental investigation of phase change materials for chilled food refrigerated cabinet applications. *Appl. Energy* **2013**, *112*, 1376–1382. [CrossRef]
70. Reyes, A.; Mahn, A.; Vásquez, F. Mushrooms dehydration in a hybrid-solar dryer, using a phase change material. *Energy Convers. Manag.* **2014**, *83*, 241–248. [CrossRef]
71. Beyhan, B.; Paksoy, H.; Daşgan, Y. Root zone temperature control with thermal energy storage in phase change materials for soilless greenhouse applications. *Energy Convers. Manag.* **2013**, *74*, 446–453. [CrossRef]
72. Shalaby, S.M.; Bek, M.A.; El-Sebaii, A.A. Solar dryers with PCM as energy storage medium: A review. *Renew. Sustain. Energy Rev.* **2014**, *33*, 110–116. [CrossRef]
73. Shalaby, S.M.; Bek, M.A. Experimental investigation of a novel indirect solar dryer implementing PCM as energy storage medium. *Energy Convers. Manag.* **2014**, *83*, 1–8. [CrossRef]
74. Bouadila, S.; Kooli, S.; Skouri, S.; Lazaar, M.; Farhat, A. Improvement of the greenhouse climate using a solar air heater with latent storage energy. *Energy* **2014**, *64*, 663–672. [CrossRef]
75. Prakash, O.; Kumar, A. Solar greenhouse drying: A review. *Renew. Sustain. Energy Rev.* **2014**, *29*, 905–910. [CrossRef]
76. Saxena, A.; Agarwal, N.; Srivastava, G. Design and performance of a solar air heater with long term heat storage. *Int. J. Heat Mass Transf.* **2013**, *60*, 8–16. [CrossRef]
77. Li, Z.-R.; Hu, N.; Fan, L.-W. Nanocomposite phase change materials for high-performance thermal energy storage: A critical review. *Energy Storage Mater.* **2023**, *55*, 727–753. [CrossRef]
78. Shah, K.W. A review on enhancement of phase change materials—A nanomaterials perspective. *Energy Build.* **2018**, *175*, 57–68. [CrossRef]
79. Ranjbar, S.G.; Roudini, G.; Barahuie, F. Fabrication and characterization of phase change material-SiO_2 nanocomposite for thermal energy storage in buildings. *J. Energy Storage* **2020**, *27*, 101168. [CrossRef]
80. Paul, J.; Kadirgama, K.; Samykano, M.; Pandey, A.K.; Tyagi, V.V. A comprehensive review on thermophysical properties and solar thermal applications of organic nano composite phase change materials. *J. Energy Storage* **2022**, *45*, 103415. [CrossRef]
81. Wang, J.; Li, Y.; Zheng, D.; Mikulčić, H.; Vujanović, M.; Sundén, B. Preparation and thermophysical property analysis of nanocomposite phase change materials for energy storage. *Renew. Sustain. Energy Rev.* **2021**, *151*, 111541. [CrossRef]
82. Rathore, P.K.S.; Shukla, S.K.; Gupta, N.K. Potential of microencapsulated PCM for energy savings in buildings: A critical review. *Sustain. Cities Soc.* **2020**, *53*, 101884. [CrossRef]
83. Rathore, P.K.S.; Shukla, S.K. Potential of macroencapsulated PCM for thermal energy storage in buildings: A comprehensive review. *Constr. Build. Mater.* **2019**, *225*, 723–744. [CrossRef]
84. Ismail, A.; Wang, J.; Salami, B.A.; Oyedele, L.O.; Otukogbe, G.K. Microencapsulated phase change materials for enhanced thermal energy storage performance in construction materials: A critical review. *Constr. Build. Mater.* **2023**, *401*, 132877. [CrossRef]
85. Konuklu, Y.; Ostry, M.; Paksoy, H.O.; Charvat, P. Review on using microencapsulated phase change materials (PCM) in building applications. *Energy Build.* **2015**, *106*, 134–155. [CrossRef]
86. Huang, X.; Zhu, C.; Lin, Y.; Fang, G. Thermal properties and applications of microencapsulated PCM for thermal energy storage: A review. *Appl. Therm. Eng.* **2019**, *147*, 841–855. [CrossRef]
87. Reyez-Araiza, J.L.; Pineda-Piñón, J.; López-Romero, J.M.; Gasca-Tirado, J.R.; Arroyo Contreras, M.; Jáuregui Correa, J.C.; Apátiga-Castro, L.M.; Rivera-Muñoz, E.M.; Velazquez-Castillo, R.R.; Pérez Bueno, J.D.J.; et al. Thermal Energy Storage by the Encapsulation of Phase Change Materials in Building Elements—A Review. *Materials* **2021**, *14*, 1420. [CrossRef]
88. Gandhi, M.; Kumar, A.; Elangovan, R.; Meena, C.S.; Kulkarni, K.S.; Kumar, A.; Bhanot, G.; Kapoor, N.R. A Review on Shape-Stabilized Phase Change Materials for Latent Energy Storage in Buildings. *Sustainability* **2020**, *12*, 9481. [CrossRef]
89. Zhu, N.; Li, S.; Hu, P.; Wei, S.; Deng, R.; Lei, F. A review on applications of shape-stabilized phase change materials embedded in building enclosure in recent ten years. *Sustain. Cities Soc.* **2018**, *43*, 251–264. [CrossRef]
90. Chinnasamy, V.; Heo, J.; Jung, S.; Lee, H.; Cho, H. Shape stabilized phase change materials based on different support structures for thermal energy storage applications—A review. *Energy* **2023**, *262*, 125463. [CrossRef]
91. Kim, H.B.; Mae, M.; Choi, Y. Application of shape-stabilized phase-change material sheets as thermal energy storage to reduce heating load in Japanese climate. *Build. Environ.* **2017**, *125*, 1–14. [CrossRef]
92. Kim, H.B.; Mae, M.; Choi, Y.; Kiyota, T. Experimental analysis of thermal performance in buildings with shape-stabilized phase change materials. *Energy Build.* **2017**, *152*, 524–533. [CrossRef]
93. Mohaisen, K.O.; Zahir, M.H.; Maslehuddin, M.; Al-Dulaijan, S.U. Development of a shape-stabilized phase change material utilizing natural and industrial byproducts for thermal energy storage in buildings. *J. Energy Storage* **2022**, *50*, 104205. [CrossRef]

94. Simonsen, G.; Ravotti, R.; O'Neill, P.; Stamatiou, A. Biobased phase change materials in energy storage and thermal management technologies. *Renew. Sustain. Energy Rev.* **2023**, *184*, 113546. [CrossRef]
95. Mehrizi, A.A.; Karimi-Maleh, H.; Naddafi, M.; Karimi, F. Application of bio-based phase change materials for effective heat management. *J. Energy Storage* **2023**, *61*, 106859. [CrossRef]
96. Baylis, C.; Cruickshank, C.A. Review of bio-based phase change materials as passive thermal storage in buildings. *Renew. Sustain. Energy Rev.* **2023**, *186*, 113690. [CrossRef]
97. Rashid, F.L.; Al-Obaidi, M.A.; Dhaidan, N.S.; Hussein, A.K.; Ali, B.; Hamida, M.B.B.; Younis, O. Bio-based phase change materials for thermal energy storage and release: A review. *J. Energy Storage* **2023**, *73*, 109219. [CrossRef]
98. Li, D.; Zhuang, B.; Chen, Y.; Li, B.; Landry, V.; Kaboorani, A.; Wu, Z.; Wang, X.A. Incorporation technology of bio-based phase change materials for building envelope: A review. *Energy Build.* **2022**, *260*, 111920. [CrossRef]
99. Jain, S.; Kumar, K.R.; Rakshit, D. Heat transfer augmentation in single and multiple (cascade) phase change materials based thermal energy storage: Research progress, challenges, and recommendations. *Sustain. Energy Technol. Assess.* **2021**, *48*, 101633. [CrossRef]
100. Dardir, M.; Panchabikesan, K.; Haghighat, F.; El Mankibi, M.; Yuan, Y. Opportunities and challenges of PCM-to-air heat exchangers (PAHXs) for building free cooling applications—A comprehensive review. *J. Energy Storage* **2019**, *22*, 157–175. [CrossRef]
101. Bondareva, N.S.; Sheremet, M.A. A Numerical Study of Heat Performance of Multi-PCM Brick in a Heat Storage Building. *Mathematics* **2023**, *11*, 2825. [CrossRef]
102. Rashid, F.L.; Al-Obaidi, M.A.; Dulaimi, A.; Mahmood, D.M.N.; Sopian, K. A Review of Recent Improvements, Developments, and Effects of Using Phase-Change Materials in Buildings to Store Thermal Energy. *Designs* **2023**, *7*, 90. [CrossRef]
103. Nair, A.M.; Wilson, C.; Huang, M.J.; Griffiths, P.; Hewitt, N. Phase change materials in building integrated space heating and domestic hot water applications: A review. *J. Energy Storage* **2022**, *54*, 105227. [CrossRef]
104. Liu, M.; Saman, W.; Bruno, F. Review on storage materials and thermal performance enhancement techniques for high temperature phase change thermal storage systems. *Renew. Sustain. Energy Rev.* **2012**, *16*, 2118–2132. [CrossRef]
105. Pomianowski, M.; Heiselberg, P.; Zhang, Y. Review of thermal energy storage technologies based on PCM application in buildings. *Energy Build.* **2013**, *67*, 56–69. [CrossRef]
106. Zahir, M.H.; Irshad, K.; Shafiullah, M.; Ibrahim, N.I.; Islam, A.K.; Mohaisen, K.O.; Sulaiman, F.A.A. Challenges of the application of PCMs to achieve zero energy buildings under hot weather conditions: A review. *J. Energy Storage* **2023**, *64*, 107156. [CrossRef]
107. Wang, X.; Li, W.; Luo, Z.; Wang, K.; Shah, S.P. A critical review on phase change materials (PCM) for sustainable and energy efficient building: Design, characteristic, performance and application. *Energy Build.* **2022**, *260*, 111923. [CrossRef]
108. Ferrer, G.; Solé, A.; Barreneche, C.; Martorell, I.; Cabeza, L.F. Review on the methodology used in thermal stability characterization of phase change materials. *Renew. Sustain. Energy Rev.* **2015**, *50*, 665–685. [CrossRef]
109. Noël, J.A.; Kahwaji, S.; Desgrosseilliers, L.; Groulx, D.; White, M.A. Chapter 13—Phase Change Materials. In *Storing Energy*; Letcher, T.M., Ed.; Elsevier: Oxford, UK, 2016; pp. 249–272. [CrossRef]
110. Sikiru, S.; Oladosu, T.L.; Amosa, T.I.; Kolawole, S.Y.; Soleimani, H. Recent advances and impact of phase change materials on solar energy: A comprehensive review. *J. Energy Storage* **2022**, *53*, 105200. [CrossRef]
111. Hu, X.; Chen, P.; Zhang, X.; Gong, X.; Shuai, Y.; Wang, Y.; Xing, X.; Zheng, X.; Wang, G. Experimental and numerical study on thermal management performance of PCM-based heat sinks with various configurations fabricated by additive manufacturing. *Renew. Energy* **2024**, *232*, 121069. [CrossRef]
112. Righetti, G.; Savio, G.; Meneghello, R.; Doretti, L.; Mancin, S. Experimental study of phase change material (PCM) embedded in 3D periodic structures realized via additive manufacturing. *Int. J. Therm. Sci.* **2020**, *153*, 106376. [CrossRef]
113. Du, K.; Calautit, J.; Wang, Z.; Wu, Y.; Liu, H. A review of the applications of phase change materials in cooling, heating and power generation in different temperature ranges. *Appl. Energy* **2018**, *220*, 242–273. [CrossRef]
114. Qiu, J.; Huo, D.; Xia, Y. Phase-Change Materials for Controlled Release and Related Applications. *Adv. Mater.* **2020**, *32*, 2000660. [CrossRef]
115. Hakami, A.; Srinivasan, S.S.; Biswas, P.K.; Krishnegowda, A.; Wallen, S.L.; Stefanakos, E.K. Review on thermochromic materials: Development, characterization, and applications. *J. Coat. Technol. Res.* **2022**, *19*, 377–402. [CrossRef]
116. Wang, X.; Narayan, S. Thermochromic Materials for Smart Windows: A State-of-Art Review. *Front. Energy Res.* **2021**, *9*, 800382. [CrossRef]
117. Nagare, S.M.; Hakami, A.; Biswas, P.K.; Stefanakos, E.K.; Srinivasan, S.S. A review of thermochromic materials for coating applications: Production, protection, and degradation of organic thermochromic materials. *J. Coat. Technol. Res.* **2024**. [CrossRef]
118. Aklujkar, P.S.; Kandasubramanian, B. A review of microencapsulated thermochromic coatings for sustainable building applications. *J. Coat. Technol. Res.* **2021**, *18*, 19–37. [CrossRef]
119. Zhao, Y.; Ji, H.; Lu, M.; Tao, J.; Ou, Y.; Wang, Y.; Chen, Y.; Huang, Y.; Wang, J.; Mao, Y. Thermochromic Smart Windows Assisted by Photothermal Nanomaterials. *Nanomaterials* **2022**, *12*, 3865. [CrossRef]
120. Zhang, Z.; Zhang, L.; Zhou, Y.; Cui, Y.; Chen, Z.; Liu, Y.; Li, J.; Long, Y.; Gao, Y. Thermochromic Energy Efficient Windows: Fundamentals, Recent Advances, and Perspectives. *Chem. Rev.* **2023**, *123*, 7025–7080. [CrossRef] [PubMed]
121. Aburas, M.; Soebarto, V.; Williamson, T.; Liang, R.; Ebendorff-Heidepriem, H.; Wu, Y. Thermochromic smart window technologies for building application: A review. *Appl. Energy* **2019**, *255*, 113522. [CrossRef]

122. Zou, X.; Ji, H.; Zhao, Y.; Lu, M.; Tao, J.; Tang, P.; Liu, B.; Yu, X.; Mao, Y. Research Progress of Photo-/Electro-Driven Thermochromic Smart Windows. *Nanomaterials* **2021**, *11*, 3335. [CrossRef] [PubMed]
123. Anayurt, R.A.; Alkan, C. Most Recent Applications of Microencapsulated Phase Change Materials. *J. New Results Sci.* **2021**, *10*, 24–37.
124. Wu, S.; Sun, H.; Duan, M.; Mao, H.; Wu, Y.; Zhao, H.; Lin, B. Applications of thermochromic and electrochromic smart windows: Materials to buildings. *Cell Rep. Phys. Sci.* **2023**, *4*, 101370. [CrossRef]
125. Kamalisarvestani, M.; Saidur, R.; Mekhilef, S.; Javadi, F.S. Performance, materials and coating technologies of thermochromic thin films on smart windows. *Renew. Sustain. Energy Rev.* **2013**, *26*, 353–364. [CrossRef]
126. Ke, Y.; Chen, J.; Lin, G.; Wang, S.; Zhou, Y.; Yin, J.; Lee, P.S.; Long, Y. Smart Windows: Electro-, Thermo-, Mechano-, Photochromics, and Beyond. *Adv. Energy Mater.* **2019**, *9*, 1902066. [CrossRef]
127. Zhai, X.Q.; Wang, X.L.; Wang, T.; Wang, R.Z. A review on phase change cold storage in air conditioning system: Materials and applications. *Renew. Sustain. Energy Rev.* **2013**, *22*, 108–120. [CrossRef]
128. Pielichowska, K.; Pielichowski, K. Phase change materials for thermal energy storage. *Prog. Mater. Sci.* **2014**, *65*, 67–123. [CrossRef]
129. Dutil, Y.; Rousse, D.R.; Salah, N.B.; Lassue, S.P.; Zalewski, L. A review on phase-change materials: Mathematical modeling and simulations. *Renew. Sustain. Energy Rev.* **2011**, *15*, 112–130. [CrossRef]
130. Zanganeh, G.; Commerford, M.; Haselbacher, A.; Pedretti, A.; Steinfeld, A. Stabilization of the Outflow Temperature of a Packed-Bed Thermal Energy Storage by Combining Rocks with Phase Change Materials. *Appl. Therm. Eng.* **2014**, *70*, 316–320. [CrossRef]
131. Jradi, M.; Gillott, M.; Riffat, S. Simulation of the transient behaviour of encapsulated organic and inorganic phase change materials for low-temperature energy storage. *Appl. Therm. Eng.* **2013**, *59*, 211–222. [CrossRef]
132. Li, T.; Lee, J.-H.; Wang, R.; Kang, Y.T. Heat transfer characteristics of phase change nanocomposite materials for thermal energy storage application. *Int. J. Heat Mass Transf.* **2014**, *75*, 1–11. [CrossRef]
133. Farid, M.M.; Khudhair, A.M.; Siddique, A.K.; Sari, A. A review on phase change energy storage: Materials and applications. *Energy Convers. Manag.* **2004**, *45*, 1597–1615. [CrossRef]
134. Dincer, I.; Rosen, M.A. *Thermal Energy Storage: Systems and Applications*; John Wiley and Sons: Hoboken, NJ, USA, 2011.
135. Agyenim, F.; Hewitt, N.; Eames, P.; Smyth, M. A review of materials, heat transfer and phase change problem formulation for latent heat thermal energy storage systems (LHTESS). *Renew. Sustain. Energy Rev.* **2010**, *14*, 615–628. [CrossRef]
136. Khodadadi, J.M.; Fan, L.; Babaei, H. Thermal conductivity enhancement of nanostructure-based colloidal suspensions utilized as phase change materials for thermal energy storage: A review. *Renew. Sustain. Energy Rev.* **2013**, *24*, 418–444. [CrossRef]
137. Rathod, M.K.; Kanzaria, H.V. A methodological concept for phase change material selection based on multiple criteria decision analysis with and without fuzzy environment. *Mater. Des.* **2011**, *32*, 3578–3585. [CrossRef]
138. Rathod, M.K.; Banerjee, J. Thermal stability of phase change materials used in latent heat energy storage systems: A review. *Renew. Sustain. Energy Rev.* **2013**, *18*, 246–258. [CrossRef]
139. Delgado, M.; Lazaro, A.; Mazo, J.; Zalba, B. Review on phase change material emulsions and microencapsulated phase change material slurries: Materials, heat transfer studies and applications. *Renew. Sustain. Energy Rev.* **2012**, *16*, 253–273. [CrossRef]
140. Sharma, A.; Tyagi, V.; Chen, C.R.; Buddhi, D. Review on thermal energy storage with phase change materials and applications. *Renew. Sustain. Energy Rev.* **2009**, *13*, 318–345. [CrossRef]
141. Kenisarin, M.M. High temperature phase change materials for thermal energy storage. *Renew. Sustain. Energy Rev.* **2010**, *14*, 955–970. [CrossRef]
142. Paksoy, H.O. Thermal Energy Storage for Sustainable Energy Consumption. In Proceedings of the NATO Advance Study Institute on Thermal Energy Storage for Sustainable Energy Consumption—Fundamentals, Case Studies and Design, Izmir, Turkey, 6–17 June 2005.
143. Nkwetta, D.N.; Vouillamoz, P.-E.; Haghighat, F.; El-Mankibi, M.; Moreau, A.; Daoud, A. Impact of phase change materials types and positioning on hot water tank thermal performance: Using measured water demand profile. *Appl. Therm. Eng.* **2014**, *67*, 460–468. [CrossRef]
144. Tchinda, R. A review of the mathematical models for predicting solar air heaters systems. *Renew. Sustain. Energy Rev.* **2009**, *13*, 1734–1759. [CrossRef]
145. Izquierdo-Barrientos, M.A.; Sobrino, C.; Almendros-Ibáñez, J.A. Thermal energy storage in a fluidized bed of PCM. *Chem. Eng. J.* **2013**, *230*, 573–583. [CrossRef]
146. Close, D.J. Solar air heaters. *Sol. Energy* **1963**, *7*, 117–129. [CrossRef]
147. El-Sawi, A.M.; Wifi, A.S.; Younan, M.Y.; Elsayed, E.A.; Basily, B.B. Application of folded sheet metal in flat bed solar air collectors. *Appl. Therm. Eng.* **2010**, *30*, 864–871. [CrossRef]
148. Alkan, C.; Kaya, K.; Sari, A. Preparation, thermal properties and thermal reliability of form-stable paraffin/polypropylene composite for thermal energy storage. *J. Polym. Environ.* **2009**, *17*, 254–258. [CrossRef]
149. Shukla, A.; Buddhi, D.; Sawhney, R.L. Thermal cycling test of few selected inorganic and organic phase change materials. *Renew. Energy* **2008**, *33*, 2606–2614. [CrossRef]
150. El-Sebaii, A.A.; Al-Amir, S.; Al-Marzouki, F.M.; Faidah, A.S.; Al-Ghamdi, A.; Al-Heniti, S. Fast thermal cycling of acetanilide and magnesium chloride hexahydrate for indoor solar cooking. *Energy Convers. Manag.* **2009**, *50*, 3104–3111. [CrossRef]
151. Sari, A. Thermal reliability test of some fatty acids as PCMs used for solar thermal latent heat storage applications. *Energy Convers. Manag.* **2003**, *44*, 2277–2287. [CrossRef]

152. Hasan, A.; Sayigh, A.A. Some fatty acids as phase-change thermal energy storage materials. *Renew. Energy* **1994**, *4*, 69–76. [CrossRef]
153. Sharma, A.; Sharma, S.D.; Buddhi, D. Accelerated thermal cycle test of acetamide, stearic acid and paraffin wax for solar thermal latent heat storage applications. *Energy Convers. Manag.* **2002**, *43*, 1923–1930. [CrossRef]
154. Kimura, H.; Kai, J. Phase change stability of $CaCl_{2.6}H_2O$. *Sol. Energy* **1984**, *33*, 49–55. [CrossRef]
155. Fellchenfeld, H.; Sarig, S. Calcium chloride hexahydrate: A phase-changing material for energy storage. *Ind. Eng. Chem. Product. Res. Dev.* **1985**, *24*, 130–133. [CrossRef]
156. Tyagi, V.V.; Buddhi, D. Thermal cycle testing of calcium chloride hexahydrate as a possible PCM for latent heat storage. *Sol. Energy Mater. Sol. Cells* **2008**, *92*, 891–899. [CrossRef]
157. Marks, S. An investigation of the thermal energy storage capacity of glauber's salt with respect to thermal cycling. *Sol. Energy* **1980**, *25*, 225–258. [CrossRef]
158. Wada, T.; Yamamoto, R.; Matsuo, Y. Heat storage capacity of sodium acetate trihydrate during thermal cycling. *Sol. Energy* **1984**, *33*, 373–375. [CrossRef]
159. Porosini, F.C. Salt hydrates used for latent heat storage: Corrosion of metals and reliability of thermal performance. *Sol. Energy* **1988**, *41*, 193–197. [CrossRef]
160. Sun, J.Q.; Zhang, R.Y.; Liu, Z.P.; Lu, G.H. Thermal reliability test of Al–34%Mg–6%Zn alloy as latent heat storage material and corrosion of metal with respect to thermal cycling. *Energy Convers. Manag.* **2007**, *48*, 619–624. [CrossRef]
161. Shilei, L.; Neng, Z.; Guohui, F. Eutectic mixture of capric acid and lauric acid applied in building wallboards for heat energy storage. *Energy Build.* **2006**, *38*, 708–711. [CrossRef]
162. Karaipekli, A.; Sari, A. Capric–mystric acid/expanded perlite composite as form-stable phase change material for latent heat thermal energy storage. *Renew. Energy* **2008**, *33*, 2599–2605. [CrossRef]
163. Karaipekli, A.; Sari, A.; Kaygusuz, K. Thermal properties and thermal reliability of capric acid/stearic acid mixture for latent heat thermal energy storage. *Energy Sources* **2009**, *31*, 199–207. [CrossRef]
164. Sari, A. Eutectic mixtures of some fatty acids for low temperature solar heating applications: Thermal properties and thermal reliability. *Appl. Therm. Eng.* **2005**, *25*, 2100–2107. [CrossRef]
165. Sari, A.; Bicer, A.; Karaipekli, A.; Alkan, C.; Karadag, A. Synthesis, thermal energy storage properties and thermal reliability of some fatty acid esters with glycerol as novel solid–liquid phase change materials. *Sol. Energy Mater. Sol. Cells* **2010**, *94*, 1711–1715. [CrossRef]
166. Nagano, K.; Takeda, S.; Mochida, T.; Shimakura, K.; Nakamura, T. Study of a floor supply air conditioning system using granular phase change material to augment building mass thermal storage—Heat response in small scale experiments. *Energy Build.* **2006**, *38*, 436–446. [CrossRef]
167. El-Sebaii, A.A.; Aboul-Eneein, S.; Ramadan, M.I.; EL-Bialy, E.E. Year round performance of double pass solar air heater with packed bed. *Energy Convers. Manag.* **2007**, *48*, 990–1003. [CrossRef]
168. de Gracia, A.; Oró, E.; Farid, M.M.; Cabeza, L.F. Thermal analysis of including phase change material in a domestic hot water cylinder. *Appl. Therm. Eng.* **2011**, *31*, 3938–3945. [CrossRef]
169. Haillot, D.; Franquet, E.; Gibout, S.; Bédécarrats, J.-P. Optimization of solar DHW system including PCM media. *Appl. Energy* **2013**, *109*, 470–475. [CrossRef]
170. International Renewable Energy Agency. *Thermal Energy Storage Technology Brief*; International Renewable Energy Agency: Masdar, United Arab Emirates, 2013.
171. Erro, I.; Aranguren, P.; Sorbet, F.J.; Bonilla-Campos, I.; Astrain, D. Installation of a Thermoelectric Heat Pump in a Thermal Energy Storage Cycle for Decentralized Energy Generation. *Appl. Therm. Eng.* **2023**, *246*, 122923. [CrossRef]
172. Biswas, K.; Lu, J.; Soroushian, P.; Shrestha, S. Combined experimental and numerical evaluation of a prototype nano-PCM enhanced wallboard. *Appl. Energy* **2014**, *131*, 517–529. [CrossRef]
173. Qiao, Y.; Liu, Y.; Yang, L.; Bao, J.; Liu, J. Experimental Investigation of PCM Wallboard in Artificial Controlled Environment with Different Climate Conditions. In Proceedings of the 11th International Symposium on Heating, Ventilation and Air Conditioning (ISHVAC 2019), Harbin, China, 12–15 July 2019; Springer: Singapore, 2020. [CrossRef]
174. Saxena, R.; Rakshi, D.; Kaushik, S.C. Experimental Assessment of Characterised PCMs for Thermal Management of Buildings in Tropical Composite Climate. In Proceedings of the 4th World Congress on Mechanical, Chemical, and Material Engineering (MCM'18), Madrid, Spain, 16–18 August 2018. [CrossRef]
175. Bal, L.M.; Sudhakar, P.; Satya, S.; Naik, N.S. Solar dryer with latent heat storage systems for drying agricultural food products. In Proceedings of the International Conference on Food Security and Environmental Sustainability, Colombo, Sri Lanka, 17–19 December 2009.
176. Cakmak, G.; Yidiz, C. The drying kinetics of seeded grape in solar dryer with PCM-based solar integrated collector. *Food Bioprod. Process.* **2011**, *89*, 103–108. [CrossRef]
177. Esakkimuthu, S.; Hassabou, A.; Palaniappan, C.; Spinnler, M.; Blumenberg, J.; Velraj, R. Experimental investigation on phase change material based thermal storage system for solar air heating applications. *Sol. Energy* **2013**, *88*, 144–153. [CrossRef]
178. Lamnatou, C.; Papanicolaou, E.; Belessiotis, V.; Kyriakis, N. Experimental investigation and thermodynamic performance analysis of a solar dryer using an evacuated-tube air collector. *Appl. Energy* **2012**, *94*, 232–243. [CrossRef]

179. Boughali, S.; Benmoussa, H. Crop drying by indirect active hybrid solar–electrical dryer. *Sol. Energy* **2009**, *83*, 2223–2232. [CrossRef]
180. Amer, B.; Hossain, M.; Gottschalk, K. Design and performance evaluation of a new hybrid solar dryer for banana. *Energy Convers. Manag.* **2010**, *51*, 813–820. [CrossRef]
181. Gao, Y.; Fan, R.; Zhang, X.Y.; An, Y.J.; Wang, M.X.; Gao, Y.K.; Yu, Y. Thermal performance and parameter analysis of a U-pipe evacuated solar tube collector. *Sol. Energy* **2014**, *107*, 714–727. [CrossRef]
182. Liang, R.; Ma, L.; Zhang, J.; Zhao, L. Performance analysis of a new-design filled-type solar collector with double U-tubes. *Energy Build.* **2013**, *57*, 220–226. [CrossRef]
183. Tiwari, A.K.; Kumar, A.; Said, Z. Applications of Nano-enhanced PCMs in Solar Energy. In *Nano Enhanced Phase Change Materials*; Said, Z., Pandey, A.K., Eds.; Materials Horizons: From Nature to Nanomaterials; Springer: Singapore, 2023.
184. Kaygusuz, K. The viability of thermal energy storage. *Energy Sources* **1999**, *21*, 745–755. [CrossRef]
185. Almeida, F. Experimental and Numerical Study on the Thermal Performance of a Vertical PCM Panel. Ph.D. Thesis, Toronto Metropolitan University, Toronto, ON, USA, 2023. [CrossRef]
186. Diaconu, B.M.; Varga, S.; Oliveira, A.C. Experimental study of natural convection heat transfer in a microencapsulated phase change material slurry. *Energy* **2010**, *35*, 2688–2693. [CrossRef]
187. Tiwari, A.K.; Kumar, A.; Said, Z. Influence of Nanoparticles on Thermophysical Properties of PCMs. In *Nano Enhanced Phase Change Materials*; Said, Z., Pandey, A.K., Eds.; Materials Horizons: From Nature to Nanomaterials; Springer: Singapore, 2023.
188. Saxena, R.; Dwivedi, C.; Dutta, V.; Kaushik, S.C.; Rakshit, D. Nano-enhanced PCMs for low-temperature thermal energy storage systems and passive conditioning applications. *Clean. Technol. Environ. Policy* **2021**, *23*, 1161–1168. [CrossRef]
189. Ren, Q.; Xu, H.; Luo, Z. PCM charging process accelerated with combination of optimized triangle fins and nanoparticles. *Int. J. Therm. Sci.* **2019**, *140*, 466–479. [CrossRef]
190. Ma, Z.; Lin, W.; Sohel, M.I. Nano-enhanced phase change materials for improved building performance. *Renew. Sustain. Energy Rev.* **2016**, *58*, 1256–1268. [CrossRef]
191. Liu, M.; Belusko, M.; Tay, N.H.S.; Bruno, F. Impact of the heat transfer fluid in a flat plate phase change thermal storage unit for concentrated solar tower plants. *Sol. Energy* **2014**, *101*, 220–231. [CrossRef]
192. Alkilani, M.M.; Sopian, K.; Mat, S.; Alghoul, M.A. Output air temperature prediction in a solar air heater integrated with phase change material. *Eur. J. Sci. Res.* **2009**, *27*, 334–434.
193. Charvát, P.; Klimeš, L.; Ostr, M. Numerical and experimental investigation of a PCM-based thermal storage unit for solar air systems. *Energy Build.* **2014**, *68*, 488–497. [CrossRef]
194. Colella, F.; Sciacovelli, A.; Verda, V. Numerical analysis of a medium scale latent energy storage unit for district heating systems. *Energy* **2012**, *45*, 397–406. [CrossRef]
195. Piroozmand, V.; Ahmadi, R. Enhancement of PCMs performance using nano-particles in horizontal triple-series shell-and-tube heat exchangers: A numerical study. *J. Energy Storage* **2024**, *85*, 111057. [CrossRef]
196. Keshteli, A.N.; Sheikholeslami, M. Nanoparticle enhanced PCM applications for intensification of thermal performance in building: A review. *J. Mol. Liq.* **2019**, *274*, 516–533. [CrossRef]
197. Yerzhan, B.; Toleukhanov, A.; Yerdesh, Y.; Seitov, A.; Aliuly, A.; Botella, O. Numerical simulation of cascade latent heat thermal energy storage device thermal performance using multiple PCMs. *AIP Conf. Proc.* **2024**, *3126*, 020001. [CrossRef]
198. Ahmed, S.S.; Radwan, A.; Fouda, M.S.; Sultan, A.A.; Abdelrehim, O. Energy assessment of a sliding window integrated with PV cell and multiple PCMs. *J. Energy Storage* **2024**, *86*, 111341. [CrossRef]
199. Ahmed, S.S.; Fouda, M.S.; Cheng, P. Power saving in a central air conditioning system by using multiple PCMs integrated with fresh air path. *Energy Build.* **2023**, *300*, 113691. [CrossRef]
200. He, W.; Su, Y.; Wang, Y.Q.; Riffat, S.B.; Ji, J. A study on incorporation of thermoelectric modules with evacuated-tube heat-pipe solar collectors. *Renew. Energy* **2012**, *37*, 142–149. [CrossRef]
201. Abed, A.M.; Mouziraji, H.R.; Bakhshi, J.; Dulaimi, A.; Mohammed, H.I.; Ibrahem, R.K.; Ben Khedher, N.; Yaïci, W.; Mahdi, J.M. Numerical analysis of the energy-storage performance of a PCM-based triplex-tube containment system equipped with arc-shaped fins. *Front. Chem.* **2022**, *10*, 1057196. [CrossRef] [PubMed]
202. Ben Khedher, N.; Biswas, N.; Togun, H.; Mohammed, H.I.; Mahdi, J.M.; Ibrahem, R.K.; Talebizadehsardari, P. Geometry modification of a vertical shell-and-tube latent heat thermal energy storage system using a framed structure with different undulated shapes for the phase change material container during the melting process. *J. Energy Storage* **2023**, *72 Pt B*, 108365. [CrossRef]
203. Oztop, H.F.; Bayrak, F.; Hepbasli, A. Energetic and exergetic aspects of solar air heating (solar collector) systems. *Renew. Sustain. Energy Rev.* **2013**, *21*, 59–83. [CrossRef]
204. Dolado, P.; Lazaro, A.; Marin, J.M.; Zalba, B. Characterization of melting and solidification in a real scale PCM-air heat exchanger: Numerical model and experimental validation. *Energy Convers. Manag.* **2011**, *52*, 1890–1907. [CrossRef]
205. Dolado, P.; Lazaro, A.; Marín, J.M.; Zalba, B. Characterization of melting and solidification in a real-scale PCM-air heat exchanger: Experimental results and empirical model. *Renew. Energy* **2011**, *36*, 2906–2917. [CrossRef]
206. Khalifa, A.J.N.; Suffer, K.H.; Mahmoud, M.S. A storage domestic solar hot water system with a back layer of phase change material. *Exp. Therm. Fluid. Sci.* **2013**, *44*, 174–181. [CrossRef]

207. Zhu, N.; Hu, P.; Xu, L.; Jiang, Z.; Lei, F. Recent research and applications of ground source heat pump integrated with thermal energy storage systems: A review. *Appl. Therm. Eng.* **2014**, *71*, 142–151. [CrossRef]
208. Tumirah, K.; Hussein, M.Z.; Zulkarnain, Z.; Rafeadah, R. Nano-encapsulated organic phase change material based on copolymer nanocomposites for thermal energy storage. *Energy* **2014**, *66*, 881–890. [CrossRef]
209. López-Navarro, A.; Biosca-Taronger, J.; Corberán, J.M.; Peñalosa, C.; Lázaro, A.; Dolado, P.; Payá, J. Performance characterization of a PCM storage tank. *Appl. Energy* **2014**, *119*, 151–162. [CrossRef]
210. Parameshwara, R.; Dhamodharana, P.; Kalaiselvam, S. Study on thermal storage properties of hybrid nanocomposite-dibasic ester as phase change material. *Thermochim. Acta* **2013**, *573*, 106–120. [CrossRef]
211. Fan, L.W.; Fang, X.; Wang, X.; Zeng, Y.; Xiao, Y.Q.; Yu, Z.T.; Xu, X.; Hu, Y.C.; Cen, K.F. Effects of various carbon nanofillers on the thermal conductivity and energy storage properties of paraffin-based nanocomposite phase change materials. *Appl. Energy* **2013**, *110*, 163–172. [CrossRef]
212. Fang, X.; Fan, L.W.; Ding, Q.; Yao, X.L.; Wu, Y.Y.; Hou, J.F.; Wang, X.; Yu, Z.T.; Cheng, G.H.; Hu, Y.C. Thermal energy storage performance of paraffin-based composite phase change materials filled with hexagonal boron nitride nanosheets. *Energy Convers. Manag.* **2014**, *80*, 103–109. [CrossRef]
213. Teng, T.-P.; Yu, C.-C. The Effect on Heating Rate for Phase Change Materials Containing MWCNTs. *Int. J. Chem. Eng. Appl.* **2012**, *3*, 340–342. [CrossRef]
214. Benita, S. *Microencapsulation: Methods and Industrial Applications*; Marcel Dekker: New York, NY, USA, 1996.
215. Ma, Z.W.; Zhang, P. Modeling the heat transfer characteristics of flow melting of phase change material slurries in the circular tubes. *Int. J. Heat Mass Transf.* **2013**, *64*, 874–881. [CrossRef]
216. Sarı, A.; Alkan, C.; Döğüşcü, D.K.; Biçer, A. Micro/nano-encapsulatedn-heptadecane with polystyrene shell for latent heat thermal energy storage. *Sol. Energy Mater. Sol. Cells* **2014**, *126*, 42–50. [CrossRef]
217. Lin, X.; Guiyin, F.; Fan, Y. Study on the preparation and performance of the microcapsule phase change material for cold storage. *Vac. Cryog.* **2006**, *12*, 153–156.
218. Alkan, C.; Sari, A.; Karaipekli, A.; Uzun, O. Preparation, characterization, and thermal properties of microencapsulated phase change material for thermal energy storage. *Sol. Energy Mater. Sol. Cells* **2009**, *93*, 143–147. [CrossRef]
219. Sari, A.; Alkan, C.; Karaipekli, A.; Uzun, O. Microencapsulated n-octacosane as phase change material for thermal energy storage. *Sol. Energy* **2009**, *83*, 1757–1763. [CrossRef]
220. Hawlader, M.N.A.; Uddin, M.S.; Zhu, H. Encapsulated phase change materials for thermal energy storage: Experiments and simulation. *Int. J. Energy Res.* **2002**, *26*, 159–171. [CrossRef]
221. Arshad, A.; Jabbal, M.; Yan, Y.; Darkwa, J. The micro-/nano-PCMs for thermal energy storage systems: A state of art review. *Int. J. Energy Res.* **2019**, *43*, 5572–5620. [CrossRef]
222. Hawlader, M.N.A.; Uddin, M.S.; Khin, M.M. Microencapsulated PCM thermal-energy storage system. *Appl. Energy* **2003**, *74*, 195–202. [CrossRef]
223. Ozonur, Y.; Mazman, M.; Paksoy, H.O.; Evliya, H. Microencapsulation of coco fatty acid mixture for thermal energy storage with phase change material. *Int. J. Energy Res.* **2006**, *30*, 741–749. [CrossRef]
224. Peng, S.; Fuchs, A.; Wirtz, R.A. Polymeric phase change composites for thermal energy storage. *J. Appl. Polym. Sci.* **2004**, *93*, 1240–1251. [CrossRef]
225. Lee, S.H.; Yoon, S.J.; Kim, Y.G.; Choi, Y.C.; Kim, J.H.; Lee, J.G. Development of building materials by using micro-encapsulated phase change material. *Korean J. Chem. Eng.* **2007**, *24*, 332–335. [CrossRef]
226. Chen, L.; Xu, L.; Shang, H.; Zhang, Z. Microencapsulation of butyl stearate as a phase change material by interfacial polycondensation in a polyurea system. *Energy Convers. Manag.* **2009**, *50*, 723–729.
227. Yu, F.; Chen, Z.H.; Zeng, X.R. Preparation, characterization, and thermal properties of microPCMs containing n-dodecanol by using different types of styrene–maleic anhydride as emulsifier. *Colloid Polym. Sci.* **2009**, *287*, 549–560. [CrossRef]
228. Fang, G.; Li, H.; Yang, F.; Liua, X.; Wu, S. Preparation and characterization of nano-encapsulated n-tetradecane as phase change material for thermal energy storage. *Chem. Eng. J.* **2009**, *153*, 217–221. [CrossRef]
229. Zhang, H.; Wang, X. Synthesis and properties of microencapsulated n-octadecane with polyurea shells containing different soft segments for heat energy storage and thermal regulation. *Sol. Energy Mater. Sol. Cells* **2009**, *93*, 1366–1376. [CrossRef]
230. Kaygusuz, K.; Sari, A. High density polyethylene/paraffin composites as form–stable phase change material for thermal energy storage. *Energy Sources Part A* **2007**, *29*, 261–270. [CrossRef]
231. Kenisarin, M.M.; Kenisarin, K.M. Form–stable phase change materials for thermal energy storage. *Renew. Sustain. Energy Rev.* **2012**, *16*, 1999–2040. [CrossRef]
232. Trigui, A.; Karkri, M.; Krupa, I. Thermal conductivity and latent heat thermal energy storage properties of LDPE/wax as a shape-stabilized composite phase change material. *Energy Convers. Manag.* **2014**, *77*, 586–596. [CrossRef]
233. Jeong, S.-G.; Lee, J.-H.; Seo, J.; Kim, S. Thermal performance evaluation of Bio-based shape stabilized PCM with boron nitride for energy saving. *Int. J. Heat Mass Transf.* **2014**, *71*, 245–250. [CrossRef]
234. Sari, A. Form-stable paraffin/high density polyethylene composites as solid-liquid phase change material for thermal energy storage: Preparation and thermal properties. *Energy Convers. Manag.* **2004**, *45*, 2033–2042. [CrossRef]
235. Alkan, C.; Sari, A.; Uzun, O. Poly(ethylene glycol)/acrylic polymer blends for latent heat thermal energy storage. *AIChE J.* **2006**, *52*, 3310–3314. [CrossRef]

236. Zhang, L.; Zhu, J.; Zhou, W.; Wang, J.; Wang, Y. Characterization of polymethyl methacrylate/polyethylene glycol/aluminum nitride composite as form-stable phase change material prepared by in situ polymerization method. *Thermochim. Acta* **2011**, *254*, 128–134. [CrossRef]
237. Sari, A.; Kaygusuz, K. Thermal conductivity and latent heat thermal energy storage characteristics of paraffin/expanded graphite composite as phase change material. *Appl. Therm. Eng.* **2007**, *27*, 1271–1277. [CrossRef]
238. Sari, A.; Kaygusuz, K. Preparation, thermal properties and thermal reliability of palmitic acid/expanded graphite composite as form-stable PCM for thermal energy storage. *Sol. Energy Mater. Sol. Cells* **2009**, *93*, 571–576. [CrossRef]
239. Behzadi, S.; Farid, M.M. Long term thermal stability of organic PCMs. *Appl. Energy* **2014**, *122*, 11–16. [CrossRef]
240. Xiao, X.; Zhang, P.; Li, M. Thermal characterization of nitrates and nitrates/expanded graphite mixture phase change materials for solar energy storage. *Energy Convers. Manag.* **2013**, *73*, 86–94. [CrossRef]
241. Tian, Y.; Zhao, C.Y. A numerical investigation of heat transfer in phase change materials (PCMs) embedded in porous metals. *Energy* **2011**, *36*, 5539–5546. [CrossRef]
242. Fukai, J.; Kanou, M.; Kodama, Y.; Miyatake, O. Thermal conductivity enhancement of energy storage media using carbon fibers. *Energy Convers. Manag.* **2000**, *41*, 1543–1556. [CrossRef]
243. Wu, S.; Wang, H.; Xiao, S.; Zhu, D. Numerical simulation on thermal energy storage behavior of Cu/paraffin nanofluids PCMs. *Procedia Eng.* **2012**, *31*, 240–244. [CrossRef]
244. Mills, A.; Farid, M.; Selman, J.R.; Al-Hallaj, S. Thermal conductivity enhancement of phase change materials using a graphite matrix. *Appl. Therm. Eng.* **2006**, *26*, 1652–1661. [CrossRef]
245. Li, M.; Wu, Z.; Tan, J. Properties of form-stable paraffin/silicon dioxide/expanded graphite phase change composites prepared by sol-gel method. *Appl. Energy* **2012**, *92*, 456–461. [CrossRef]
246. Chieruzzi, M.; Cerritelli, G.F.; Miliozzi, A.; Kenny, J.M. Effect of nanoparticles on heat capacity of nanofluids based on molten salts as PCM for thermal energy storage. *Nanoscale Res. Lett.* **2013**, *8*, 448. [CrossRef]
247. Ismail, K.A.R.; Henrı´quez, J.R. Numerical and experimental study of spherical capsules packed bed latent heat storage system. *Appl. Therm. Eng.* **2002**, *22*, 1705–1716. [CrossRef]
248. Pisello, A.L. State of the art on the development of cool coatings for buildings and cities. *Sol. Energy* **2017**, *144*, 660–680. [CrossRef]
249. Lima, O., Jr.; Freitas, E.; Cardoso, P.; Segundo, I.R.; Margalho, É.; Moreira, L.; Nascimento, J.H.O.; Landi, S., Jr.; Carneiro, J. Mitigation of Urban Heat Island Effects by Thermochromic Asphalt Pavement. *Coatings* **2023**, *13*, 35. [CrossRef]
250. Zhang, Y.; Wang, X.; Wu, D. Design and fabrication of dual-functional microcapsules containing phase change material core and zirconium oxide shell with fluorescent characteristics. *Sol. Energy Mater. Sol. Cells* **2015**, *133*, 56–68. [CrossRef]
251. Geng, X.; Li, W.; Wang, Y.; Lu, J.; Wang, J.; Wang, N.; Li, J.; Zhang, X. Design and fabrication of reversible thermochromic microencapsulated phase change materials for thermal energy storage and its antibacterial activity. *Energy* **2018**, *159*, 857–869. [CrossRef]
252. Abdollahi, A.; Roghani-Mamaqani, H.; Razavi, B. Stimuli-chromism of photoswitches in smart polymers: Recent advances and applications as chemosensors. *Prog. Polym. Sci.* **2019**, *98*, 101149. [CrossRef]
253. Vassalini, I.; Alessandri, I.; de Ceglia, D. Stimuli-Responsive Phase Change Materials: Optical and Optoelectronic Applications. *Materials* **2021**, *14*, 3396. [CrossRef]
254. Zhang, Y.; Chou, J.B.; Li, J.; Li, H.; Du, Q.; Yadav, A.; Zhou, S.; Shalaginov, M.Y.; Fang, Z.; Zhong, H.; et al. Broadband transparent optical phase change materials for high-performance nonvolatile photonics. *Nat. Commun.* **2019**, *10*, 4279. [CrossRef]
255. Yuan, A.; Zhao, S.; Liu, T.; Zhao, Y.; Jiang, L.; Lei, J. Temperature-Responsive Thermochromic Phase Change Materials with Tunable Photothermal Conversion Efficiency Toward Solar Energy Storage. *Adv. Mater. Technol.* **2022**, *7*, 2200226. [CrossRef]
256. Cacciarini, M.; Skov, A.B.; Jevric, M.; Hansen, A.S.; Elm, J.; Kjaergaard, H.G.; Mikkelsen, K.V.; Brøndsted Nielsen, M. Towards Solar Energy Storage in the Photochromic Dihydroazulene-Vinylheptafulvene System. *Chem.–A Eur. J.* **2015**, *21*, 7454–7461. [CrossRef]
257. IEA. Technology Roadmap: *Solar Photovoltaic Energy and Technology Roadmap: Solar Thermal* Electricity; IEA: Paris, France, 2014.
258. Amberger, B.; Savji, N. Thermochromism of Transition Metal Compounds. 2008. Available online: https://web.archive.org/web/20090531010111/http://www3.amherst.edu/~thoughts/contents/amberger-thermochromism.html (accessed on 30 September 2024).
259. Bukleski, M.; Petruševski, V.M. Preparation and Properties of a Spectacular Thermochromic Solid. *J. Chem. Educ.* **2009**, *86*, 30. [CrossRef]
260. Bamfield, P.; Hutchings, M.G. *Chromic Phenomena: Technological Applications of Colour Chemistry*; Royal Society of Chemistry: London, UK, 2010.
261. Miller, B.J. How smart windows save energy. *Knowable Magazine*, 6 August 2022. [CrossRef]
262. Seeboth, A.; Lötzsch, D. *Thermochromic and Thermotropic Materials*; CRC Press: Boca Raton, FL, USA, 2013.
263. Kitsopoulou, A.; Bellos, E.; Sammoutos, C.; Lykas, P.; Vrachopoulos, M.G.; Tzivanidis, C. A detailed investigation of thermochromic dye-based roof coatings for Greek climatic conditions. *J. Build. Eng.* **2024**, *84*, 108570. [CrossRef]
264. Garshasbi, S.; Santamouris, M. Using advanced thermochromic technologies in the built environment: Recent development and potential to decrease the energy consumption and fight urban overheating. *Sol. Energy Mater. Sol. Cells* **2019**, *191*, 21–32. [CrossRef]

265. Santamouris, M. On the energy impact of urban heat island and global warming on buildings. *Energy Build.* **2014**, *82*, 100–113. [CrossRef]
266. Yang, Q.; Xiong, J.; Mao, G.; Zhang, Y. A Novel Molecular PCM Wall with Inorganic Composite: Dynamic Thermal Analysis and Optimization in Charge–Discharge Cycles. *Materials* **2023**, *16*, 5955. [CrossRef] [PubMed]
267. Kalair, A.R.; Seyedmahmoudian, M.; Mekhilef, S.; Stojcevski, A. Dynamic Analysis of Solar Heat Stimulated Residential Absorption Cooling with Integrated Thermal Wall for Space Heating. In Proceedings of the 2022 7th International Conference on Smart and Sustainable Technologies (SpliTech), Split/Bol, Croatia, 5–8 July 2022; pp. 1–5. [CrossRef]
268. Kalair, A.R. Synergetic Electrothermal Storage Integrated Trigeneration Nanogrid. Ph.D. Thesis, Swinburne University of Technology, Melbourne, Australia, 2024. [CrossRef]
269. Al-Absi, Z.A.; Isa, M.H.M.; Ismail, M. Phase Change Materials (PCMs) and Their Optimum Position in Building Walls. *Sustainability* **2020**, *12*, 1294. [CrossRef]

energies

MDPI

Article

Effect of Macrocapsule Geometry on PCM Performance for Thermal Regulation in Buildings

Margarida Gonçalves [1,*], António Figueiredo [1], German Vela [1], Filipe Rebelo [1], Ricardo M. S. F. Almeida [2,3], Mónica S. A. Oliveira [4,5] and Romeu Vicente [1]

1 CERIS-Civil Engineering Research and Innovation for Sustainability, Civil Engineering Department, University of Aveiro, Campus Universitário de Santiago, 3810-193 Aveiro, Portugal; ajfigueiredo@ua.pt (A.F.); german@ua.pt (G.V.); filiperebelo@ua.pt (F.R.); romvic@ua.pt (R.V.)
2 Department of Civil Engineering, Polytechnic Institute of Viseu, School of Technology and Management, Campus Politécnico de Repeses, 3504-510 Viseu, Portugal; ralmeida@estgv.ipv.pt
3 Institute of R&D in Structures and Construction-LFC (CONSTRUCT-LFC), Faculty of Engineering (FEUP), University of Porto, Rua Dr. Roberto Frias s/n, 4200-465 Porto, Portugal
4 TEMA-Centre for Mechanical Technology and Automation, Department of Mechanical Engineering, University of Aveiro, Campus Universitário de Santiago, 3810-193 Aveiro, Portugal; monica.oliveira@ua.pt
5 LASI-Intelligent Systems Associate Laboratory, 4800-058 Guimarães, Portugal
* Correspondence: margaridagoncalves@ua.pt

Abstract: The integration of phase-change materials (PCMs) into thermal energy storage systems offers significant potential for reducing energy consumption and improving thermal comfort, crucial issues for achieving sustainable building stocks. Nevertheless, the performance of PCM-based systems is strongly influenced by the container geometry. Among the various forms of incorporating PCMs into building applications, macroencapsulation is the most versatile and is, therefore, widely used. Herewith, this paper analyzes the impact of macrocapsule geometry on PCM thermal performance. Thermal properties of the material were first tested using Differential Scanning Calorimetry at five heating/cooling rates to evaluate its influence on phase-change temperatures and enthalpy. Then, an experimental setup evaluated four macrocapsule geometries on the enclosed PCM behavior during charging and discharging processes. The PCM characterization revealed that the slowest-tested rate minimized the supercooling effect. Analysis across different macrocapsule geometries showed that sectioning the contact surface improved heat transfer efficiency by fully mobilizing the PCM and reducing phase-change times. Conversely, double-layered geometry designs hindered heat transfer, presenting challenges in completing PCM charging and discharging. These findings suggest that optimizing its performance is a necessary direction for further research, which may include adjusting the PCM operating temperature range across layers or redesigning the geometry to misalign contact surfaces.

Keywords: phase-change materials; thermal energy storage; Differential Scanning Calorimetry; macrocapsule geometry; melting and solidification

1. Introduction

Current architectural trends increasingly emphasize the use of large glazed areas [1], particularly through a higher Window-to-Wall Ratio (WWR), to maximize natural lighting and to enhance the aesthetic appeal [2]. However, this design strategy significantly increases solar gains, which can have a major impact on both energy consumption and thermal comfort. Several studies estimate that glazed areas can account for approximately 25% to 40% of a building's energy needs [3]. While increased solar gain may reduce heating demands in colder months, they often introduce significant challenges during warmer seasons by raising cooling loads and, therefore, increasing the overheating risk [4]. As awareness of climate change grows, balancing these effects requires careful designing of glazing and shading solutions, prioritizing the decarbonization of the building stock, and improving its thermal and energy performance.

Solar energy is widely recognized as a promising renewable resource due to its abundance, environmental sustainability, accessibility, and direct usability [5,6]. In this context, integrating Thermal Energy Storage (TES) systems into building components or solutions has proven its significant potential for enhancing energy efficiency by increasing the share of renewable energy and reducing both energy demand and peak loads for heating and cooling [7]. Among TES systems, latent heat storage utilizing phase-change materials (PCMs) is particularly notable. These systems can store solar thermal energy during the material's melting phase and release it during solidification, maintaining an approximately constant temperature throughout the cycle [8,9]. In recent years, research on the integration of PCMs into building envelopes has gained notable attention in scientific circles due to their remarkable potential in various heat transfer and energy storage applications. However, it is important to note that the efficiency of PCM-based systems is largely determined by the material's ability to completely undergo rapid and reliable melting and solidification processes. Herewith, the PCM response time plays a crucial role, indicating how quickly it transitions between its liquid and solid phases in response to environmental temperature fluctuations [10].

Various methods are available for integrating PCMs into building applications, including direct incorporation, impregnation, encapsulation, immersion, and shape stabilization [11,12]. Among these techniques, encapsulation is widely explored due to its ability to enhance the surface area interaction between the PCM and the thermal load [13]. Within the encapsulation techniques, macroencapsulation stands out, as it is considered a valuable option for applications in the building sector given its cost-effectiveness, enhanced mechanical stability, and greater surface area [14]. Nevertheless, the design of PCM-based systems is significantly influenced by the geometry of the PCM container, including its shape, structure, and thickness. Selecting an appropriate PCM container geometry is crucial for increasing PCM performance by ensuring efficient heat transfer, as well as effective storage and release of thermal energy [15]. Different geometries and surface areas impact heat transfer rates and thermal efficiency in distinct ways [16].

To date, extensive research has been conducted on various PCMs and container designs encompassing different shapes, sizes, and materials, mainly using numerical methods. Park et al. [17] investigated flexible PCM containers—elliptical PCM vessels—and compared their total melting times with those of traditional circular containers. They identified an optimal axis ratio (relation between the vertical and longitudinal axes) in the range of 0.05–0.20 for effective heat transfer enhancement in PCM configurations. Hekmat et al. [18] evaluated different geometries for PCM containers, including circular, horizontal elliptical, vertical elliptical, square, triangular, and downward/upward trapezoidal shapes. Their analysis of melting and solidification rates indicated that the downward trapezoidal and horizontal elliptical containers exhibited the best performance. Container geometry influenced melting and solidification times by up to 57% and 38%, respectively. Similarly, Qin et al. [19] simulated PCM storage containers with identical cross-sectional areas but varying configurations, such as rectangular shapes, concave folded sidewalls, and protruding folded sidewalls. They reported a 25% reduction in melting time for containers with protruding-shaped sidewalls. Huang et al. [20] examined an optimized PCM container with an elongated and curved shape. By simulating the convection and melting behavior of the PCM and comparing it to a rectangular container, they observed a reduction in total melting time by over 20%. Manikandan et al. [15] tested four distinct container configurations—vertical rectangle, horizontal rectangle, triangle, and cylindrical—alongside solid and hollow fins. Numerical simulations showed that horizontal rectangular containers exhibited the fastest melting and solidification rates, whereas cylindrical shapes resulted in longer phase-change durations. The introduction of fins significantly decreased both the melting and solidification times. Liu et al. [21] also examined the effect of fin geometry in a numerical study on rectangular containers aimed at optimizing the PCM melting process and enhancing heat transfer. They analyzed rectangular, trapezoidal, and triangular longitudinal fins with varying thicknesses. The results showed that triangular fins reduced the melting completion time by 16% and increased the heat storage rate by 19%, improving their effectiveness as the fin length increased. In their experimental study, Zhang et al. [22] investigated six shapes of PCM capsules: sphere, tall cylinder, short cylinder, pyramid, tetrahedron, and red-blood-cell shape. The outcomes revealed that geometry significantly impacts the phase-change time. The shortest cylindrical capsule with a horizontal axis had the longest phase-change duration, whereas the pyramidal capsule with a horizontal base exhibited the shortest duration. Furthermore, pyramidal and tetrahedral capsules enhanced heat transfer rates due to their larger surface areas. More recently, Issa and Thirunavukkarasu [23] conducted experimental comparisons among copper cylindrical, stainless-steel cylindrical, and stainless-steel spherical capsules. Their findings emphasized the significant influence of geometry, which was found to have a greater impact than the container material. The cylindrical configurations demonstrated a distinct advantage, achieving a 58% reduction in discharge time and storage efficiency of 80–82%, outperforming spherical containers. Erlbeck et al. [24] experimentally and numerically investigated the impact of an encapsulated PCM design and its positioning on the thermal behavior of a concrete block, aiming to modify and enhance the heat transfer area. The study considered various configurations: a cuboid PCM package (reference case), two superimposed longitudinally distributed PCM plates, three thinner elongated longitudinal plates, and eight thicker transversally distributed plates. The longitudinal plates exhibited a slight increase in melting and solidification times due to a thermal shading effect, which isolated individual PCM elements from the heat source, reducing the heat flux and extending the heating and cooling periods. In contrast, the transversally distributed plates provided a larger heat transfer area, increased thickness for natural convection, and aligned all the PCM packages in a row, mitigating the thermal shading effect. This configuration resulted in significantly shorter phase-change times.

From the literature, it is observed that numerous studies confirm the importance of container geometries on PCM melting and solidification processes. However, experimental studies on the impact of macrocapsule geometry on the PCM's thermal performance remain limited. Furthermore, a research gap has been identified concerning the analysis of identical geometries subdivided into smaller segments, which involves exploring the relationship between surface area and PCM efficiency. Therefore, this study aims to investigate the effects of container geometry on the melting and solidification rates of PCMs, as well as its impact on thermal management performance, specifically when positioned in front of a glazed window. Experimental tests were conducted on various container configurations to evaluate the impact of distinct contact surface areas, focusing on four distinct geometries of squared shapes. The comprehensive analysis yielded results on the evolution of PCM temperature profiles and response times, providing valuable insights into optimal container designs for enhanced performance in building applications.

2. Materials and Methods

2.1. Materials

To investigate the impact of different geometries on the thermal performance of PCMs, novel-shaped macrocapsules were developed and analyzed in this study. These macrocapsules were fabricated using a transparent PolyLactic Acid (PLA) filament through 3D printing (*UltiMaker B.V.* S5, Geldermalsen, Netherlands; Figure 1) and conceptualized based on a traditional commercial solution (rectangular or square geometry). The potential for the expansion of 3D printing in building applications is vast, providing new opportunities for innovative and customized designs and reducing material waste [25]. The macrocapsules' dimensions are 200 mm in their width and length and a thickness of 30 mm.

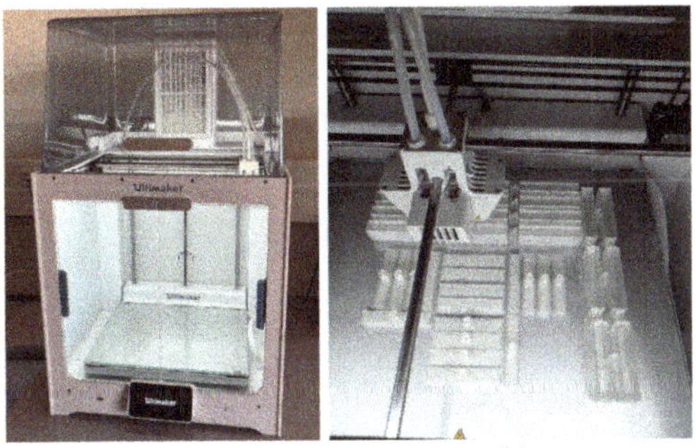

(a) (b)

Figure 1. Manufacture of the macrocapsules: (**a**) *UltiMaker S5* equipment; (**b**) 3D printing process.

The selected PCM was a paraffin from *Rubitherm*, Berlin, Germany (reference RT25HC) [26], with a melting temperature range of 22–26 °C, according to the manufacturer's datasheet. Table 1 provides the thermophysical properties of the materials.

Table 1. Main thermophysical properties of the materials [26–28].

PLA Macrocapsules	
Maximum operating temperature (°C)	59
Impact strength (kJ m^{-2})	4
Thermal conductivity (W m^{-1} $°C^{-1}$)	0.13
Density (kg m^{-3})	1240
PCM	
Phase-change temperature range (°C)	22 to 26
Specific heat capacity (kJ kg^{-1} $°C^{-1}$)	2
Latent heat (kJ kg^{-1})	230
Thermal conductivity (W m^{-1} $°C^{-1}$)	0.20
Density (kg m^{-3})	880 (solid); 770 (liquid)

2.2. Macrocapsule Geometries

The experimental campaign involved four square geometries with varying contact areas to evaluate their impact on PCM behavior during the charging and discharging processes, as well as the overall heat transfer performance of each macrocapsule. The contact area between the PCM and the container surfaces, and consequently the thermal environment, was adjusted to study these effects. Each macrocapsule contained a PCM volume of 450 mL. The geometry, dimensions (in millimeters, mm), and surface contact areas for each macrocapsule (A_{cont}) are illustrated in Figure 2, featuring a wall thickness of 1.5 mm and a stopper thickness of 3 mm. Moreover, Figure 3 presents a real view of the developed containers. It is important to emphasize that the joints connecting the macrocapsule sections (H1, H2, and H3) and the two layers of the double-layered solutions (H1 and H3) were designed to be solid and filled with PLA, i.e., without incorporating PCMs.

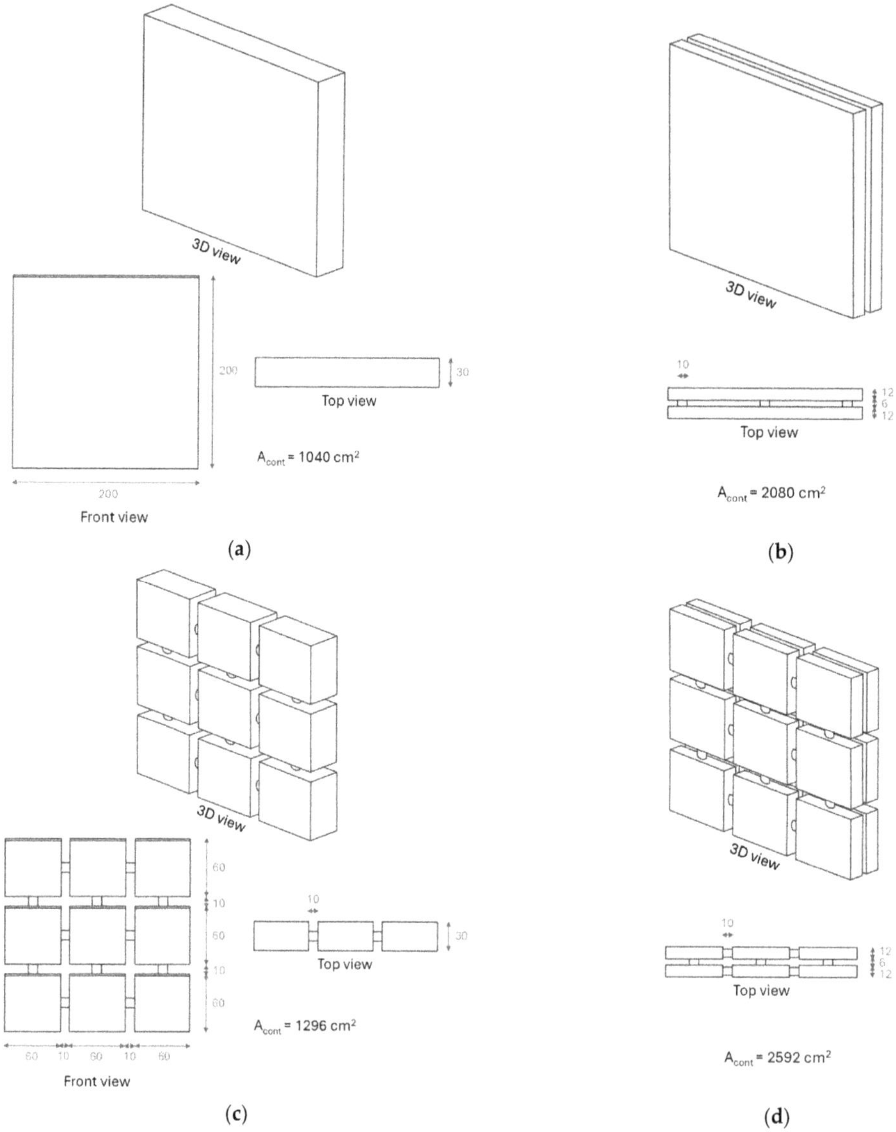

Figure 2. PCM macrocapsule geometries and their corresponding dimensions (in mm): (**a**) Reference macrocapsule (REF); (**b**) H1 macrocapsule; (**c**) H2 macrocapsule; (**d**) H3 macrocapsule.

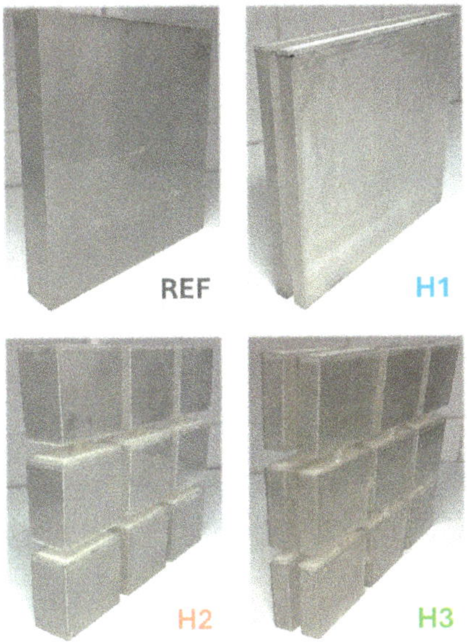

Figure 3. PCM macrocapsules: real view.

2.3. Test Procedures

2.3.1. PCM Thermal Characterization: Differential Scanning Calorimetry

To compare the experimentally obtained thermal properties with those provided by the manufacturer, the selected PCM was tested using the Differential Scanning Calorimetry (DSC) method. DSC analysis involves subjecting a sample to controlled temperature variations at specific heating and cooling rates while measuring the associated heat flows [29,30].

Since the test rates are crucial for accurate DSC measurements [31], various heating and cooling rates were tested to assess their impact on the thermal properties of the selected PCM. Seven PCM samples, each weighing approximately 6–8.5 mg, were analyzed using calibrated DSC 250 equipment from *TA Instruments*, New Castle, DE, USA (Figure 4).

(a) (b)

Figure 4. DSC analysis: (**a**) *TA Instruments* DSC 250 equipment and (**b**) the crucible used.

The temperature range for testing was consistent across all samples and aligned with the recommendation by Christy et al. [32], which suggests an optimal range of ±20 °C around the phase-change temperature. Accordingly, the heating process was conducted from 5 °C to 45 °C, with the reverse range used for cooling.

Lastly, five different decreasing heating/cooling rates were selected to evaluate their influence on the PCM's enthalpy, onset, peak, and endset temperatures: 4 °C min^{-1}, 2 °C min^{-1}, 1 °C min^{-1}, 0.5 °C min^{-1}, and 0.25 °C min^{-1}. The higher rates were chosen because they are commonly used in the literature, while the slower rates were included to better represent PCM behavior in real-scale building applications. Moreover, testing these slower rates is crucial for comparison with the experimental performance of the PCM, as detailed in Section 2.3.2.

2.3.2. Macrocapsule Geometries: Analysis and Comparison of the PCM Thermal Performance

In order to evaluate the effect of different macrocapsule geometries on the thermal performance and PCM behavior during charging and discharging cycles, two chambers were used. Each chamber has internal dimensions of 800 mm × 650 mm × 600 mm (height × width × length) and are separated by a mounting ring where the test sample is installed.

One chamber (marked in red in Figure 5a,b) functioned as a climatic chamber to input a thermal load, simulating the external environment and promoting the PCM phase change. This chamber was heated using a 175 W infrared lamp to promote heat transfer and to simulate the average daily solar radiation peak during the summer season: 700 W m^{-2} (global solar irradiance, monitored by a local weather station). Herewith, the irradiance value in the mounting ring was preliminary verified using a pyranometer (SR05-D2A2; *Hukseflux*, Delft, Netherlands). During the discharging phase, the chamber was cooled to a setpoint temperature of 10 °C. The experimental protocol was structured as follows:

1. Stabilization of the PCM temperature at approximately 15 °C for 1 h (to ensure the complete solid state of the whole PCM mass);
2. Beginning of the heating cycle by turning on the infrared lamp for 2.5 h;
3. Beginning of the cooling cycle by setting the climatic chamber at 10 °C for 12 h.

The second chamber (highlighted in blue in Figure 5a,d) was kept in passive operation under free-running conditions. Within the mounting ring, which has internal dimensions of 800 mm × 650 mm × 390 mm (height × width × length), an extruded polystyrene (XPS) contour frame was installed to ensure adiabatic boundary conditions. The frame featured a central opening measuring 200 mm × 200 mm. The test sample consisted of a 6 mm glass pane fixed within this opening in contact with the climatic chamber, representing a simple glazing system, with the various macrocapsules positioned approximately 2 cm from the glass (see Figure 5c).

Indoor air temperatures in the chambers were monitored using PT100 sensors—one in the climatic chamber and three in the free-running chamber, as shown in Figure 3. Additionally, K-type thermocouples (calibrated under laboratory conditions using the stirred liquid bath method; operating range: −100 °C to 450 °C; accuracy: ±0.5 °C) were used to measure surface temperatures across all layers of the macrocapsules, the PCM temperature inside the macrocapsules (see Figure 5a), and the outer surface temperature of the glass (on the side facing the climatic chamber). The thermocouples embedded in the PCM were positioned at the midpoint of each layer's thickness and at the center of the macrocapsule's filled height (in the H2 and H3 solutions, the sensors were in the central sections of the macrocapsule). All sensors were connected to a datalogger system (CR1000X; *Campbell Scientific*, Logan, UT, USA), with measurements recorded continuously at 5 min intervals.

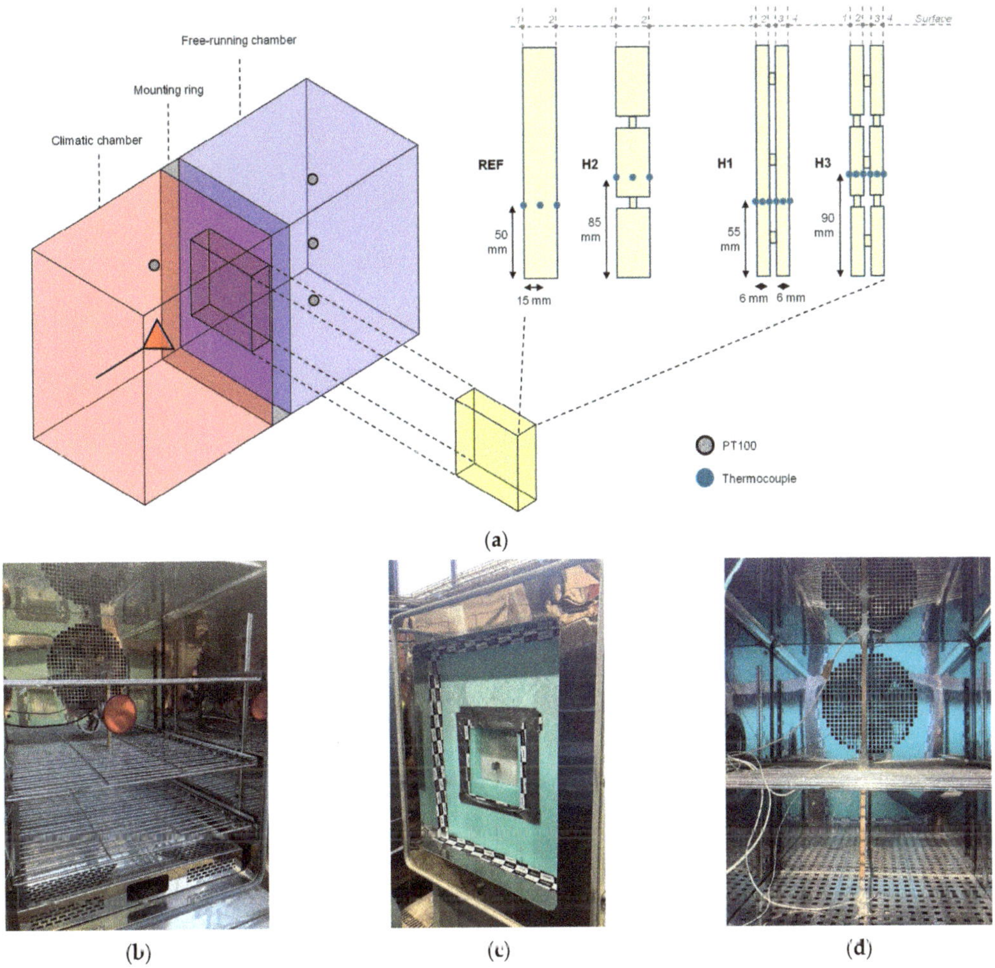

Figure 5. Experimental setup: (**a**) Schematic depiction of the climatic and free-running chambers and the mounting ring, and the layout of sensors positioned inside the chambers and on the macrocapsule layers; (**b**) Climatic chamber, (**c**) Mounting ring with a test sample; (**d**) Free-running chamber.

3. Results

3.1. PCM: Differential Scanning Calorimetry Analysis

Thermal characterization is essential for understanding the behavior of heat storage materials; specifically, to determine the phase-change temperature range and enthalpy variation during both the melting and solidification processes. As previously noted, DSC tests were conducted on various PCM samples at different heating and cooling rates: 4 °C min^{-1}; 2 °C min^{-1}; 1 °C min^{-1}; 0.5 °C min^{-1}; and 0.25 °C min^{-1}. The experimental test results are presented in Figure 6, and Table 2 depicts a summary of the phase-change temperatures and enthalpies. The plots in Figure 6b–e also show the percentage differences in the peak temperatures and supercooling effect, which are always relative to the previously tested rate.

4 °C min^{-1}

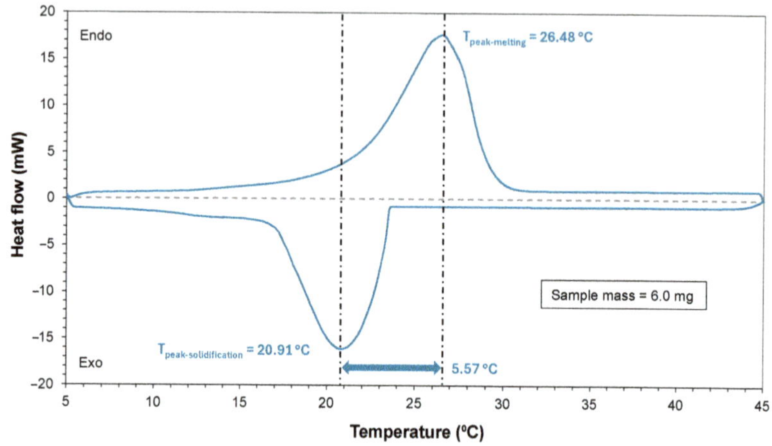

(a)

2 °C min^{-1}

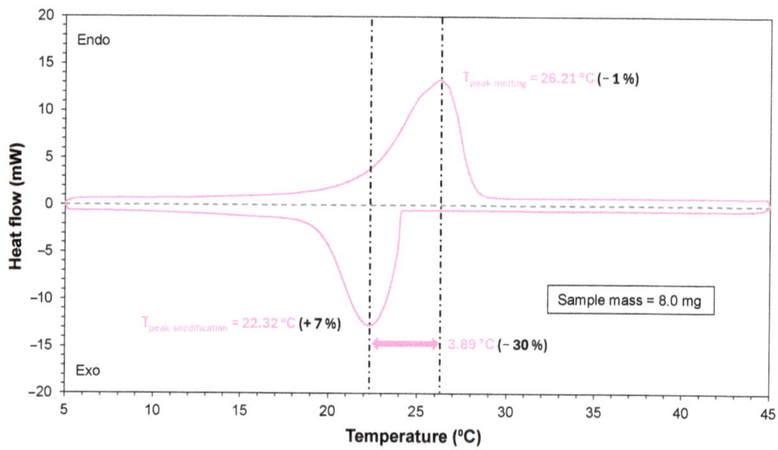

(b)

1 °C min^{-1}

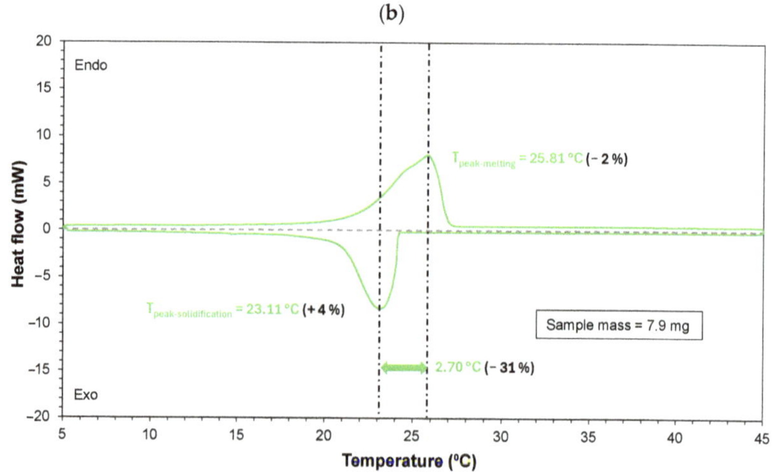

(c)

Figure 6. *Cont.*

0.5 °C min-1

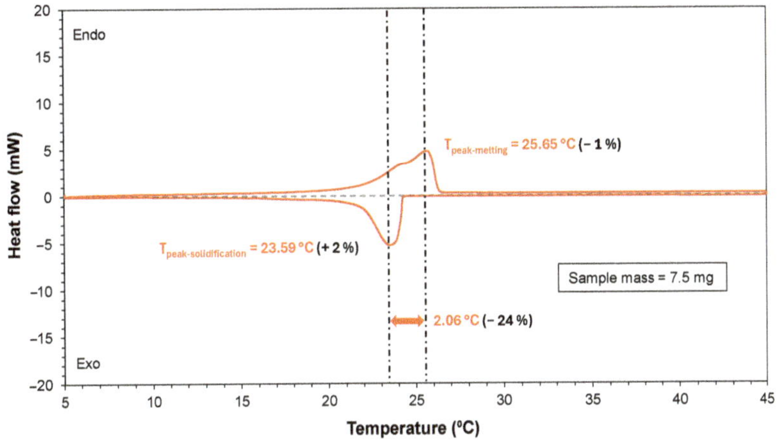

(d)

0.25 °C min-1

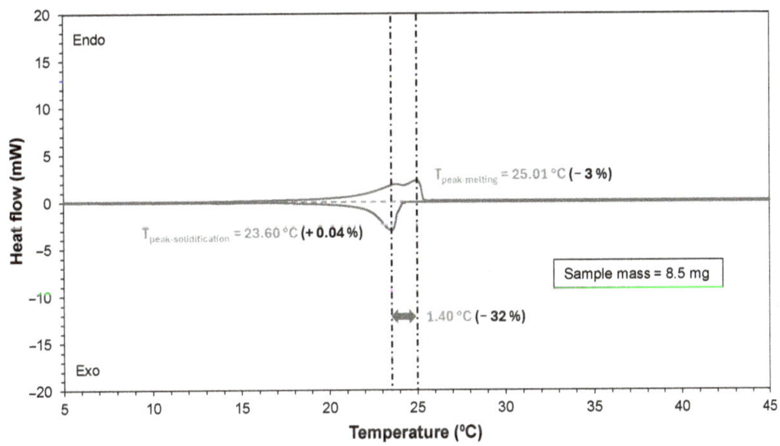

(e)

Figure 6. DSC results for the tested heating/cooling rates and analysis of the differences between the peak melting and solidification temperatures and the supercooling degree: (**a**) 4 °C min^{-1}; (**b**) 2 °C min^{-1}; (**c**) 1 °C min^{-1}; (**d**) 0.5 °C min^{-1}; (**e**) 0.25 °C min^{-1}.

Table 2. PCM RT25HC: Summary of onset temperature (T_{on}), endset temperature (T_{end}), peak temperature (T_{peak}), and phase-change enthalpy (ΔH) for the tested DSC heating/cooling rates.

		4 °C min^{-1}	**2 °C min^{-1}**	**1 °C min^{-1}**	**0.5 °C min^{-1}**	**0.25 °C min^{-1}**
Melting	**T_{on} (°C)**	18.07	18.95	18.76	19.50	20.00
	T_{end} (°C)	30.54	28.81	27.31	26.62	25.60
	T_{peak} (°C)	26.48	26.21	25.81	25.65	25.01
	ΔH (J g^{-1})	200.29	181.01	181.00	182.59	155.99
Solidification	**T_{on} (°C)**	23.52	24.06	24.22	24.25	24.59
	T_{end} (°C)	16.19	16.81	17.31	18.82	19.61
	T_{peak} (°C)	20.91	22.32	23.11	23.59	23.60
	ΔH (J g^{-1})	146.29	143.18	142.80	143.47	130.89

It was observed that the DSC curve is influenced by the heating and cooling rates; lower heating rates during the test resulted in a narrower phase-change range and reduced heat flow through the sample.

Figure 6 illustrates that reducing the heating rate by half caused the melting peak temperature to decrease by 1% for rates of 2 °C min^{-1} and 0.5 °C min^{-1}, by 2% at a rate of 1 °C min^{-1}, and by 3% at a rate of 0.25 °C min^{-1}. Similarly, proportional reductions in the cooling rate led to an increase in the solidification peak temperature, ranging from 0.04% at 0.25 °C min^{-1} to 7% at 2 °C min^{-1}. This trend has been observed in previous studies [29,33] and can be attributed to differences in the thermal gradient within the sample. Higher test rates reduce the time available for uniform heat distribution and phase change, leading to variations in thermal behavior.

The effect of sample mass on the variability of thermal results in the DSC analysis, often referred to in the literature [32,34], was evaluated using a heating/cooling rate of 0.25 °C min^{-1}. This rate closely replicates the material's behavior in building components under real conditions. To assess this effect, the test was repeated with three different samples of varying masses while maintaining a constant test rate. For a sample mass variation of 7.7 ± 0.8 mg, the results showed that the melting and solidification peak temperatures changed by 2%. These changes remained within the 5% variability threshold reported in the literature [35], indicating that sample mass does not significantly influence the DSC test results.

Figure 6 also revealed the presence of a supercooling effect in the PCM sample, characterized by the difference between the melting and solidification peak temperatures. The supercooling phenomenon occurs when there is a significant delay in crystal formation, resulting in a lower observed solidification peak temperature during testing [36]. To clearly illustrate the evolution of the supercooling degree observed in the DSC tests, the results for the different heating and cooling rates are presented in Figure 7.

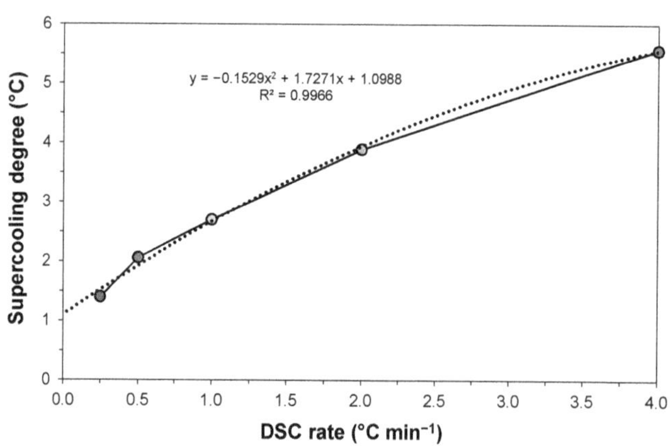

Figure 7. Evolution of the supercooling degree under the tested heating/cooling rates.

Similarly, variations in the heating and cooling rates demonstrated a significant impact on the supercooling degree. The highest tested rate (4 °C min^{-1}) resulted in a supercooling degree of 5.57 °C. Reducing the heating/cooling rates by half led to an approximate 30% decrease in the difference between the melting and solidification peak temperatures. The smallest reduction was observed at a rate of 0.5 °C min^{-1}, with a 24% decrease, while the most favorable performance was achieved at the lowest tested rate (0.25 °C min^{-1}), yielding a supercooling degree of 1.40 °C, 32% lower than the value obtained at the preceding rate.

The results obtained indicate an excellent R^2 value, demonstrating a strong fit with the experimental data. The second-order polynomial regression curve also indicated that a DSC rate below 0.25 °C min^{-1} would have no significant influence on supercooling. Thus, the degree of supercooling exhibits a practically negligible decrease for rates below this threshold.

The onset temperature (T_{on}) and endset temperature (T_{end}) also exhibit a narrowing trend with decreasing test rates, consistent with the behavior observed in the evolution of phase-change peak temperatures, as indicated in Table 2. Generally, during both the heating and cooling cycles, T_{on} increases while T_{end} decreases at lower DSC rates. The most pronounced narrowing was observed at the lowest rate, 0.25 °C min^{-1}, with a difference of 5.6 °C between T_{on} and T_{end} during the heating phase and 4.98 °C during the cooling phase.

However, at lower heating and cooling rates, the phase-change enthalpy values (ΔH) were reduced, as indicated in Table 2. A similar trend was reported by Castellón et al. [37], who found that increasing the heating and cooling rates improves the accuracy of enthalpy determination but also introduces greater temperature uncertainty. The ΔH values were also notably different from those provided in the manufacturer's datasheet. During the heating process, discrepancies ranged from −15% to −47%, while during the cooling process, the differences were even more pronounced, varying from −57% to −76%. Additionally, the DSC results showed slightly higher enthalpy values during the melting process, a phenomenon in line with findings in the literature [38,39].

Based on these results, the testing rate of 0.25 °C min^{-1} demonstrates the lowest degree of supercooling and bears a closer resemblance to the manufacturer-reported phase-change temperatures. Furthermore, given that this application is designed for passive use with a daily cycle, this rate more accurately reflects the PCM's charging and discharging behavior under real-world conditions when applied to building components.

3.2. Thermal Performance Analysis and Comparison of Different Macrocapsule Geometries

3.2.1. Thermal Performance Tests

This section presents and analyzes the results of the thermal performance tests conducted on the developed macrocapsules, providing a comparative evaluation of surface temperatures during both the heating and cooling phases. Figure 8 illustrates the environmental temperatures inside the climatic and free-running chambers, the external glass surface temperature, the surface temperatures of each macrocapsule, and the PCM temperature within the studied containers.

The analysis of temperature distributions across the cross-sections of the tested solutions revealed significant differences between the inner (surface 1) and outer surfaces (surface 2 for REF and H2, surface 4 for H1 and H3) during the melting phase, except for solution H2. The recorded thermal amplitudes, considering the maximum surface temperature, were 11 °C for the REF solution, 26 °C for the H1 macrocapsule, and 15 °C for the H3 macrocapsule. In contrast, the variation in minimum surface temperatures for these solutions was approximately 3 °C and 2 °C, respectively. Conversely, solution H2 exhibited similar behavior during both the heating and cooling phases, with temperature differences between its surfaces limited to 2 °C for both the maximum and minimum temperatures.

A detailed analysis of Figure 8 shows that in the double-layered solutions H1 and H3, where the macrocapsule features thickness sectioning, the temperature difference between the inner and outer surfaces was significantly higher. This suggests that such geometric configurations influence heat transfer through these types of macrocapsules. Nevertheless, in the H2 solution, where the contact surface with the thermal environment is sectioned, an improvement in heat transfer efficiency through the macrocapsule is observed, as evidenced by a notably reduced temperature difference between the inner and outer surfaces.

A preliminary analysis of the PCM behavior in the different tested solutions indicates a considerable reduction in the time required to complete both the melting and solidification phases compared with the performance observed in the REF solution. However, in the double-layered solutions (H1 and H3), this reduction is evident only in the first layer of the macrocapsule, which is closest to the climatic chamber. For H1 (Figure 8b), the first layer required approximately 70 min for complete melting and 460 min for the PCM to completely discharge. In contrast, the second layer did not achieve complete melting. Solution H3 (Figure 8d) performed better, with the first layer achieving complete melting and solidification within 50 min and 310 min, respectively. However, the second layer required an additional 75 min to 90 min to fully mobilize the PCM. These findings underscore the need for further optimization of double-layered macroencapsulated solutions to enhance PCM performance during the charging and discharging processes.

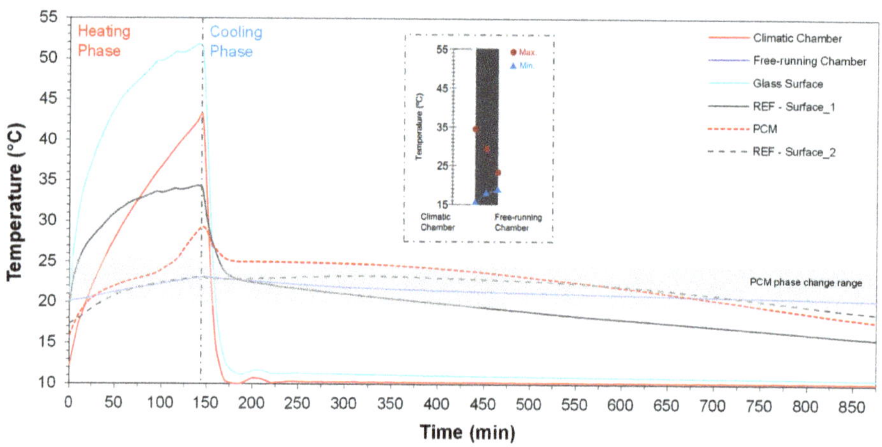

(**a**)

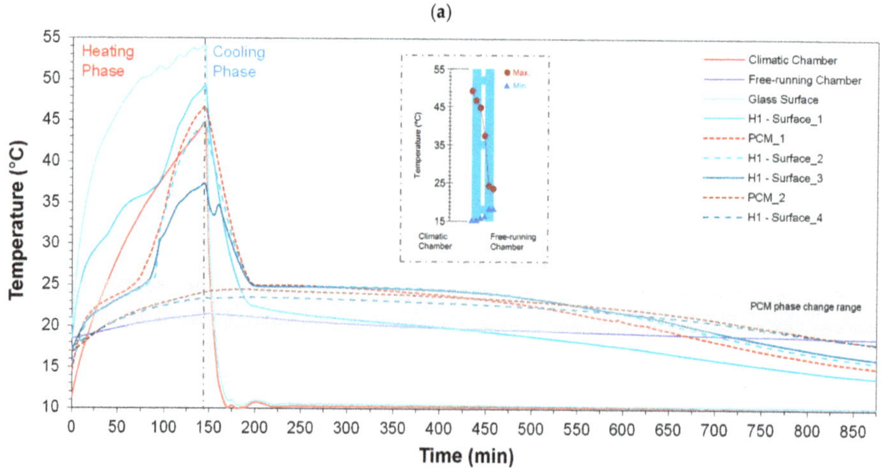

(**b**)

Figure 8. *Cont.*

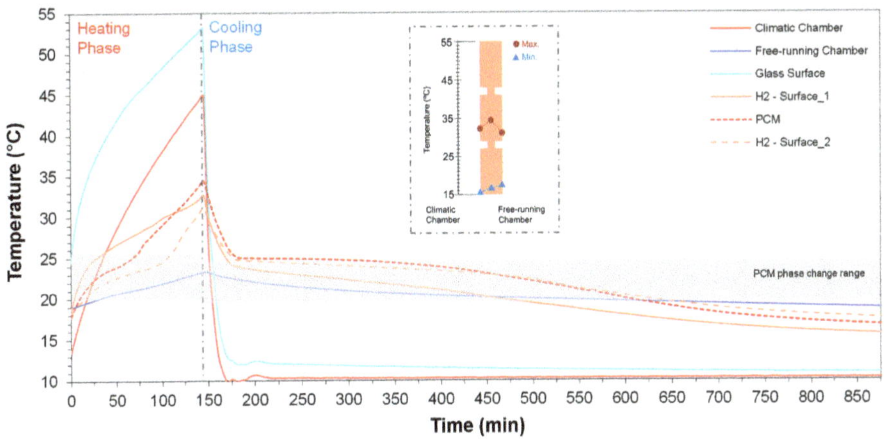

(**c**)

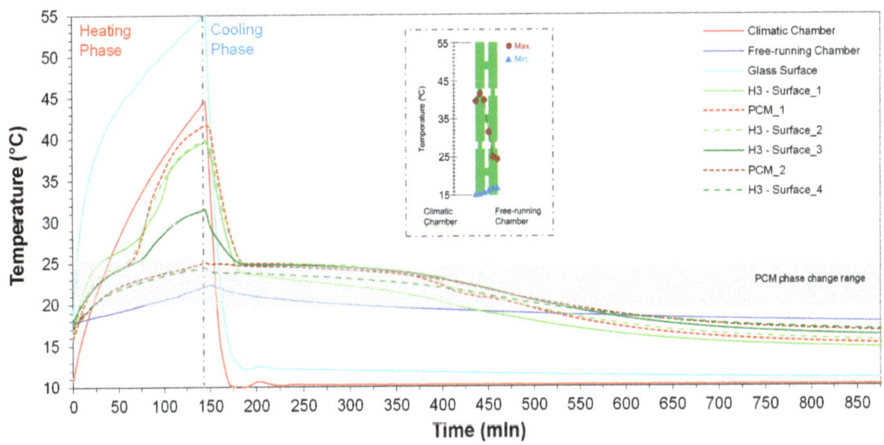

(**d**)

Figure 8. Temperature profiles and temperature distributions in cross-sections of the macrocapsule solutions during the heating and cooling cycles for (**a**) REF, (**b**) H1, (**c**) H2, and (**d**) H3.

3.2.2. PCM Performance Evaluation

Examining the performance of the PCM within the four tested macrocapsule configurations requires a detailed evaluation of its behavior during both the charging and discharging phases. This assessment was conducted by measuring the PCM temperature within the container and by analyzing and comparing the heating rates (Figure 9a) and melting times (Figure 10a) during the charging phase, as well as the cooling rates (Figure 9b) and solidification times (Figure 10b) during the discharging process. These evaluations considered the monitored PCM temperatures and the thermal characterization results of the material presented in Section 3.1. The results highlighted that the heating and cooling inputs were maintained at a constant intensity throughout the test.

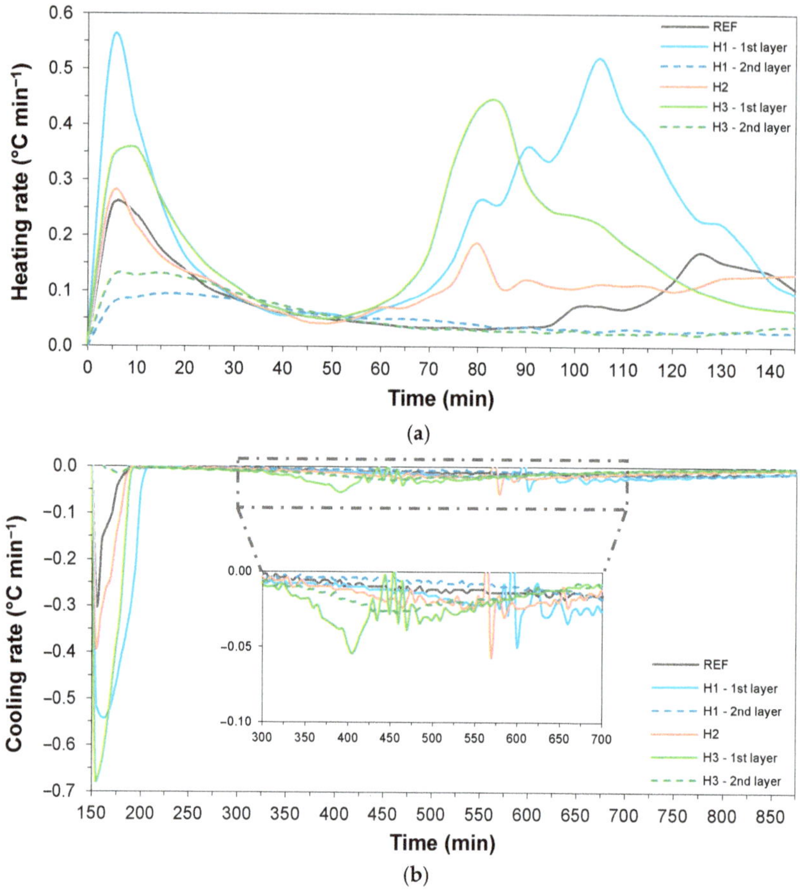

Figure 9. PCM melting and solidification behaviors of the PCM for each tested macrocapsule: (**a**) Heating rate; (**b**) Cooling rate.

Analyzing the single-layer solutions, the results indicate that the PCM within the H2 macrocapsule exhibits slightly higher heating and cooling rates (heating rate: 0.2–0.3 °C min^{-1}; cooling rate: 0.4 °C min^{-1}) compared with the REF macrocapsule (heating rate: 0.2–0.25 °C min^{-1}; cooling rate: 0.3 °C min^{-1}) during the material's sensible phase. This difference can be attributed to the geometry of the H2 macrocapsule, which promotes more efficient heat transfer.

In the double-layer solutions, the behavior of the PCM during the sensible phase is notably different: the PCM layer closer to the climatic chamber exhibits significantly higher heating and cooling rates (H1: heating and cooling rate of 0.55 °C min^{-1}; H3: heating rate of 0.35–0.45 °C min^{-1} and cooling rate of 0.7 °C min^{-1}), whereas the PCM layer closer to the free-running chamber shows a much slower response during both the heating and cooling phases (H1 and H3: heating rate of 0.1 °C min^{-1} and cooling rate of 0.02 °C min^{-1}).

The latent phase of the PCM is clearly observed in the plots presented in Figure 9. During the phase-change process, the material does not present a significant variation in its temperature, as reflected in the reduction and stabilization of the rates. This observation leads to the preliminary conclusion that, for all the tested macrocapsules, the solidification phase (Figure 9b) is significantly slower than the melting phase (Figure 9a).

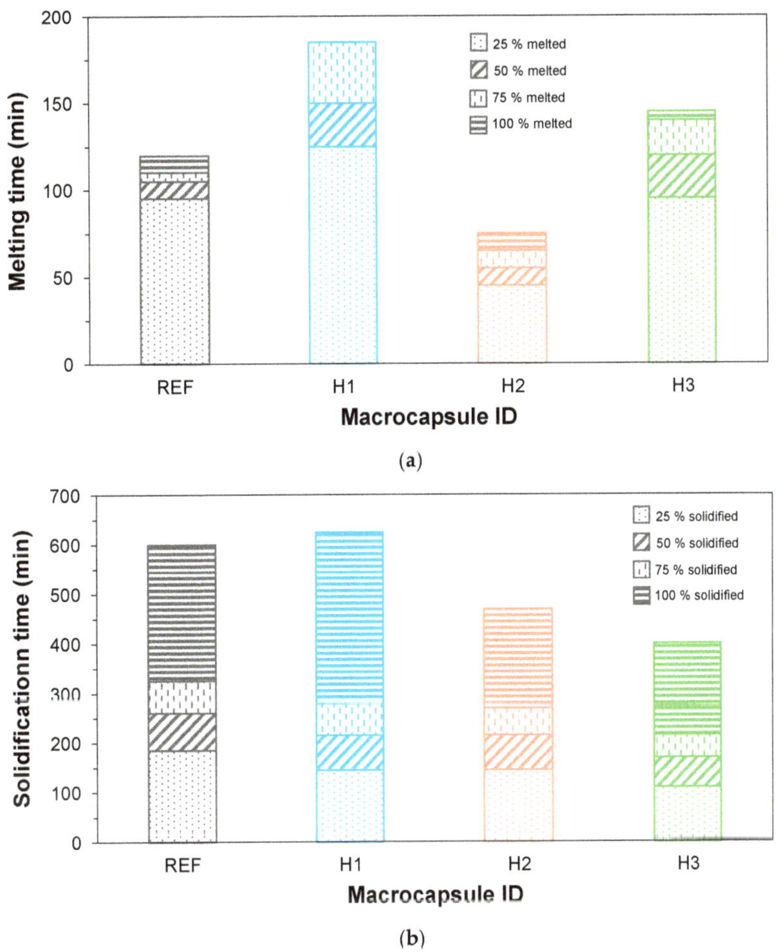

Figure 10. Melting (**a**) and solidification times (**b**) of the PCM for each tested macrocapsule.

When comparing these rates to those obtained from the DSC tests (Section 3.1), the values fall within a similar range (0–0.7 °C min^{-1}). This confirms that, in real-scenario building applications, PCMs experience slower heating and cooling processes than those materials typically employed in material characterization tests.

To complement the previous analysis and to accurately characterize the thermal performance of the PCM [40], the evolution of the liquid fraction of the PCM (f_{LIQUID}) was determined using the analytical model proposed by Thonon et al. [41], as described in Equation (1):

$$f_{\mathrm{LIQUID}} = \begin{cases} \frac{\sigma_S}{\sigma_S+\sigma_L}\left[\mathrm{erf}\left(\frac{T-T_{PC}}{\sigma_S}\right)+1\right], & T \leq T_{PC} \\ \frac{1}{\sigma_S+\sigma_L}\left[\sigma_L.\mathrm{erf}\left(\frac{T-T_{PC}}{\sigma_L}\right)+\sigma_S\right], & T > T_{PC} \end{cases}, \quad (1)$$

The coefficients σ_S (Equation (2)) and σ_L (Equation (3)) ensure that the liquid fraction approaches 0 when the PCM is fully solid and approaches 1 when the PCM is fully melted.

$$\sigma_S = \frac{\sqrt{2}}{4}(T_{PC} - T_S), \quad (2)$$

$$\sigma_L = \frac{\sqrt{2}}{4}(T_L - T_{PC}), \tag{3}$$

Based on the DSC results obtained at a rate of 0.25 °C min^{-1}, the following parameters are defined: T_{PC} = 25 °C (peak phase-change temperature during melting), T_S = 19.61 °C (temperature at which the PCM transitions into its solid state), T_L = 25.6 °C (temperature at which the PCM transitions into its liquid state), while T represents the instantaneous temperature of the PCM (°C). Using this model, the liquid fraction (f_{LIQUID}) of the PCM was quantified, and the solid fraction was subsequently determined as the difference between 1 and the instantaneous liquid fraction. Figure 10 illustrates the melting and solidification times for each of the tested macrocapsule solutions, considering thresholds of 25%, 50%, 75%, and 100% of the PCM's charge and discharge potential.

The analysis of the phase-change time, detailed for different percentages, reinforces the earlier observation; for this case study, the complete solidification of the PCM was approximately three to six times slower than complete melting, although the heating and cooling inputs are different.

Figure 10a highlights that during the heating process, all solutions require more time to achieve a 25% melted fraction of the PCM, whereas the melting of the remaining 75% occurs significantly faster. Notably, the H1 macrocapsule is the only solution that did not reach a fully melted state (100%) of the PCM. This is essentially due to its unsectioned contact surface and its double-layered structure. The first layer of H1, being slightly thinner than half the thickness of REF, facilitates rapid PCM melting (as shown in Figure 8b). On the other hand, the second layer experiences greater difficulty in this process. Thus, this solution attained only approximately melting of 75% the total PCM volume. Among the tested solutions, H2 exhibited the best performance during heating, achieving full melting within 75 min, owing to its sectioned contact surface. This design was followed by REF, which fully melted within 120 min. The H3 solution demonstrated slower PCM melting, requiring 145 min, mainly due to delays in melting the PCM in the second layer of the macrocapsules. Lastly, H1 displayed the least effective performance, achieving only a 75% melted fraction, primarily due to its more complex geometry.

During the solidification process, the behavior of the solutions differed slightly from that observed during melting, although all achieved a fully solidified state. The macrocapsules exhibited a higher discharge capacity for 75% of the PCM volume, while the remaining 25% required a significantly longer discharge time. Among the tested solutions, H3 delivered the best discharge performance, with a total solidification time of 400 min. This performance can be attributed to its sectioned geometry and the fact that the temperature of the PCM in the second layer of the macrocapsule barely exceeded the melting endset temperature (Figure 8d). The H2 design exhibited the second-best discharge performance (470 min), attributable to its geometric design and the higher temperatures attained by the PCM during the heating phase (Figure 8c). REF followed, with a solidification time of 600 min. As in the heating phase, solution H1 required the longest time to complete solidification (625 min) due to the same geometric limitations previously noted.

In summary, the sectioned geometry of the macrocapsule substantially improved heat transfer, enhancing the PCM's performance in retaining and dissipating thermal energy. Conversely, the double-layered geometry appeared to slow down the mobilization of the PCM in the second layer, negatively impacting the overall performance of the solution.

4. Conclusions

Selecting an appropriate container geometry is a key factor in ensuring efficient heat transfer, storage, and thermal energy release in PCM-based solutions. The present research focuses on the thermal characterization of PCMs and the experimental evaluation of four container geometries for enhanced thermal regulation in buildings, with the aim of assessing their impact on thermal performance and to compare the melting and solidification behaviors of the PCM

The DSC analysis was performed at five decreasing heating and cooling rates (4 °C min^{-1}, 2 °C min^{-1}, 1 °C min^{-1}, 0.5 °C min^{-1}, and 0.25 °C min^{-1}) to evaluate their influence on the PCM's thermal properties. The lowest tested rate significantly reduced the degree of supercooling by 75%, lowered the melting peak temperature by 6%, and increased the solidification peak temperature by 13%, showing a closer resemblance with the properties reported in the manufacturer's datasheet.

The experimental tests on the four macrocapsule geometries revealed notable differences in thermal performance. Configurations with sectioned contact surfaces, particularly the H2 solution, demonstrated superior heat transfer efficiency compared with the REF solution, as evidenced by a significantly reduced temperature gradient between the inner and outer surfaces. The H2 solution also reduced the melting and solidification times by 38% and 22%, respectively. Conversely, double-layered geometries (H1 and H3) faced challenges in mobilizing the PCM within the second layer, negatively affecting the overall performance. The H1 solution was unable to fully mobilize the volume of the PCM during the melting phase and exhibited a solidification time that was 4% longer than that of the REF solution. The H3 solution showed a slight improvement during the solidification phase, reducing the time by 33%, although it required 21% more time to fully melt compared with the REF solution.

Thus, a promising direction for further investigation and optimization of the developed solutions could involve using PCMs with different phase-change ranges (i.e., different operating temperature ranges) for the first and second layers of the macrocapsule to maximize the latent heat potential. This solution also offers an interesting advantage due to the dual functionality of the macrocapsule. The first layer could act as thermal insulation, reducing heat transfer between outdoor and indoor environments, while the second layer could serve as a thermal buffer to regulate indoor temperature fluctuations. Another effective strategy might involve redesigning the geometry by misaligning the contact surfaces between the first and second layers, thereby enhancing the charging and discharging processes and improving the thermal exchange efficiency. Additionally, establishing a thermal bond between the layers by modifying the connection joints to be hollow and incorporating the PCM could improve heat transfer within the PCM, thereby optimizing the overall macrocapsule thermal performance.

Author Contributions: Conceptualization, A.F.; methodology, M.G., A.F., G.V., F.R., R.M.S.F.A., M.S.A.O. and R.V.; validation, M.G., A.F., G.V., F.R., R.M.S.F.A., M.S.A.O. and R.V.; investigation, M.G., A.F., G.V. and F.R.; resources, M.S.A.O. and R.V.; data curation, M.G., A.F., G.V. and F.R.; writing—original draft preparation, M.G., A.F., G.V. and F.R.; writing—review and editing, M.G., A.F., G.V., F.R., R.M.S.F.A., M.S.A.O. and R.V.; visualization, M.G., A.F., G.V. and F.R.; supervision, R.M.S.F.A., M.S.A.O. and R.V.; project administration, R.V.; funding acquisition, R.V. All authors have read and agreed to the published version of the manuscript.

Funding: This research was supported by the Foundation for Science and Technology (FCT)—PhD Grant Ref. 2022.12799.BD. This research was also supported by RISCO—The Aveiro Research Centre for Risks and Sustainability in Construction, University of Aveiro, Portugal (FCT/UIDB/ECI/04450/2020

and DOI identifier 10.54499/UIDB/04450/2020), and by the Center for Mechanical Technology and Automation (TEMA) under project codes UIDB/00481/2020 and UIDP/00481/2020.

Data Availability Statement: The raw data supporting the conclusions of this article will be made available by the authors on request.

Conflicts of Interest: The authors declare no conflicts of interest.

References

1. Moghaddam, S.A.; Serra, C.; da Silva, M.G.; Simões, N. Comprehensive Review and Analysis of Glazing Systems towards Nearly Zero-Energy Buildings: Energy Performance, Thermal Comfort, Cost-Effectiveness, and Environmental Impact Perspectives. *Energies* **2023**, *16*, 6283. [CrossRef]
2. Xie, X.; Xu, B.; Fei, Y.; Chen, X.-N.; Pei, G.; Ji, J. Passive energy-saving design strategy and realization on high window-wall ratio buildings in subtropical regions. *Renew. Energy* **2024**, *229*, 120709. [CrossRef]
3. Feng, F.; Kunwar, N.; Cetin, K.; O'Neill, Z. A critical review of fenestration/window system design methods for high performance buildings. *Energy Build.* **2021**, *248*, 111184. [CrossRef]
4. Hwang, R.-L.; Chen, W.-A. Creating glazed facades performance map based on energy and thermal comfort perspective for office building design strategies in Asian hot-humid climate zone. *Appl. Energy* **2022**, *311*, 118689. [CrossRef]
5. Reddy, V.J.; Hariram, N.P.; Ghazali, M.F.; Kumarasamy, S. Pathway to Sustainability: An Overview of Renewable Energy Integration in Building Systems. *Sustainability* **2024**, *16*, 638. [CrossRef]
6. Irfeey, A.M.M.; Jamei, E.; Chau, H.-W.; Ramasubramanian, B. Enhancing Occupants' Thermal Comfort in Buildings by Applying Solar-Powered Techniques. *Architecture* **2023**, *3*, 213–233. [CrossRef]
7. Heier, J.; Bales, C.; Martin, V. Combining thermal energy storage with buildings—A review. *Renew. Sustain. Energy Rev.* **2015**, *42*, 1305–1325. [CrossRef]
8. Al-Yasiri, Q.; Szabó, M. Incorporation of phase change materials into building envelope for thermal comfort and energy saving: A comprehensive analysis. *J. Build. Eng.* **2021**, *36*, 102122. [CrossRef]
9. Reddy, K.; Mudgal, V.; Mallick, T. Review of latent heat thermal energy storage for improved material stability and effective load management. *J. Energy Storage* **2018**, *15*, 205–227. [CrossRef]
10. Al-Musaedi, S.; Asadi, Z.; Ganji, D.; Javidan, M. Simulation of melting and solidification processes of organic phase change materials used in storing heat energy in multi-layers heat exchanger. *J. Energy Storage* **2022**, *50*, 104687. [CrossRef]
11. Zhou, D.; Zhao, C.Y.; Tian, Y. Review on thermal energy storage with phase change materials (PCMs) in building applications. *Appl. Energy* **2012**, *92*, 593–605. [CrossRef]
12. Liu, Z.; Yu, Z.; Yang, T.; Qin, D.; Li, S.; Zhang, G.; Haghighat, F.; Joybari, M.M. A review on macro-encapsulated phase change material for building envelope applications. *Build. Environ.* **2018**, *144*, 281–294. [CrossRef]
13. Palacios, A.; Navarro-Rivero, M.; Zou, B.; Jiang, Z.; Harrison, M.; Ding, Y. A perspective on Phase Change Material encapsulation: Guidance for encapsulation design methodology from low to high-temperature thermal energy storage applications. *J. Energy Storage* **2023**, *72*, 108597. [CrossRef]
14. Yang, L.; Villalobos, U.; Akhmetov, B.; Gil, A.; Khor, J.O.; Palacios, A.; Li, Y.; Ding, Y.; Cabeza, L.F.; Tan, W.L.; et al. A comprehensive review on sub-zero temperature cold thermal energy storage materials, technologies, and applications: State of the art and recent developments. *Appl. Energy* **2021**, *288*, 116555. [CrossRef]
15. Manikandan, S.; Muthuvairavan, G.; Samykano, M.; Natarajan, S.K. Numerical simulation of various PCM container configurations for solar dryer application. *J. Energy Storage* **2024**, *86*, 111294. [CrossRef]
16. Punniakodi, B.M.S.; Senthil, R. A review on container geometry and orientations of phase change materials for solar thermal systems. *J. Energy Storage* **2021**, *36*, 102452. [CrossRef]
17. Park, J.; Shin, D.H.; Shin, Y.; Karng, S.W. Analysis of heat transfer in latent heat thermal energy storage using a flexible PCM container. *Heat Mass Transf.* **2019**, *55*, 1571–1581. [CrossRef]
18. Hekmat, M.H.; Haghani, M.H.K.; Izadpanah, E.; Sadeghi, H. The influence of energy storage container geometry on the melting and solidification of PCM. *Int. Commun. Heat Mass Transf.* **2022**, *137*, 106237. [CrossRef]
19. Qin, Z.; Ji, C.; Low, Z.H.; Tong, W.; Wu, C.; Duan, F. Geometry effect of phase change material container on waste heat recovery enhancement. *Appl. Energy* **2022**, *327*, 120108. [CrossRef]
20. Huang, B.; Yang, S.; Wang, J.; Lund, P.D. Optimizing the shape of PCM container to enhance the melting process. *Oxf. Open Energy* **2022**, *1*, oiab006. [CrossRef]
21. Liu, H.-Y.; Qu, B.-C.; Wu, C.-M.; Li, Y.-R. Numerical study of PCM melting performance in a rectangular container with various longitudinal fin structures. *J. Energy Storage* **2024**, *94*, 112529. [CrossRef]

22. Zhang, Y.; Bozorg, M.V.; Torres, J.F.; Zhao, Y.; Wang, X. Dynamic melting of encapsulated PCM in various geometries driven by natural convection of surrounding air: A modelling-based parametric study. *J. Energy Storage* **2022**, *48*, 103975. [CrossRef]
23. Issa, O.O.; Thirunavukkarasu, V. Experimental study on charging and discharging behavior of PCM encapsulations for thermal energy storage of concentrating solar power system. *J. Energy Storage* **2024**, *85*, 111071. [CrossRef]
24. Erlbeck, L.; Schreiner, P.; Schlachter, K.; Dörnhofer, P.; Fasel, F.; Methner, F.-J.; Rädle, M. Adjustment of thermal behavior by changing the shape of PCM inclusions in concrete blocks. *Energy Convers. Manag.* **2018**, *158*, 256–265. [CrossRef]
25. Tu, H.; Wei, Z.; Bahrami, A.; Ben Kahla, N.; Ahmad, A.; Özkılıç, Y.O. Recent advancements and future trends in 3D concrete printing using waste materials. *Dev. Built Environ.* **2023**, *16*, 100187. [CrossRef]
26. RUBITHERM®. Data Sheet—RT25HC. 2020. Available online: https://www.rubitherm.eu/media/products/datasheets/Techdata_-RT25HC_EN_09102020.PDF (accessed on 15 October 2024).
27. Islam, S.; Bhat, G.; Sikdar, P. Thermal and acoustic performance evaluation of 3D-Printable PLA materials. *J. Build. Eng.* **2023**, *67*, 105979. [CrossRef]
28. UltiMaker. PLA for S Series. 2024. Available online: https://ultimaker.com/materials/s-series-pla/ (accessed on 15 October 2024).
29. Kheradmand, M.; Azenha, M.; de Aguiar, J.L.B.; Krakowiak, K.J. Thermal behavior of cement based plastering mortar containing hybrid microencapsulated phase change materials. *Energy Build.* **2014**, *84*, 526–536. [CrossRef]
30. Rebelo, F.; Figueiredo, A.; Vicente, R.; Ferreira, V. Study of a thermally enhanced mortar incorporating phase change materials for overheating reduction in buildings. *J. Energy Storage* **2022**, *46*, 103876. [CrossRef]
31. Iten, M.; Liu, S.; Shukla, A.; Silva, P. Investigating the impact of C_p-T values determined by DSC on the PCM-CFD model. *Appl. Therm. Eng.* **2017**, *117*, 65–75. [CrossRef]
32. Christy, J.V.; Balwani, A.; Mehling, H.; Agrawal, N. Optimization of DSC measurements for organic phase change materials. *J. Energy Storage* **2023**, *73*, 109032. [CrossRef]
33. Kousksou, T.; Jamil, A.; Zeraouli, Y. Enthalpy and apparent specific heat capacity of the binary solution during the melting process: DSC modeling. *Thermochim. Acta* **2012**, *541*, 31–41. [CrossRef]
34. Feng, G.; Huang, K.; Xie, H.; Li, H.; Liu, X.; Liu, S.; Cao, C. DSC test error of phase change material (PCM) and its influence on the simulation of the PCM floor. *Renew. Energy* **2016**, *87*, 1148–1153. [CrossRef]
35. Cook, D.; Julias, M.; Nauman, E. Biological variability in biomechanical engineering research: Significance and meta-analysis of current modeling practices. *J. Biomech.* **2014**, *47*, 1241–1250. [CrossRef]
36. Kośny, J.; Thakkar, J.; Kamidollayev, T.; Sobkowicz, M.J.; Trelles, J.P.; Schmid, C.; Phan, S.; Annavajjala, S.; Horwath, P. Dynamic Thermal Performance Analysis of PCM Products Used for Energy Efficiency and Internal Climate Control in Buildings. *Buildings* **2023**, *13*, 1516. [CrossRef]
37. Castellón, C.; Günther, E.; Mehling, H.; Hiebler, S.; Cabeza, L.F. Determination of the enthalpy of PCM as a function of temperature using a heat-flux DSC-A study of different measurement procedures and their accuracy. *Int. J. Energy Res.* **2008**, *32*, 1258–1265. [CrossRef]
38. Ukrainczyk, N.; Kurajica, S.; Šipušić, J. Thermophysical Comparison of Five Commercial Paraffin Waxes as Latent Heat Storage Materials. *Chem. Biochem. Eng. Q.* **2010**, *24*, 55015.
39. Soares, N.; Matias, T.; Durães, L.; Simões, P.; Costa, J. Thermophysical characterization of paraffin-based PCMs for low temperature thermal energy storage applications for buildings. *Energy* **2023**, *269*, 126745. [CrossRef]
40. Gonçalves, M.; Figueiredo, A.; Almeida, R.M.; Vicente, R.; Samagaio, A.; Kośny, J. Thermal performance evaluation of a light steel framing building with macroencapsulated phase change materials in a Mediterranean climate. *Energy Build.* **2024**, *323*, 114837. [CrossRef]
41. Thonon, M.; Fraisse, G.; Zalewski, L.; Pailha, M. Towards a better analytical modelling of the thermodynamic behaviour of phase change materials. *J. Energy Storage* **2020**, *32*, 101826. [CrossRef]

energies

MDPI

Article

From Waste to Functional Material—Carbon Aerogels from Citrus Biomass Infiltrated with Phase Change Materials for Possible Application in Solar-Thermal Energy Conversion and Storage

Katarzyna Suchorowiec [1], Martyna Bieda [2], Martyna Szatkowska [1], Małgorzata Sieradzka [3], Monika Kuźnia [3], Magdalena Ziąbka [4] and Kinga Pielichowska [1,*]

1 Department of Glass Technology and Amorphous Coatings, Faculty of Materials Science and Ceramics, AGH University of Krakow, Al. Mickiewicza 30, 30-059 Krakow, Poland; suchorowiec@agh.edu.pl (K.S.); szatkowska@agh.edu.pl (M.S.)

2 Department of Biomaterials and Composites, Faculty of Materials Science and Ceramics, AGH University of Krakow, Al. Mickiewicza 30, 30-059 Krakow, Poland; mbieda@student.agh.edu.pl

3 Department of Heat Engineering & Environment Protection, Faculty of Metals Engineering and Industrial Computer Science, AGH University of Krakow, Al. Mickiewicza 30, 30-059 Krakow, Poland; msieradz@agh.edu.pl (M.S.); kuznia@agh.edu.pl (M.K.)

4 Department of Ceramics and Refractories, Faculty of Materials Science and Ceramics, AGH University of Krakow, Al. Mickiewicza 30, 30-059 Krakow, Poland; ziabka@agh.edu.pl

* Correspondence: kingapie@agh.edu.pl

Abstract: Green energy conversion and storage materials have become a focal point of research in recent times, especially in energy-consuming buildings. Phase change materials (PCMs) have gained more and more attention not only for energy storage but also in composites for solar energy conversion. This research investigates a sustainable method for converting orange biomass waste (OBW) into advanced porous carbon aerogel (PCA) composites, designed for solar-thermal energy harvesting and storage in building applications. Using potato starch as a binder, the research develops a process for producing a uniform and lightweight carbon matrix that could be scalable. The best results were found for PCA obtained with 2.5% starch, where the lowest mass loss (8.2, 0.5, 11.2% pt) was observed during the leakage test. This study highlights the suitability of OBW-derived aerogels as effective matrices for PCM impregnation and shape stabilization, indicating their future potential in solar-thermal energy conversion and storage and potentially lowering energy consumption in buildings. By repurposing agricultural waste, this work contributes to sustainable material development and advances the application of renewable energy technologies.

Keywords: biomass; phase change materials; solar energy harvesting; biomass waste; citrus waste; carbon aerogels

1. Introduction

Energy consumption has increased rapidly over the last decades [1]. This increase in energy consumption has led to extensive usage of fossil fuels, followed by an increase in greenhouse gas emissions because of the still-increasing and dominant nonrenewable energy sources share. To overcome those problems, researchers aim to develop new technologies that could help with harvesting green energy and develop more sustainable methods of energy generation and storage. Researchers' attention has shifted to renewable energy sources like solar, wind, hydropower, biogas or geothermal [2]. Despite numerous works related to those fields, there is still a need for further improvement to eliminate the mismatch between the amount of energy generated and its demand in time and space.

According to the Global Status Report for Buildings and Construction (Buildings-GSR) [3] in 2022, the buildings sector consumed 30% of the final energy demand, mainly for cooling and heating. Additionally, four percent more was absorbed to produce construction materials. In general, the energy demand in this sector increased by one percent only in one year. Najjar et al. [4] investigated heat energy loss in buildings in three different climates based on climate classification (tropical climates, dry climates, and moist subtropical mid-latitude climates). The data obtained indicated that the buildings lose energy in two ways: ventilation heat loss, or fabric heat loss. However, only 19% of said losses are due to ventilation heat loss; 81% are due to fabric heat loss. The loss resulting from fabrics can be related to doors and windows, at 51.50%, and floors, exterior walls, and roofs, at 15.75%, 19.87%, and 12.88%, respectively [4]. Energy storage is becoming an essential element for maximizing the effective utilization of energy [5]. Energy losses can be reduced by using PCMs. PCMs are materials that store energy while heating, and release it when cooling [6]. This process is related to phase changes, in solid–liquid, solid–solid or liquid–gas changes [7]. In general, PCMs can be divided into three main groups: organic (paraffin compounds, non-paraffin compounds), inorganic (hydrated salt, metallic), and eutectic (organic–organic, organic–inorganic, inorganic–inorganic) [5].

The literature indicates numerous examples of PCMs being used in building materials, such as walls, ceilings or insulating walls [8]. There is still room to improve the performance of PCMs used in industry, and implementation of those materials in different building elements, thus, in the research and development sections of many studies, is carried out in the building industry, especially as regards energy-related materials. Hence, Li et al. [9] proposed the use of paraffin-based PCMs—such as MG29—as a modifier to triple-pane and double-pane windows. The results show that the composition of the triple-pane window's PCM was the best at reducing energy consumption and temperature shifts. During sunny days, the interior surface temperature was 2.7 °C, which was 5.5 °C lower than for the double-pane PCM window and triple pane window. The heat transfer of triple-pane windows with PCMs is reduced by 14.7% compared to double-pane windows + PCM, and is increased by 4.5% compared to standard triple-pane windows [9]. Kishore et al. [10] proposed dynamic insulation materials and system-wide thermal resistance based on indoor and outdoor conditions. The mathematical analysis performed shows that the proposed system provides a 7–38% reduction in annual heat loss and a 15–72% reduction in annual heat gain. Walls facing different directions receive varying amounts of solar heat, which affects heat gain and loss throughout the day.

Phase change materials can also be applied in the production of intelligent paint. Qin et al. [11] produced materials from thermoplastic acrylic resin, radiative binders, SiO_2-PCM (paraffin) microcapsules, sunlight-scattered particles, IR emitters, and heat containers. The obtained materials were characterized by radiative cooling and their thermal storage abilities for passive cooling [11]. An interesting solution was proposed by Yaxuan et al. [12] for composite building for heating and cooling. They produced a material from semi-coke ash and industrial $NaNO_3$ as a PCM. The team used the common waste from the fabric, which is popular, to obtain shape-stable PCMs. The thermal energy storage (TES) capacity was 325.91 J/g and the phase transition range was from 100 °C to 380 °C [12].

The global economy is turning more towards sustainable solutions and development based on the circular economy model. The aim of it is to create a sustainable economic system in terms of the optimalization of natural resource utilization, as well as minimizing waste from production process, for general environmental and social well-being. But, the core of the principle, generally, is the 'restorative use' of resources, such that raw materials do not become discarded waste [13]. The use of materials from natural sources and modifications to them have been gaining a lot of attention in recent decades, but, according to the circular economy conceptualization, there should also be a focus on the utilization and management of waste through its reprocessing [14]. Thus, biomass waste (BW) recycling and its incorporation into materials' design processes should be developed to close the loop in the production process and propose sustainable materials solutions [15].

The main problem connected with use of PCM is the solid–liquid transition, which causes a tendency for volume to change or leak, during the phase transition processes. A solution to this is the use of porous materials, including carbonized biomass [16]. Combining BW and PCMs could lead to the obtaining of a green and sustainable composite in both material sourcing and performance classification. There are several types of biomass that are used for BW-PCMs, such as wood [17–19], fruits [20–24], and vegetables [25–28], along with others, like walnut [29,30], cotton fiber [31], etc. To reduce the costs of production, BW can also be carbonized and used as an alternative for carbon materials with complex preparation processes, like graphene, graphene oxide, or carbon nanotubes. However, to obtain an affordable carbon material from the carbonization of BW, the BW should be abundant in source, cost-effective, renewable, and environmentally friendly [32].

Yearly, 75 Mt of oranges is produced (Figure 1) [33]; approximately 50–60% of these fruits are predicted to turn into waste, which adds up to ~35 Mt of biomass waste per year from one type of fruit. Orange peel is a material rich in cellulose and lignin, which is beneficial in that, through the carbonization process, it turns into a highly porous carbon microstructure; additionally, orange peels have a proper pore microstructure that can retain its properties after carbonization and be beneficial in terms of PCM loading and its further performance due to maintained structure [34]. Converting orange peel waste into valuable carbon-based products promotes the principles of a circular economy by minimizing dependence on finite resources and opening new opportunities for economic growth within sustainable, bio-based industries.

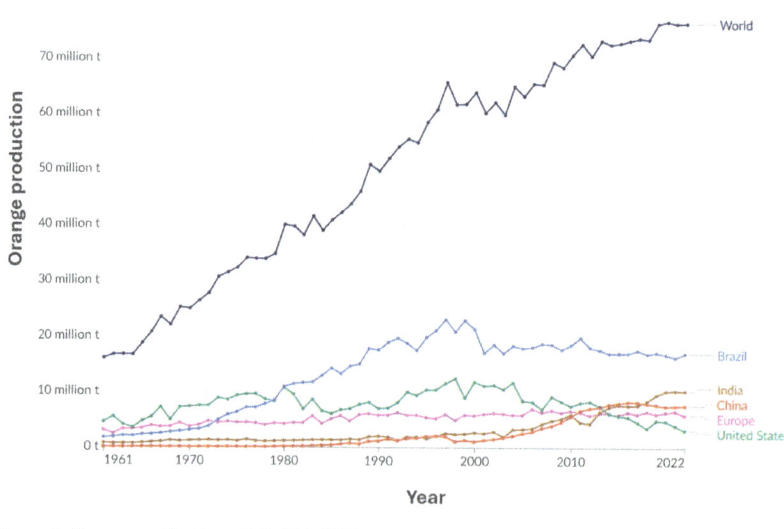

Figure 1. Orange production, 1961–2022 [33].

Carbonized biomass can be a good enhancer of PCM performance, especially in improving the thermal conductivity of organic PCMs, and it can be a shape stabilizer. The combination of a 3D carbon matrix with PCMs can provide solar-thermal conversion functionality due to the combination of optical absorption, energy transfer mechanisms, and thermal dissipation [35–37].

Sheng et al. [24] proposed research on MXene-wrapped, bio-based carbonized pomelo peel foam/poly(ethylene glycol) (PEG) composites. The pore size and MXene nanosheets determine the absorption ability of the PEG. The composites exhibited high PCM loading (96.2 wt%) and enthalpy efficiency, after the 20th thermal cycle, of 95%. Another study, presented by Ji et al. [22], discusses ultralow-density 3D porous carbon aerogels derived from waste pomelo peels impregnated with paraffin wax (PW), with an impregnation loading ratio of ~95% and a comprehensive strength of 466 kPa at 60% compression. The thermal performance of the composites reached a latent thermal storage capacity of 159.9 J/g with the solar-to-thermal energy conversion efficiency of 85.1%. However, a review of the literature proved that there is not a uniform way to process citrus biomass in such a way that it could be scalable to the industry level.

Therefore, this article aims to present a novel approach to obtain a repetitive, porous, light-carbon matrix delivered from orange biomass waste (OBW), with potato starch as a binder, to uniformize the preparation process and make it scalable. Our research includes a study on different binder contents in the OBW precursor for carbonization. Obtained carbon aerogels were further impregnated with three different PCMs (fatty acid, paraffin wax, and PEG) to evaluate the aerogel performance in composites in terms of impregnation and loading, thermal properties, shape stability, and leakage.

2. Materials and Methods

2.1. Materials

The orange biomass waste (leftovers) was collected from local cafes. Ethanol (99.9%wt) and Starch (ST) (99.9%, POCH) were acquired from Avantor Performance Materials, Poland, S.A. (Gliwice, Poland) PEG 6000 (99.9%, Sigma Aldrich, Darmstadt, Germany), palmitic acid (99.0%, Chempur, Piekary Śląskie, Poland), and Octacosane (99+% Thermo Scientific Chemicals, Bremen, Germany) were used as PCMs to infiltrate carbon aerogels.

2.2. Preparation of Porous Carbon Aerogels (PCA)

The procedure for the preparation of PCA was designed according to the literature [38–40] and TGA results for raw biomass. The orange biomass waste (OBW) was washed once with deionized water, once with ethanol, and then with deionized water again to ensure the removal of impurities. Then, the OBW was blended to obtain similarly sized small pieces (~1 mm^3) of biomass, and was then dried for ~30 min in the oven at 90 °C. Next, the ST binder was prepared in 0.8 g of deionized water with ST to obtain 2.5, 5, 10, and 20 wt.% ST in the samples; later, the proper amount of blended OBW was added to the ST gel. The mixtures were kept in an ultrasonic bath at 80 °C for 10 min, then transferred to the cylindrical containers. The samples were frozen at −80 °C for 24 h and subsequently freeze-dried for 48 h (temp −40 °C and 0.055 mbar/7 Pa). The lyophilized samples were next subjected to a four-step carbonization processes in a tube furnace (Czylok, Jastrzębie-Zdrój, Poland) and N_2 atmosphere, involving the following: I—heating to 200 °C with a heating rate of 2 °C/min; II—maintenance of temperature at 200 °C for 1.5 h; III—heating to 600 °C with a heating rate 5 °C/min; IV—maintenance of temperature at 200 °C for 1.5 h. Then, the samples were left to cool to room temperature in the furnace.

2.3. Preparation of PCM-PCA Composites

The PCM-PCA composites were prepared by vacuum assisted impregnation. The PCMs were melted at 80 °C in a vacuum oven in glass containers. Then, the PCA was inserted into the completely melted PCM. The samples were impregnated in a vacuum for 1 h. The vacuum dryer had previously been purged with dry nitrogen. After the vacuum assisted (500 mbar) impregnation, the samples were transferred to filter paper, to remove of excess PCM (on the outside of the PCA), and left for cooling. Preparation steps for PCM-PCA composites are presented in Figure 2.

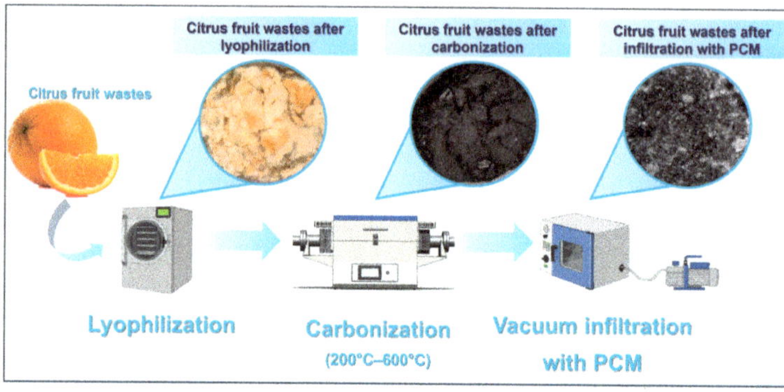

Figure 2. Preparation scheme for PCM-PCA composites.

2.4. Characterization

The CHN elemental analysis was performed using LECO CHN 628 (St. Joseph, MI, USA) to evaluate the carbon element content, especially regarding carbon content before and after the carbonization process, and to compare it with the samples with different ST contents. Samples after the carbonization process were weighed and measured to determine the density and theoretical porosity of the PCA obtained. The density ($\rho_{aerogel}$) was calculated according to Equation (1), where m is the weight of the PCA, and V is volume of the PCA. Porosity (ε) of the PCA was determined according to Equation (2), where $\rho_{aerogel}$ is from Equation (1) [41] and $\rho_{theoretical}$ is the theoretical density of the aerogel substrate, which has a theoretical carbon element density of 2.2 g/cm^3 [42].

$$\rho_{aerogel} = \frac{m}{V} \tag{1}$$

$$\varepsilon = \rho_{aerogel}\left(\frac{1}{\rho_{aerogel}} - \frac{1}{\rho_{theoretical}}\right) \times 100\% \tag{2}$$

The lyophilised samples were examined by thermogravimetric analysis (TGA) (TA Instruments TGA 550, New Castle, DE, USA) and evaluated using Trios software (Version 4.3.1 39215) The temperature range was 40–600 °C, with a heating rate of 10 °C/min and a nitrogen atmosphere (30 mL/min); in open platinum pans, the sample mass was ~5 mg. an analysis was performed to evaluate the thermal decomposition of samples that may occur during the carbonization process and adjust the temperature resulting from the carbonization steps. The same parameters were used to perform the thermal stability evaluation of the PCA and PCM-PCA composite samples.

The microstructurea of the lyophilised PCA and PCM-PCA composites samples were studied using scanning electron microscopy (SEM) (Apreo 2S low vac., Waltham, MA, USA) to evaluate the changes in microstructure before and after carbonization, and additionally evaluate the impregnation quality of composites. Observations were performed in high vacuum conditions (50 Pa) using a secondary-electron ETD detector with an acceleration voltage of 10 HV. Image J software (Version 1.52v) function *Analyze Particles*) was used to determine the average pore size (μm^2) and the percentage of pore area on the image (%).

Optical microscopy, performed on a Keyence VHX-900F (Mechelen, Belgium), was used to evaluate the incorporation of PCM into the PCA, and after the leakage and shape stability tests, was performed on the PCM-PCA composites.

The thermal properties of PCM-PCA composites were examined by differential scanning calorimetry (DSC), using the Melteler Toledo DSC 1 STARe System (Greifensee, Switzerland). Measurements were performed in a temperature range of −40–200 °C, with heating and cooling rates of 10 °C/min and an atmosphere of N_2 with 30 mL/min flow. Samples with mass ca. 6 mg were placed in pierced aluminum pans. The supercooling of PCM-PCA composites, T_s, was calculated according to Equation (3) [43], where T_m is the temperature of the melting peak and T_c is the temperature of the crystallization peak.

$$T_s = T_m - T_c \tag{3}$$

The data obtained from the analysis were also used to calculate the encapsulation yield (λ) of the PCA impregnated with PCMs according to Equation (4) [44], where $\Delta H_{m\ PCM-PCA}$ is the melting enthalpy of the PCM-PCA composite sample measured with DSC, and $\Delta H_{m\ PCM}$ is the melting enthalpy of the pure PCM used for the impregnation. The relative heat storage efficiency (η) was calculated according to Equation (6) [45], using the DSC results and loading rate (ω) obtained from Equation (5) [46], where m_c is the weight of PCM-PCA composites after the impregnation and m is the weight of PCA before impregnation.

$$\lambda = \frac{\Delta H_{m\ PCM-PCA}}{\Delta H_{m\ PCM}} \cdot 100\% \tag{4}$$

$$\omega = \frac{m_c - m}{m} \tag{5}$$

$$\eta = \frac{\Delta H_{m\ PCM-PCA}}{\omega \cdot \Delta H_{m\ PCM}} \cdot 100\% \tag{6}$$

The shape stability and leakage tests were performed on a heated plate placed in the oven at 80 °C. Digital photographs of PCM-PCA composite samples were taken at 0, 2, 7, 15, 30, and 60 min. According to the images and analysis performed using ImageJ software of the mean diameter, the leakage was calculated. Additionally, the loading rate of the PCM-PCA composite samples was calculated according to Equation (5), using the mass of samples after the leakage test as the m_c value.

3. Results and Discussion

3.1. SEM

The morphology of PCA and the PCM-PCA samples were recorded by SEM. The morphology of PCA samples differ from each other in regard to the base of the ST content in non-carbonized samples; nevertheless, all of them have highly porous 3D structures, which may result in high adsorption and impregnation ratios. The PCA obtained from pure OBW is less porous, with a wider specific surface area. The pore area is 19.8% (total image area), as seen in Figure 3A. This could be beneficial in terms of enhancing solar-thermal conversion in the final composite, but at the same time limits the loading capacity of the matrix. Some small pores could be distinguished within the ~15 μm mean diameter range; additionally, in Figure 3A, some closed pores could be noted. The samples with the addition of ST exhibit well-distributed pores within their structures (Figure 3B–E). A further increase in ST content in the sample results in pore increases, and in the case of the 2.5% ST PCA sample, the mean diameter is ~30 μm with 45.2% (total image) area, similar to the 5% ST (45.17% area). In case of the 20% ST PCA sample, the microstructure is less uniform, and the pore area is only 23.8% (total image area), which is less in comparison to other samples with ST, for example in the case of 10% ST, for which the area is 32.51%. Additionally, the higher ST content led to the creation of ST beans (average diameter ~20 μm), most of which lasted through the carbonization process and remained in the form of carbon beans, and some of which broke, which may be good for further PCM incorporation and increased active surface areas (Figure 3E, yellow arrow); in contrast, the sample with lower ST content does not exhibit bean-like microstructures, but the ST precursor left more flake-like carbon matrix fragments. The 2.5% ST PCA sample exhibits a more regular open-pore structure, which may be beneficial in terms of loading and incorporation of PCMs. The pore presence increases as the ST content gets higher, but the 20% ST sample represents a more closed structure due to the ST beans. The morphology of PCM-PCA (Figure 3F–J) indicates that the impregnation procedure was effective, leading to a good distribution of PCMs within the matrix. The micrographs of the cross section of PCM-PCA indicate that PCA pores were almost filled with PEG materials, and the same applies to the rest of the PCMs used in this research. Only in the cases of the 2.5% and 10% ST PEG-PCA samples were some pores present; in the 10% ST sample, this could be the result of a more open, highly porous PCA structure (the highest among all PCA samples). The 20% ST sample showed some starch-origin beans, which were closed during the carbonization process and could not be impregnated with PCM material. This indicates that the proper proportion of binder being added to the biomass waste in the precursor stage is crucial to obtain a material with good porosity and open infrastructure pathways (an open porous network) to enable the effective impregnation, and in the same way prevent leakage, during melting phase changes in PCM materials from the PCA. Compared to other, similar works, such as the SEM results presented by Xie et al. [23,47] on carbonized lemon and pitaya peals and PEG, the evaluation of materials after impregnation reveals fewer empty pores and a more uniform structure, indicating better impregnation quality of ST-based PCA-PEG samples (Figure 3F–J).

3.2. Results for Precursor and PCA

After the lyophilization process, the samples (precursors) were examined with TGA within the carbonization temperature range. The DTG curve indicates five main mass-loss stages (Figure 4B), at 40~120, 121~249, 250~287, 288~388, and 389~600 °C, which are attributed to the degradation of the following main components of orange peels: (1) water and soluble components, (2) pectin, (3) hemicellulose, (4) cellulose, and (5) lignin [48–50]. The fourth peek, at 288~388 (331) °C, increases with the starch content in the sample, which is attributed to the starch degradation that occurs around 320 °C [51].

According to the CHN analysis, the overall percentage of carbon in the PCA samples was almost two times higher than in the lyophilized one, reaching values from ~82–88% (Figure 4C). The carbon percentage in the PCA samples increases with the higher starch content in the precursor lyophilized samples. This is an indication that the samples turned out highly carbonized and that they could represent a promising material in terms of their potential in solar energy harvesting.

Additionally, the hydrogen percentage in all the samples before and after carbonization oscillate within the same value range (~7%), and after the carbonization process the values dropped from ~7 to 3% (Figure 4D). This is the result of the degradation processes of cellulose, hemicellulose, pectin, and other ingredients in orange peels. The percentage of nitrogen in the samples may slightly increase due to the nonvolatile nature of bonded nitrogen (compared to nitrogen and carbon in volatile compounds, such as CO_2, CO, CH_4 and H_2O) and what remains from the degradation of nitrogen-containing compounds like proteins or amino acids (Figure 4E). The amount of nitrogen is not correlated with the starch content in the precursors, and the range before the carbonization process is ~1%, and after carbonization is 1.5%. The lyophilized precursor samples exhibited a decreasing tendency for nitrogen content, which was observed along with an increase in ST content (less proteins, etc., from OBW). The highest percentage of nitrogen occurs in the matrix with 0% starch content, and the lowest in the lyophilized product with 20% starch content.

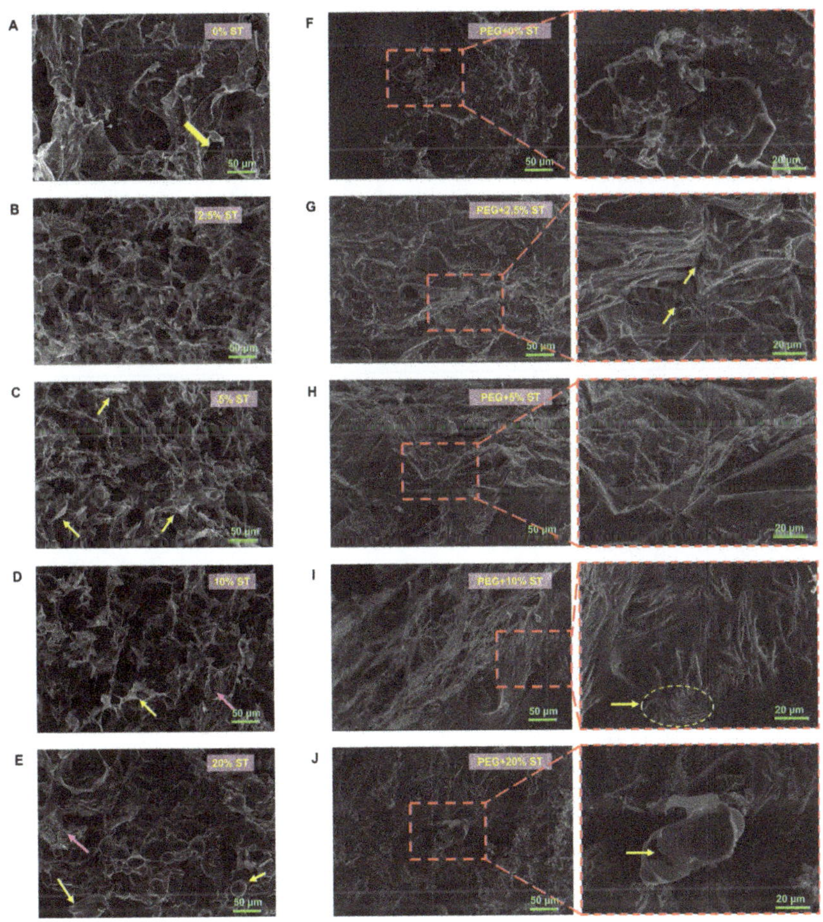

Figure 3. SEM micrographs of PCA with (**A**) 0% ST, (**B**) 2,5% ST, (**C**) 5% ST, (**D**) 10% ST, and (**E**) 20% ST, and PCM-PCA composites infiltrated with PEG on the PCA with (**F**) 0% ST, (**G**) 2,5% ST, (**H**) 5% ST, (**I**) 10% ST, and (**J**) 20% ST. (yellow arrows/circles—empty pores, pink arrows—ST sheets).

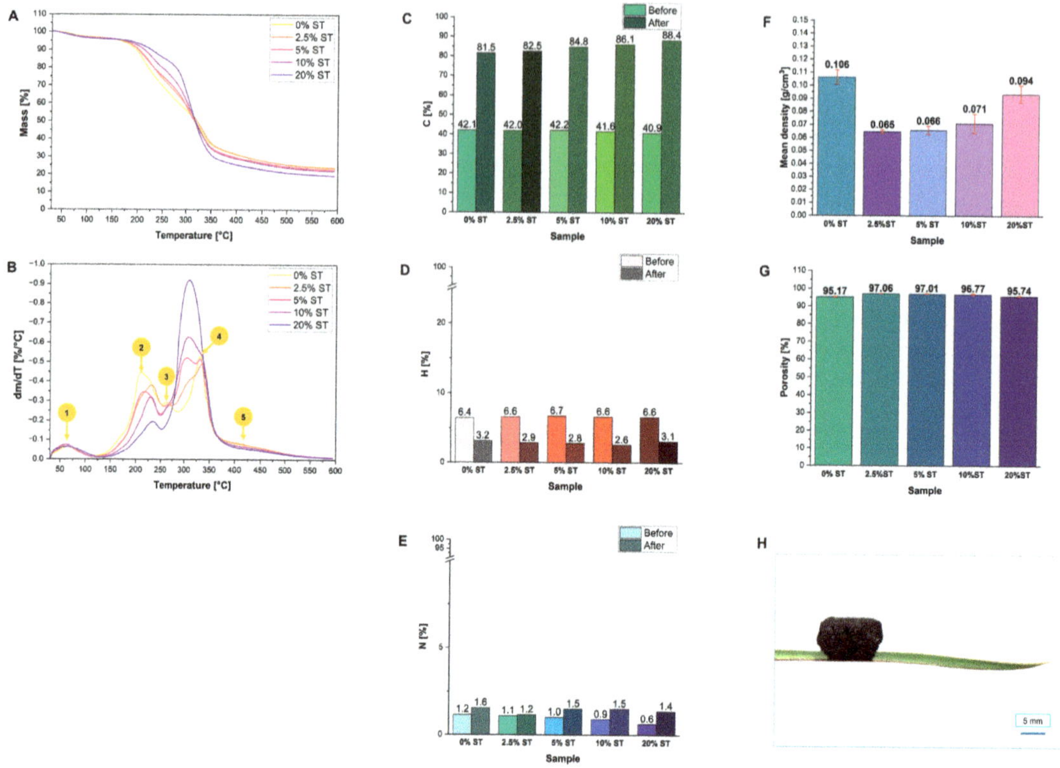

Figure 4. TG (**A**) and DTG (**B**) curves of lyophilized samples; elemental analysis results before and after carbonization for (**C**) carbon, (**D**) hydrogen, and (**E**) nitrogen elements; (**F**) mean density of PCA; (**G**) porosity of PCA; (**H**) digital photo of carbonized OBW on dragon tree leaf.

The PCA, after the carbonization process, exhibits a mean density in the range of 0.065 to 0.106 g/cm^3, resulting in ultra-light material (Figure 4H). The highest density is observed for the pure-orange sample; with the addition of the ST binder, the density has decreased (Figure 4F). The densities of the PCA samples with lower ST content at the precursor stage were lower because of the less packed structure; at the opposite site, higher starch concentrations led to the filling of the empty spaces within the OBW, making the structure denser.

The theoretical porosity of PCA (Figure 4G) is high in all the PCA samples above 95%, which is characteristic for the carbon aerogels [52]. The high porosity is beneficial for combining the porous carbon aerogels with PCM to create a PCA-PCM composite with a high impregnation ratio.

3.3. TGA Results (Thermal Stability of PCM-PCA)

The thermal properties of PCM-PCA were evaluated using TGA. The results of the analysis and the temperatures found for 1%, 3%, 5%, 10%, and 50% mass loss are displayed in Table 1. The changes in mass of the samples come mostly from the PCM that has been impregnated into the PCA, since the carbon matrix does not degrade in the examination temperature range. The PCA samples impregnated with PEG as a PCM exhibited the highest thermal stability in comparison to the pure PEG. Additionally, matrices with increased ST content (at the precursor stage) delay the thermal degradation process. This could be a cause the of well-distributed pores and the active surface, which is made possible by physical interactions and the strong adhesion of PEG to the pore walls. The first significant change in mass for the PEG-PCA composites was observed above ~350 °C. All composites show significantly higher thermal stability than pure PEG. The composite with 5% ST content has the lowest stability; the biggest mass loss occurs at 400 °C, when compared to the other composites (above 410 °C). Analyzing the TG and DTG curves for composites with PEG (Figure 5A–F), a shift in the maximum mass loss of composites towards higher temperatures was observed. The sample with 2.5% ST content had the highest thermal stability, with mass loss occurring at the highest temperature, close to 420 °C.

Table 1. TGA results for PCM-PCA samples.

PCM	Matrix	$T_{1\%}$ [°C]	$T_{3\%}$ [°C]	$T_{5\%}$ [°C]	$T_{10\%}$ [°C]	$T_{50\%}$ [°C]	$T_{DTG\ max}$ [°C]	Char Residue [%]
PEG 6000	-	211	233	249	286	380	392	0.233
	0% ST	330	369	378	387	410	410	9.922
	2.5% ST	350	373	380	389	412	416	6.691
	5% ST	322	361	371	380	400	415	6.23
	10% ST	353	374	381	389	411	414	6.489
	20% ST	349	371	379	388	411	415	8.908
PA	-	165	178	185	197	229	245	0.248
	0% ST	161	177	184	195	228	236	12.547
	2.5% ST	159	174	182	193	224	240	6.105
	5% ST	161	178	186	199	235	236	6.297
	10% ST	157	174	183	194	227	247	6.654
	20% ST	161	177	184	195	228	237	12.115
Octacosane	-	189	204	212	225	264	282	0.333
	0% ST	187	217	229	245	287	295	11.76
	2.5% ST	186	203	211	224	281	298	4.723
	5% ST	186	203	214	228	274	290	6.385
	10% ST	181	200	210	226	277	295	6.039
	20% ST	185	203	213	227	268	284	10.088

The PA-PCA samples are thermally stable to 130 °C. In comparison with the PEG-PCA, the PA exhibited no significant changes in thermal stability when impregnated into the PCA matrices. Among all samples, the 10% ST PCA sample exhibited the highest temperature for the main mass loss, at ~250 °C. The rest of the samples had a temperature lower than 240 °C which is lower than a pure PA degradation temperature (245 °C).

The pure OC sample exhibited two main peaks associated with the degradation of two different components; the DTG curves were not symmetrical, and shifted to the left. A hidden peak could be observed around 250 °C. The peaks emerge even more in the sample that contains more ST in the precursor stage. The beginning of the degradation could be observed at ~150 °C for all samples. For a 2.5% ST sample, the first peak of mass loss occurred at 230 °C, while the maximum mass loss was observed at 300 °C, showing the highest thermal stability among OC-PCA samples. The sample with a 20% ST content had its first significant mass loss at ~250 °C; the most significant mass loss occurred at a lower temperature (285 °C) than for the other composite samples (above 290 °C). Although there was no direct tendency for the thermal stability of the OC-PCA or the starch content at the precursor matrix stage, conclusions could still be made according to the morphology of PCA based on SEM images. The carbon matrix could act as a thermal buffer, delaying the degradation of octadecane. It could also restrict the migration of volatile degradation products, possibly influencing the degradation mechanism. It follows that the thermal behavior of the OC-PCA materials can be affected by the porosity and pore distribution of the PCA. The thermal stability results correspond to the SEM results. When pores are well distributed with similar sizes (like in the 2.5% ST sample), the thermal stability is higher (due to restriction of the migration of volatile degradation products); when the pores are less evenly distributed, the thermal degradation process can occur more easily at lower temperatures.

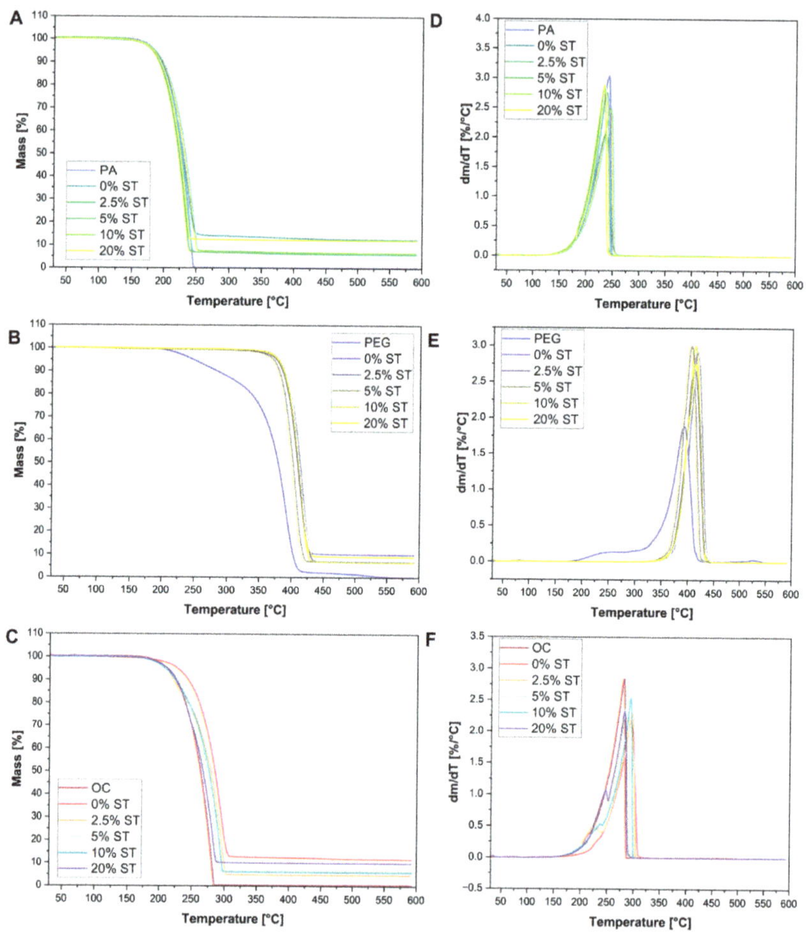

Figure 5. Results for PCM-PCA composites from TGA: TG curves of (**A**) PA, (**B**) PEG, and (**C**) OC impregnated PCA; DTG curves of (**D**) PA, (**E**) PEG, and (**F**) OC impregnated PCA.

All examined samples exhibited no thermal degradation or changes in mass fraction during heating in the range of their PCM phase transition, which could be beneficial in the case of long-lifetime materials and their performance.

3.4. DSC Results for PCM-PCA Composites

The thermal properties of PCM-PCA composites were evaluated using the DSC method. Temperatures for the beginning (T_b), peak (T_p), and end (T_e) of phase transitions (melting and crystallization), together with the heat of phase transition, were summarized in Table 2.

Table 2. Characteristic temperatures for heating and cooling cycles in PCM-PCA materials.

Phase Change	PCM	Matrix	T_b [°C]	T_p [°C]	T_e [°C]	Heat of Phase Transition [J/g]
Heating	PEG 6000	-	50	65	78	184.4
		0% ST	45	69	82	163.8
		2.5% ST	43	69	85	172.5
		5% ST	40	75	84	173.1
		10% ST	47	71	82	164.7
		20% ST	48	72	84	164.8
	PA	-	52	66	83	195.7
		0% ST	51	73	86	173.6
		2.5% ST	53	72	83	188.6
		5% ST	53	74	86	184.9
		10% ST	46	75	87	186.5
		20% ST	46	75	93	194.1
	Octacosane	-	52	64	72	248.7
		0% ST	48	74	90	214.6
		2.5% ST	51	75	90	239.0
		5% ST	53	69	86	230.8
		10% ST	55	74	88	224.7
		20% ST	54	67	86	228.4
Cooling	PEG 6000	-	49	36	25	161.5
		0% ST	48	40	26	154.5
		2.5% ST	46	40	24	160.8
		5% ST	46	41	27	159.7
		10% ST	45	40	26	156.9
		20% ST	46	40	25	156.8
	PA	-	67	59	39	196.5
		0% ST	60	60	42	171.1
		2.5% ST	61	56	47	188.3
		5% ST	61	60	43	183.8
		10% ST	66	60	37	184.5
		20% ST	62	59	39	140.2
	Octacosane	-	61	60/54	47	246.3
		0% ST	61	60/54	40	212.4
		2.5% ST	61	59/54	44	234.8
		5% ST	61	59/54	42	226.1
		10% ST	61	59/54	42	223.0
		20% ST	61	57/54	44	226.3

The results obtained for PEG-PCA composites present some tendencies. The melting T_b is lower than that of pure PEG, due to the influence of the carbon matrix in increasing heat conductivity; thus, the melting process starts faster. Nevertheless, the T_p and T_e of the melting phase transition are shifted to higher temperatures because of the PEG encapsulation within the pores, and, consequently, the PCM material entrapped within the pores requires higher temperatures to reach it and cause the phase transition. In all cases (as could be observed in Figure 6B), the melting process is elongated compared to pure PEG. The enthalpy of PEG-PC composites is lower than that of pure PCM because of the incorporation of the matrix, which does not show any phase transition in the measurement temperature range. The PEG-PCA heat flow curves obtained from the test were not as symmetrical as the pure PEG curve, and, additionally, as could be seen in Figure 6B, some double peaks may appear, as was the case for 10% ST content. Additionally, the slopes of the left curves tend to flatten out toward higher temperatures. This is caused by the better pore distribution seen as the ST content in the sample increases (Figure 3). The melting temperature range for all PEG-PCM samples starts at 40 and reaches 80 °C.

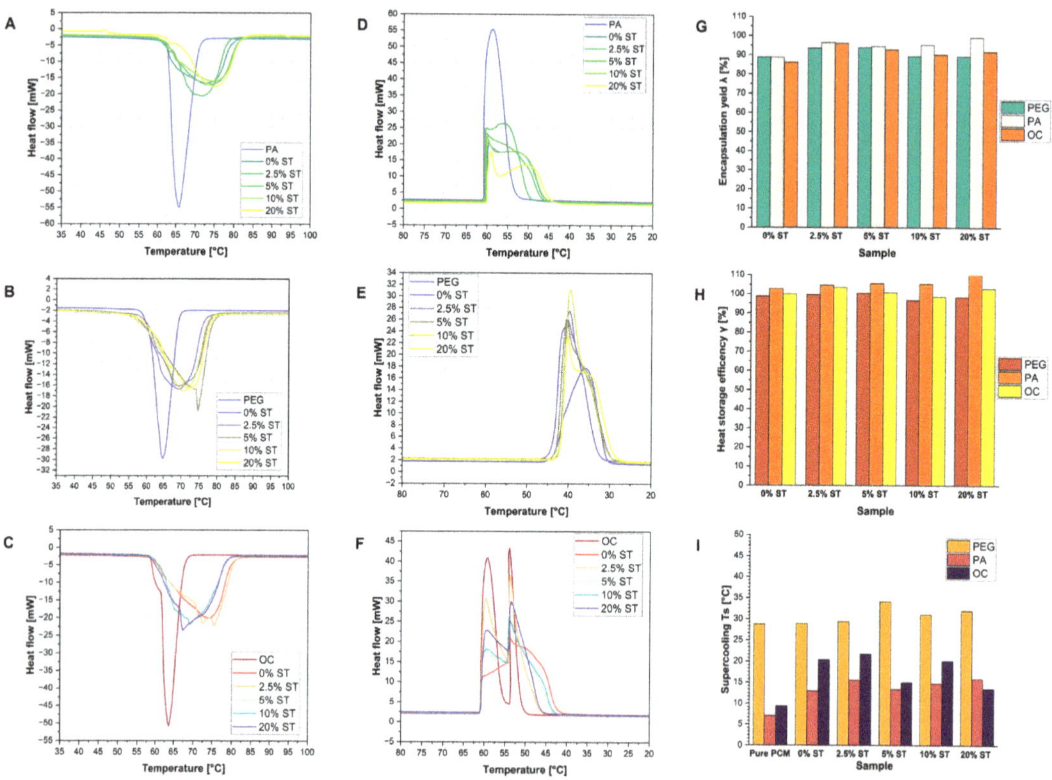

Figure 6. Results for PCM-PCA composites from DSC: (**A**) PA, (**B**) PEG, (**C**) OC impregnated PCA; for DSC cooling cycle, (**D**) PA, (**E**) PEG, (**F**) OC impregnated PCA; (**G**) encapsulation yield; (**H**) heat storage efficiency; (**I**) supercooling.

The crystallization process in PEG-PCA composites started at lower temperatures compared to pure PEG. Additionally, the peaks are less flattened; there are still some examples of double peaks that are sharper, considering the melting process. This could also be a result of leakage and crystallization processes that occur outside and inside the composite samples. The main peaks are shifted to higher temperatures compared to those of pure PEG, which may be the result of a crystallization nucleus being more likely to form in the carbon matrix. Also, among all the samples, PEG-PCA exhibits the highest supercooling (Figure 6I).

The results for the PA-PCA revealed similar tendencies as the PEG samples in terms of the characteristic temperatures for the melting process, with lower T_b and higher T_p and

T_e values than the reference pure PA sample. The heating curves of the PA-PCA samples were flattened, and the phase transition temperature range was larger by ~15 °C. The cooling curves of PA-PCA samples exhibited a wider double crystallization peak. The crystallization process also exhibited a wider temperature range than pure PA (~7 °C), which is beneficial in terms of long-term heat exchange. The double peak phase change may have been caused by the leakage of the PA from the matrix, or, alternatively, could be associated with different crystalline phase changes.

The paraffin-based OC composite samples exhibited T_b, higher T_p, and lower T_e than the reference pure OC sample, with a tendency for the phase change to be shifted to higher temperatures within wider temperatures ranges. There are some indications of two energy-related processes, related to the nonsymmetrical curves. The curves are also flattened, and the phase transition temperature range is larger by ~15 °C (Figure 6C). In Figure 6F, two peaks (especially for pure OC) in the cooling curves can be distinguished. The first one is ascribed to the crystallization from liquid to solid, and the second one is a solid–solid phase transition occurring within the crystalline phase of OC because of the multiple crystalline phases (monoclinic and orthorhombic and rotator) with different molecular arrangements [53,54].

When OC is incorporated into carbon matrices, this effect is covered by the nonequal melting processes occurring within pores, and peaks are present but not as strong. Similar to the melting process, and to other sample types, the phase transition temperature range is larger by ~10 °C, leading to a longer energy release process. The supercooling is lower than in the PEG-PCA samples, reaching a maximum value of 20 °C for 2.5% ST samples, which is two times more than the pure OC.

The encapsulation yield in all types of samples is relatively high, reaching above 85%. The highest values may be ascribed to PA-based composites with values of ~90% or more (Figure 6G). The heat storage efficiency indicates that the incorporation of PCA does not have a significant impact on the thermal performance of composites compared to that of pure PCM. Values in all cases are higher than 95% (Figure 6H), which additionally indicates an improvement in performance for PCM incorporated into PCA. Taking into consideration the encapsulation yield and heat storage efficiency, the best performance among all types of infiltrated PCMs could be ascribed to the composite that was prepared with 2.5% ST matrices.

3.5. Leakage Tests and Mass Fraction of PCM-PCA Composites

The leakage test (Figure 7A–C) showed that the best performance could be observed in PEG-PCA samples, where the leakage is, respectively, small compared to PA-PCA and OC-PCA, because of the leakage diameter, which is one-fourth the size (Figure 7G–I) (25 mm compared to >100 mm).

This result is mainly related to the viscosity of the proposed PCMs; for example, at 60 °C, the viscosity of PEG 6000 is 450–600 cP; OC is 100–200 cP; and PA is 10–50 cP. This indicates the extent of the leakage observed during the test. The more viscous the PCM, the less leakage could be observed, and the more the PCM could be entrapped within the pores of the matrix. For the PEG samples, the difference between the mass ratio before and after the test is ~1 p.%, compared to PA 5% ST or OC 10% ST, where the difference reaches more than 30% (Figure 7D–F).

A difference could also be observed on the microstructures of the samples before and after the test, as evaluated with a digital optical microscope (Figure 8). After the leakage test, a significant amount of PCM could be observed, especially for PEG-impregnated aerogels. For PA and OC, the difference between the sample before and after the test is more notable. Some open, empty pores, after the test, could also be observed for 10% ST and 20% ST samples. The type of matrix plays a very important role. As mentioned before, the starch content determines the microstructure of the carbon matrices. The best performance among all types of composites can be ascribed to samples based on 2.5% ST. The difference between the mass ratio of PCM before and after the leakage test is relatively low compared to other composites.

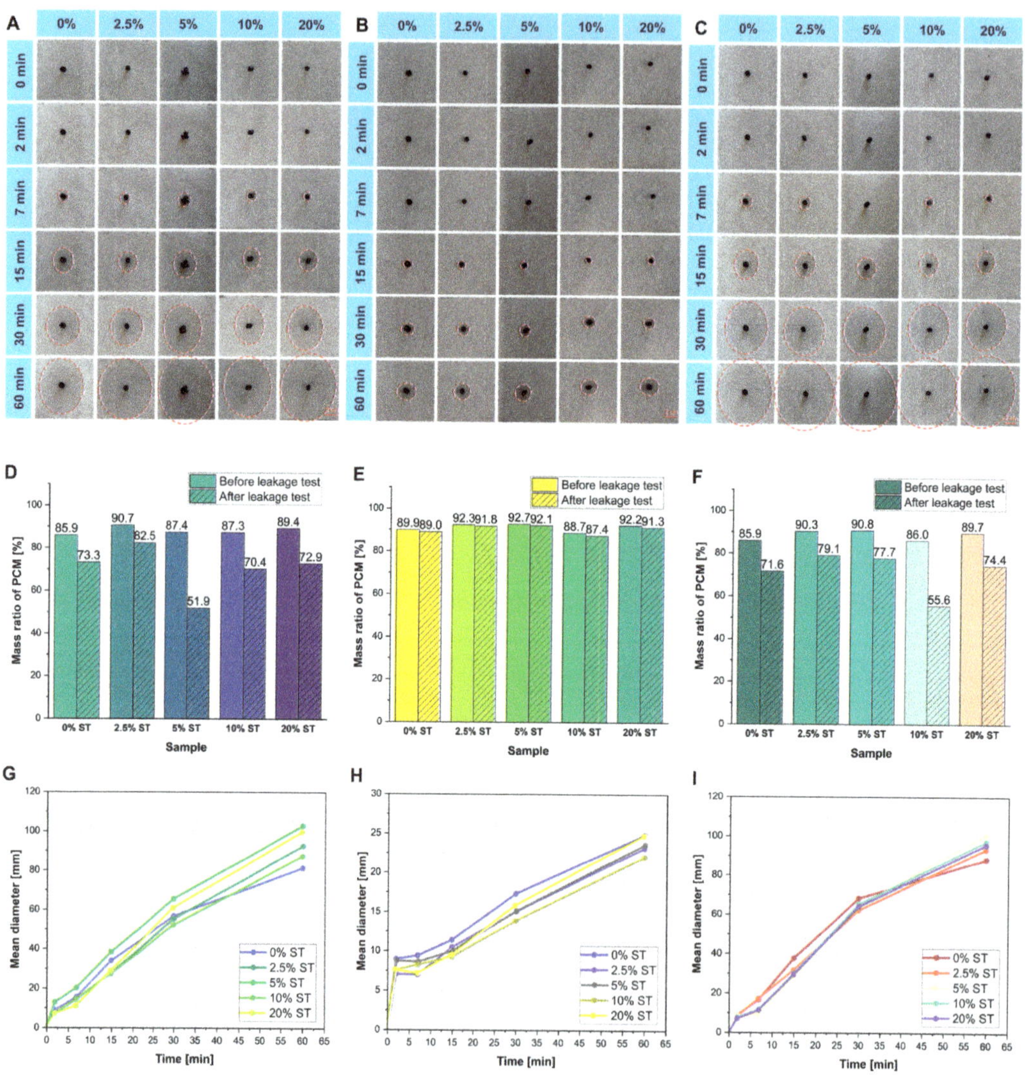

Figure 7. Digital photographs of leakage test results: (**A**) PA, (**B**) PEG, and (**C**) OC impregnated PCA; mass ratio of PCM in the (**D**) PA, (**E**) PEG, and (**F**) OC impregnated PCA composites; mean diameter of leakage of (**G**) PA, (**H**) PEG, and (**I**) OC impregnated PCA.

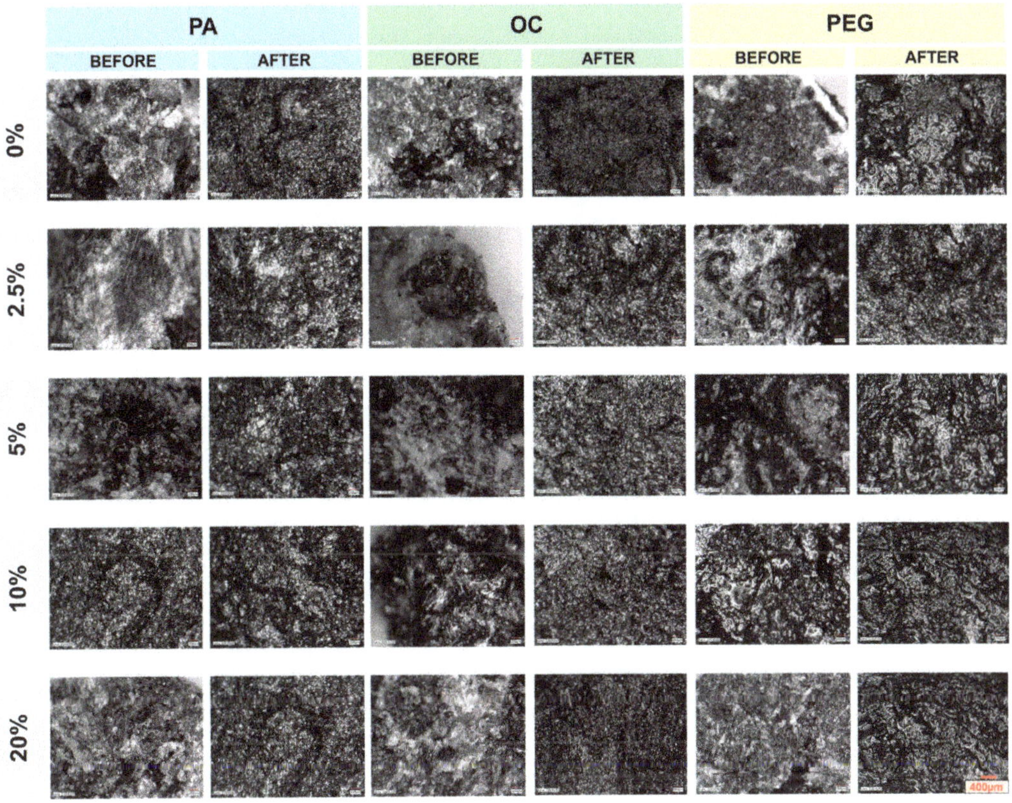

Figure 8. Digital microscope photographs of the leakage test results for composite microstructures, taken before and after testing.

4. Conclusions

In this study, PCA-PCM composites for solar-thermal energy conversion were designed and prepared. The preparation method included carbonization of orange biomass waste with a starch binder.

The following conclusions were reached:

(1) The carbonization process significantly increased the carbon content (from ~40% to ~82–88%), confirming the successful conversion to highly carbonized material. The increased carbon content correlates with higher starch content in the precursors, making the material promising for applications such as solar energy harvesting.

(2) The addition of starch (ST) significantly influences the morphology of PCA samples. Lower ST content (e.g., 2.5%) results in more regular and open-pored structures, which are ideal for loading and incorporating PCMs. A higher ST content (e.g., 20%) leads to less uniform structures with starch-derived beans, reducing overall porosity and limiting PCM impregnation.

(3) PCM impregnation successfully fills the pores of PCA samples, particularly in cases with optimal porosity and open structures, such as the 2.5% and 5% ST samples.

(4) The proper ratio of binder (ST) to biomass waste is critical in the precursor stage to achieve materials with balanced porous networks. This balance supports effective PCM impregnation while minimizing leakage during phase transitions, enhancing the

functionality of PCA-based thermal energy storage systems. The best performances of ST as a binder (considering the microstructures of aerogels obtained and PCA-PCM performance) are 2.5% and 10%.

(5) None of the tested samples experienced thermal degradation or mass changes within the temperature range corresponding to their PCM phase transitions. This property suggests their potential with regard to long-term durability and consistent performance.

(6) Among the composites tested, the 2.5% starch-based matrices showed the best overall performance, with the lowest mass loss during the leakage test. This finding underscores the importance of optimizing matrix composition to balance PCM retention and structural integrity, particularly for applications requiring minimal leakage.

Our research and analysis have proven the good quality of the PCA-PCM composites. The obtained OBW carbon aerogels could also be used in other fields, in applications such as water evaporation and distillation systems, energy storage, catalysts, gas storage, etc. This study can additionally be used as an introduction for further investigation of the presented materials and the evaluation of long-term solar-thermal conversion efficiency and overall material performance.

Author Contributions: Conceptualization, K.S. and K.P.; methodology, K.S., M.B. and M.S. (Martyna Sieradzka); validation, K.S. and K.P.; formal analysis, K.S. and K.P.; investigation, K.S. and M.B.; resources, M.K., M.Z. and K.P.; data curation, K.S.; writing—original draft preparation, K.S. and M.S. (Martyna Szatkowska); writing—review and editing, K.S. and K.P.; visualization, K.S.; supervision, K.P.; project administration, K.P.; funding acquisition, K.P. All authors have read and agreed to the published version of the manuscript.

Funding: This work was supported by a subsidy from the Ministry of Science and Higher Education for the AGH University of Krakow (Project No. 16.16.160.557) and the Program "Excellence Initiative—Research University" for the AGH University of Krakow.

Data Availability Statement: The original contributions presented in this study are included in the article. Further inquiries can be directed to the corresponding author.

Conflicts of Interest: The authors declare no conflicts of interest.

Abbreviations

DDI	Distilled water
DIMS	Dynamic insulation material and system
DSC	Differential scanning calorimetry
DW	Double-pane window
FHL	Fabric heat loss
HEL	Heat energy loss
IR	Infrared
OBW	Orange biomass waste
PCA	Porous carbon aerogel
PCM	Phase change material
SEM	Scanning electron microscopy
ST	Starch
TES	Thermal energy storage
TGA	Thermogravimetry analysis
TW	Triple-pane window
VHL	Ventilation heat loss

References

1. Ritchie, H.; Rosado, P.; Roser, M. Energy Production and Consumption. Our World in Data. 2024. Available online: https://ourworldindata.org/energy-production-consumption (accessed on 1 June 2024).
2. Chaturvedi, R.; Islam, A.; Sharma, K. A Review on the Applications of PCM in Thermal Storage of Solar Energy. *Mater. Today Proc.* **2021**, *43*, 293–297. [CrossRef]
3. United Nations Environment Programme. *2023 Global Status Report for Buildings and Construction: Beyond Foundations—Mainstreaming Sustainable Solutions to Cut Emissions from the Buildings Sector*; United Nations Environment Programme: Washington, DC, USA, 2024; ISBN 978-92-807-4131-5.
4. Najjar, M.K.; Figueiredo, K.; Hammad, A.W.A.; Tam, V.W.Y.; Evangelista, A.C.J.; Haddad, A. A Framework to Estimate Heat Energy Loss in Building Operation. *J. Clean. Prod.* **2019**, *235*, 789–800. [CrossRef]
5. Ismail, K.A.R.; Lino, F.A.M.; Teggar, M.; Arıcı, M.; Machado, P.L.O.; Alves, T.A.; De Paula, A.C.O.; Benhorma, A. A Comprehensive Review on Phase Change Materials and Applications in Buildings and Components. *ASME Open J. Eng.* **2022**, *1*, 011049. [CrossRef]
6. Pielichowska, K.; Pielichowski, K. Phase Change Materials for Thermal Energy Storage. *Prog. Mater. Sci.* **2014**, *65*, 67–123. [CrossRef]
7. Liu, K.; Tian, Z. Advances in Phase-Change Materials. *J. Appl. Phys.* **2021**, *130*, 070401. [CrossRef]
8. Kalnæs, S.E.; Jelle, B.P. Phase Change Materials and Products for Building Applications: A State-of-the-Art Review and Future Research Opportunities. *Energy Build.* **2015**, *94*, 150–176. [CrossRef]
9. Li, S.; Sun, G.; Zou, K.; Zhang, X. Experimental Research on the Dynamic Thermal Performance of a Novel Triple-Pane Building Window Filled with PCM. *Sustain. Cities Soc.* **2016**, *27*, 15–22. [CrossRef]
10. Kishore, R.A.; Bianchi, M.V.A.; Booten, C.; Vidal, J.; Jackson, R. Enhancing Building Energy Performance by Effectively Using Phase Change Material and Dynamic Insulation in Walls. *Appl. Energy* **2021**, *283*, 116306. [CrossRef]
11. Qin, M.; Xiong, F.; Aftab, W.; Shi, J.; Han, H.; Zou, R. Phase-Change Materials Reinforced Intelligent Paint for Efficient Daytime Radiative Cooling. *iScience* **2022**, *25*, 104584. [CrossRef]
12. Yaxuan, X.; Chenhua, Y.; Jing, R.; Yuting, W.; Qian, X.; Binjian, N.; Chuan, L.; Yulong, D. Waste Semicoke Ash Utilized to Fabricate Shape-Stable Phase Change Composites for Building Heating and Cooling. *Constr. Build. Mater.* **2022**, *361*, 129638. [CrossRef]
13. Geisendorf, S.; Pietrulla, F. The Circular Economy and Circular Economic Concepts—A Literature Analysis and Redefinition. *Thunderbird Int. Bus. Rev.* **2018**, *60*, 771–782. [CrossRef]
14. Zhou, C.; Wang, Y. Recent Progress in the Conversion of Biomass Wastes into Functional Materials for Value-Added Applications. *Sci. Technol. Adv. Mater.* **2020**, *21*, 787–804. [CrossRef]
15. Tiwari, S.K.; Bystrzejewski, M.; De Adhikari, A.; Huczko, A.; Wang, N. Methods for the Conversion of Biomass Waste into Value-Added Carbon Nanomaterials: Recent Progress and Applications. *Prog. Energy Combust. Sci.* **2022**, *92*, 101023. [CrossRef]
16. Liu, H.; Qian, Z.; Wang, Q.; Wu, D.; Wang, X. Development of Renewable Biomass-Derived Carbonaceous Aerogel/Mannitol Phase Change Composites for High Thermal-Energy-Release Efficiency and Shape Stabilization. *ACS Appl. Energy Mater.* **2021**, *4*, 1714–1730. [CrossRef]
17. Cheng, L.; Feng, J. Form-Stable Phase Change Materials Based on Delignified Wood Flour for Thermal Management of Buildings. *Compos. Part Appl. Sci. Manuf.* **2020**, *129*, 105690. [CrossRef]
18. Yang, H.; Wang, S.; Wang, X.; Chao, W.; Wang, N.; Ding, X.; Liu, F.; Yu, Q.; Yang, T.; Yang, Z.; et al. Wood-Based Composite Phase Change Materials with Self-Cleaning Superhydrophobic Surface for Thermal Energy Storage. *Appl. Energy* **2020**, *261*, 114481. [CrossRef]
19. Yang, Z.; Deng, Y.; Li, J. Preparation of Porous Carbonized Woods Impregnated with Lauric Acid as Shape-Stable Composite Phase Change Materials. *Appl. Therm. Eng.* **2019**, *150*, 967–976. [CrossRef]
20. Liu, H.; Zheng, Z.; Qian, Z.; Wang, Q.; Wu, D.; Wang, X. Lamellar-Structured Phase Change Composites Based on Biomass-Derived Carbonaceous Sheets and Sodium Acetate Trihydrate for High-Efficient Solar Photothermal Energy Harvest. *Sol. Energy Mater. Sol. Cells* **2021**, *229*, 111140. [CrossRef]
21. Yu, K.; Liu, Y.; Jia, M.; Yang, Y. Bio-Based Dual-Functionalized Phase Change Composite: Ultrafast Solar-to-Thermal Conversion and Reinforced Heat Storage Capacity. *Energy Fuels* **2021**, *35*, 16162–16173. [CrossRef]
22. Ji, Z.; Abdalkarim, S.Y.H.; Li, H.; Asad, R.A.M.; Yu, H.-Y. Waste Pomelo Peels-Derived Ultralow Density 3D-Porous Carbon Aerogels: Mechanisms of "Soft-Rigid" Structural Formation and Solar-Thermal Energy Storage Conversion. *Sol. Energy Mater. Sol. Cells* **2023**, *259*, 112453. [CrossRef]
23. Xie, Y.; Zou, M.; Xiao, S.; Chen, W.; Liu, Y.; Hu, X.; Jiang, X.; He, Y.; Li, Q. MXene-Modified Bio-Based Pitaya Peel Foam/Polyethylene Glycol Composite Phase Change Material with Excellent Photo-Thermal Conversion Efficiency, Thermal Energy Storage Capacity and Thermal Conductivity. *J. Energy Storage* **2024**, *78*, 110089. [CrossRef]

24. Sheng, X.; Dong, D.; Lu, X.; Zhang, L.; Chen, Y. MXene-Wrapped Bio-Based Pomelo Peel Foam/Polyethylene Glycol Composite Phase Change Material with Enhanced Light-to-Thermal Conversion Efficiency, Thermal Energy Storage Capability and Thermal Conductivity. *Compos. Part Appl. Sci. Manuf.* **2020**, *138*, 106067. [CrossRef]
25. Zhao, Y.; Min, X.; Huang, Z.; Liu, Y.; Wu, X.; Fang, M. Honeycomb-like Structured Biological Porous Carbon Encapsulating PEG: A Shape-Stable Phase Change Material with Enhanced Thermal Conductivity for Thermal Energy Storage. *Energy Build.* **2018**, *158*, 1049–1062. [CrossRef]
26. Fang, Y.; Liu, S.; Li, X.; Hu, X.; Wu, H.; Lu, X.; Qu, J. Biomass Porous Potatoes/MXene Encapsulated PEG-Based PCMs with Improved Photo-to-Thermal Conversion Capability. *Sol. Energy Mater. Sol. Cells* **2022**, *237*, 111559. [CrossRef]
27. Xie, Y.; Li, W.; Huang, H.; Dong, D.; Zhang, X.; Zhang, L.; Chen, Y.; Sheng, X.; Lu, X. Bio-Based Radish@PDA/PEG Sandwich Composite with High Efficiency Solar Thermal Energy Storage. *ACS Sustain. Chem. Eng.* **2020**, *8*, 8448–8457. [CrossRef]
28. Li, Y.; Huang, X.; Li, Y.; Xi, Z.; Hai, G.; Tao, Z.; Wang, G. Shape-Stabilized Phase-Change Materials Supported by Eggplant-Derived Porous Carbon for Efficient Solar-to-Thermal Energy Conversion and Storage. *Sustain. Energy Fuels* **2020**, *4*, 1764–1772. [CrossRef]
29. Hekimoğlu, G.; Sarı, A.; Kar, T.; Keleş, S.; Kaygusuz, K.; Tyagi, V.V.; Sharma, R.K.; Al-Ahmed, A.; Al-Sulaiman, F.A.; Saleh, T.A. Walnut Shell Derived Bio-Carbon/Methyl Palmitate as Novel Composite Phase Change Material with Enhanced Thermal Energy Storage Properties. *J. Energy Storage* **2021**, *35*, 102288. [CrossRef]
30. Zhao, P.-P.; Deng, C.; Zhao, Z.-Y.; Lu, P.; He, S.; Wang, Y.-Z. Hypophosphite Tailored Graphitized Hierarchical Porous Biochar toward Highly Efficient Solar Thermal Energy Harvesting and Stable Storage/Release. *Chem. Eng. J.* **2021**, *420*, 129942. [CrossRef]
31. Song, S.; Ai, H.; Zhu, W.; Lv, L.; Feng, R.; Dong, L. Carbon Aerogel Based Composite Phase Change Material Derived from Kapok Fiber: Exceptional Microwave Absorbility and Efficient Solar/Magnetic to Thermal Energy Storage Performance. *Compos. Part B Eng.* **2021**, *226*, 109330. [CrossRef]
32. Zhang, W.; Qiu, X.; Wang, C.; Zhong, L.; Fu, F.; Zhu, J.; Zhang, Z.; Qin, Y.; Yang, D.; Xu, C.C. Lignin Derived Carbon Materials: Current Status and Future Trends. *Carbon Res.* **2022**, *1*, 14. [CrossRef]
33. Orange Production. Available online: https://ourworldindata.org/grapher/orange-production (accessed on 6 December 2024).
34. Xiao, K.; Liu, H.; Li, Y.; Yi, L.; Zhang, X.; Hu, H.; Yao, H. Correlations between Hydrochar Properties and Chemical Constitution of Orange Peel Waste during Hydrothermal Carbonization. *Bioresour. Technol.* **2018**, *265*, 432–436. [CrossRef]
35. Singh, P.; Sharma, R.K.; Khalid, M.; Goyal, R.; Sarı, A.; Tyagi, V.V. Evaluation of Carbon Based-Supporting Materials for Developing Form-Stable Organic Phase Change Materials for Thermal Energy Storage: A Review. *Sol. Energy Mater. Sol. Cells* **2022**, *246*, 111896. [CrossRef]
36. Shi, L.; Wang, X.; Hu, Y.; He, Y.; Yan, Y. Solar-Thermal Conversion and Steam Generation: A Review. *Appl. Therm. Eng.* **2020**, *179*, 115691. [CrossRef]
37. Suchorowiec, K.; Paprota, N.; Pielichowska, K. Aerogels for Phase-Change Materials in Functional and Multifunctional Composites: A Review. *Materials* **2024**, *17*, 4405. [CrossRef] [PubMed]
38. Xu, Y.; Zhang, X.; Wu, B.; Xu, Y.; Wen, R.; Liu, Y.; Fang, M.; Wu, X.; Min, X.; Huang, Z. Preparation and Performance of Shape-Stable Phase Change Materials Based on Carbonized-Abandoned Orange Peel and Paraffin. *Fuller. Nanotub. Carbon Nanostruct.* **2019**, *27*, 289–298. [CrossRef]
39. Gou, H.; He, J.; Zhao, G.; Zhang, L.; Yang, C.; Rao, H. Porous Nitrogen-Doped Carbon Networks Derived from Orange Peel for High-Performance Supercapacitors. *Ionics* **2019**, *25*, 4371–4380. [CrossRef]
40. Liu, H.; Li, A.; Liu, Z.; Tao, Q.; Li, J.; Peng, J.; Liu, Y. Preparation of Lightweight and Hydrophobic Natural Biomass-Based Carbon Aerogels for Adsorption Oils and Organic Solvents. *J. Porous Mater.* **2022**, *29*, 1001–1009. [CrossRef]
41. Py, X.; Olives, R.; Mauran, S. Paraffin/Porous-Graphite-Matrix Composite as a High and Constant Power Thermal Storage Material. *Int. J. Heat Mass Transf.* **2001**, *44*, 2727–2737. [CrossRef]
42. Density of Elements Chart—Angstrom Sciences Elements Density Table. Available online: https://www.angstromsciences.com/density-elements-chart (accessed on 27 February 2024).
43. Zahir, M.H.; Rahman, M.M.; Mohaisen, K.O.; Helal, A.; Shaikh, M.N.; Rahman, M.M. Clarification of the Supercooling and Heat Storage Efficiency Calculation Formula for Shape-Stabilized Phase Change Materials. *ACS Omega* **2022**, *7*, 41096–41099. [CrossRef] [PubMed]
44. Yue, H.; Ou, Y.; Wang, J.; Wang, H.; Du, Z.; Du, X.; Cheng, X. Ti3C2Tx MXene/Delignified Wood Supported Flame-Retardant Phase-Change Composites with Superior Solar-Thermal Conversion Efficiency and Highly Electromagnetic Interference Shielding for Efficient Thermal Management. *Energy* **2024**, *286*, 129441. [CrossRef]
45. Lu, X.; Huang, J.; Wong, W.-Y.; Qu, J. A Novel Bio-Based Polyurethane/Wood Powder Composite as Shape-Stable Phase Change Material with High Relative Enthalpy Efficiency for Solar Thermal Energy Storage. *Sol. Energy Mater. Sol. Cells* **2019**, *200*, 109987. [CrossRef]

46. Xiao, S.; Zou, M.; Xie, Y.; Chen, W.; Hu, X.; Ma, Y.; Zu, S.; Che, Y.; Jiang, X. Nanosilver Modified Navel Orange Peel Foam/Polyethylene Glycol Composite Phase Change Materials with Improved Thermal Conductivity and Photo-Thermal Conversion Efficiency. *J. Energy Storage* **2022**, *56*, 105976. [CrossRef]
47. Xie, Y.; Zou, M.; Liu, Y.; Chen, W.; Hu, X.; Jiang, L.; Luo, L.; Luo, W.; Ma, Y.; Jiang, X.; et al. MXene-Modified Lemon Peel-Based Composite Phase Change Material with Excellent Photo-Thermal Conversion Efficiency, Thermal Storage Capacity and Thermal Conductivity for Thermal Management of Electronic Components. *J. Energy Storage* **2024**, *90*, 111971. [CrossRef]
48. Miranda, R.; Bustos-Martinez, D.; Blanco, C.S.; Villarreal, M.H.G.; Cantú, M.E.R. Pyrolysis of Sweet Orange (*Citrus Sinensis*) Dry Peel. *J. Anal. Appl. Pyrolysis* **2009**, *86*, 245–251. [CrossRef]
49. Aguiar, L.; Márquez-Montesinos, F.; Gonzalo, A.; Sánchez, J.L.; Arauzo, J. Influence of Temperature and Particle Size on the Fixed Bed Pyrolysis of Orange Peel Residues. *J. Anal. Appl. Pyrolysis* **2008**, *83*, 124–130. [CrossRef]
50. Kim, Y.-M.; Lee, H.W.; Kim, S.; Watanabe, C.; Park, Y.-K. Non-Isothermal Pyrolysis of Citrus Unshiu Peel. *BioEnergy Res.* **2015**, *8*, 431–439. [CrossRef]
51. Altayan, M.M.; Al Darouich, T.; Karabet, F. On the Plasticization Process of Potato Starch: Preparation and Characterization. *Food Biophys.* **2017**, *12*, 397–403. [CrossRef]
52. Fairén-Jiménez, D.; Carrasco-Marín, F.; Moreno-Castilla, C. Inter- and Intra-Primary-Particle Structure of Monolithic Carbon Aerogels Obtained with Varying Solvents. *Langmuir* **2008**, *24*, 2820–2825. [CrossRef] [PubMed]
53. Rajabalee, F.; Métivaud, V.; Mondieig, D.; Haget, Y.; Cuevas-Diarte, M.A. New Insights on the Crystalline Forms in Binary Systems of *n* -Alkanes: Characterization of the Solid Ordered Phases in the Phase Diagram Tricosane + Pentacosane. *J. Mater. Res.* **1999**, *14*, 2644–2654. [CrossRef]
54. Mondieig, D.; Rajabalee, F.; Metivaud, V.; Oonk, H.A.J.; Cuevas-Diarte, M.A. *N*-Alkane Binary Molecular Alloys. *Chem. Mater.* **2004**, *16*, 786–798. [CrossRef]

MDPI AG
Grosspeteranlage 5
4052 Basel
Switzerland
Tel.: +41 61 683 77 34

Energies Editorial Office
E-mail: energies@mdpi.com
www.mdpi.com/journal/energies

www.ingramcontent.com/pod-product-compliance
Lightning Source LLC
Chambersburg PA
CBHW061446210926
51850CB00008B/1626
9783725847594